AF722626

Volume 2, 1995

Telecommunications & Space Journal

Law - Economics - Public Policy

Annual Edition

Edited by
Professor Lucien Rapp
University of Toulouse I
(France)

SERDI Publishing Compagny
29 rue de Saint Petersbourg
75008 Paris, France

© L. Rapp, 1995

All rights reserved. No part of this publication may be reproduced, stored in a retrieval system, or transmitted in any form or by any means, electronic, mechanical, photocopying, recording or otherwise without the prior permission of SERDI Publishing Company.

Published by SERDI, 29 rue de Saint Petersbourg, 75008 Paris France

Printed in France - I.S.S.N. 1254-7875

TABLE OF CONTENTS

Acknowledgements

This edition would not have come into being without the friendly collaboration of the various authors, all prestigious specialists known worldwide in the area of space or telecommunications law. They accepted, out of friendship, to devote a couple of moments of their precious time to complete their contributions. Heartfelt thanks are extended to them !

The collaboration of SERDI Publishers (Editions SERDI), and especially Ms C. Giry and Ms R. Prat , was particularly useful in relation to establishing contacts with the authors and for the compilation of the final manuscript of the publication. Sincere thanks are extended to them.

This edition also is indebted to the companies which accepted to sponsor it. Our thanks to all of them for their very kind support.

All those who have participated in the completion of this edition of the Telecommunications and Space Journal represent, together, the hope of having shown the way and of being joined very quickly by those, in particular, the future authors, who in one capacity or another, wish to contribute to the success of the following editions. They are thanked in advance.

L. R.

Alphabetic List of the Contributing Authors

F. Busso, *Negri, Teijeiro and Incera, Buenos Aires, Argentina.*

A. Cerisola, *Ferrère Lamaison, Montevideo, Uruguay.*

M. Davies, *Nabarro Nathanson, Warsaw, Poland.*

I. De Vries, *Brussels, Belgium.*

E. Fasan, *Rechtsanwälte, Neunkirchen, Austria.*

J. Gavan, *Center for technological education, Holon, Israel.*

C. Golda, *University of Genoa, Genoa, Italy.*

S. Gorove, *University of Mississippi Law Center, Mississipi, USA.*

K. Harvey, *The George Washington University, Washington DC, USA.*

T. Johnson, *S.J. Berwin & Co, London, U.K..*

T. Kelly, *I.T.U., Genève, Switzerland.*

R. Kröner, *Trenité Van Doorne, Amsterdam, The Netherlands.*

D. Maclean, *I.T.U., Genève, Switzerland.*

J. J. Negri, *Negri, Teijeiro and Incera, Buenos Aires, Argentina.*

E. Paul, *S.J.Berwin and Co., London, U.K..*

L. Rapp, *University of Toulouse I, Toulouse, France.*

J. Scherer, *Döser Amereller Noack. Frankfurt, Germany..*

M. Schmittmann, *Heuking Kühn Kunz Wojtek, Dusseldorf, Germany.*

R. Schultz, *Mac Gill University, Montreal, Canada.*

S. Sin Han, *SJ.Berwin and Co., London, U.K..*

A. Spadoni, *University of Milan, Milano, Italy.*

P. Tarjanne, *I.T.U., Genève, Switzerland.*

F. Von der Dunk, *International Institute of Air & Space Law, Leiden, The Netherlands.*

S. White, *Ashurst, Morris Crisp, London, U.K..*

FOREWORD

PEKKA TARJANNE

Secretary-General, International Telecommunication Union, Place des Nations, 1211 Genève 20, Suisse.

We are witnessing today an extraordinary media attention to the issues of telecommunication. The media interest though new and unprecedented in its magnitude is a legitimate one, because telecommunication is at the heart of industrial and economic policy-making and has political, social and cultural ramifications. All these issues have been firmly placed on the global political agenda by the recent discussion of information superhighways.

The International Telecommunication Union (ITU) is carefully following this great upsurge of interest involving lawyers, policy analysts, sociologists, and politicians. The ITU applauds the establishment of the ***Telecommunication and Space Journal*** *as a welcome initiative in promoting discussion and dissemination of informed opinion based on dispassionate academic inquiry.*

As the oldest inter-governmental organization, the ITU has been involved in the evolution of the global telecommunication network. Since 1865 it has contributed to the promotion of inter-connectivity and inter-operability of telecommunication systems to allow people and nations to communicate with each other.

The advent of the Information Age has led the ITU to adapt itself to the new challenges demanded by globalization and pace of technology. This was reflected in a two-fold approach: structural and functional. The objective has been to promote responsiveness in decision-making by reforming the organizational structures in terms of shorter cycles of conferences; greater clarity in defining organizational roles and an increased commitment to enhance the participation of the private sector in the Unions activities. The Additional Plenipotentiary Conference (APP), Geneva, 1992, accomplished these objectives by putting in place the new organizational structures.

Two and a half years later in Kyoto, the ITU Plenipotentiary Conference focussed on the functional approach to telecommunication issues, namely ITU missions. Among these missions, the Kyoto Plenipotentiary Conference acknowledged the policy mission: of harmonizing ITU members policies and promoting a global approach to telecommunications as opposed to dealing with discrete technical issues.

It is my conviction that amidst refining organizational structures and elaborating new missions, we should not lose sight of the basic objective: that of ensuring that the benefits of telecommunication technology are extended to all. I would go so far as to suggest that in the information society the right to communicate, understood as the access to universal service, has to be viewed as a basic human right failing which the Information Age would have by-passed the majority of humankind.

Genève, February 1995,

Pekka Tarjanne

- I -

ANNUAL REVIEW

TELECOMMUNICATIONS & SPACE ANNUAL REVIEW, 1994

LUCIEN RAPP

Professeur Agrégé des Facultés de Droit, Directeur de l'Institut d'Etudes Internationales de l'Université de Toulouse I, Place Anatole France, 31042 Toulouse Cédex, France Tel. (33) 6175-9463 ; Fax (33) 6175-9415. Avocat au Barreau de Paris, Associé Gérant de Serra Michaud & Associés, 2 Rue de la Baume 75008 Paris, France Tel. (33). 1. 4421-9797 ; Fax : (33) 1. 4289-5790.

Abstract *: Important events have characterized the year 1994 : (1) the promising development of the Chinese market, and, beyond China, of the Asian countries markets, (2) the officialization and generalization of techniques of resale of capacity over leased lines to monopolistic operators, simultaneously in Europe and Japan, which announces, more surely than policies decided in high places, the end of public monopolies, (3) the technical and economic revolution of cordless communications, assisted by the progress of space communications, particularly the development of numerization and compression techniques, and the projects for implementation of satellite channels in low orbit, (4) the forthcoming statutory changes in international operating cooperatives like INTELSAT or INMARSAT, whose privatization is under way, (5) the appearance of initial international investment funds designed to cover the considerable financial needs of telecommunications infrastructures, in times of scarce resources and severe budgetary tensions...The present article, in particular, deals with such events and several others with the objective of providing for the benefit of all policy makers such as network operators or service providers, the basic trends which characterize telecommunications and space sectors, at the end of this century.*

Key Words *: Information Superhighways, Technological Innovations, Professional Fairs, Mobile Telecommunications, Social Transformations, Teleworking, International Agreements, Industrial Alliances, Industrial Strategies, Regional Policies, Economic Privatizations, Emerging Markets, Multimedia Services, Peripheral Sectors, Space Activities, Financial Services.*

INTRODUCTION.

Has the year 1994 been a sham in the telecommunications and space sectors ? One of false revolutions proclaimed to the grand fracas of official declarations and of real trends which produce more certainly than the former, a profound transformation in social habits. More than ever, the counsel of the German philosopher Leibnitz imposes itself on the exegete of events which have distinguished it :

"Sous la paille des mots, retrouve le grain des choses". Under the straw of events of 1994 is found the grain of substantial movements contained therein and which sketch the future of both sectors of activities in the middle of the last decade of the 20th century.

This being said, the year 1994 may appear to be dominated by the implementation of President Clinton and Vice-President Gore's project on super electronic highways. It has succeeded in capturing everybody's attention in the United States, of course, but also in Europe or Japan. It has captured the attention of governments but also of telecommunications or computer network operators and service providers. They have all been firmly involved in what has been presented to them, here or there, as a new challenge to take up, a "new frontier" to conquer. It has even been hailed as the equivalent of the Apolo space programme launched in the past by President Kennedy. Nobody wanted to be left out for fear of letting a new and apparently promising opportunity slip by. In reality, nobody knew for sure what such ambitious programmes whose financing is still largely unknown, cover. At present, there exists as many interpretations as there are questions on the subject. A quip of Edward G. Markey, the outgoing President of the Committee responsible for telecommunications in the American Chamber of Representatives, springs to mind : "Today, I have two pieces of news from Washington, one good and one bad. The good news is that each member of Congress is very actively supporting Vice-President Gore's project of super electronic highways. The bad news is that ... nobody has the slightest idea of what it entails".

Such programmes have succeeded in making us overlook other equally important events which have characterized the year 1994 : (1) the promising development of the Chinese market, and, beyond China, of the Asian countries markets, (2) the officialization and generalization of techniques of resale of capacity over leased lines to monopolistic operators, simultaneously in Europe and Japan, which announces, more surely than policies decided in high places, the end of public monopolies, (3) the technical and economic revolution of cordless communications, assisted by the progress of space communications, particularly the development of numerization and compression techniques, and the projects for implementation of satellite channels in low orbit, (4) the forthcoming statutory changes in international operating cooperatives like INTELSAT or INMARSAT, whose privatization is under way, (5) the appearance of initial international investment funds designed to cover the considerable financial needs of telecommunications infrastructures, in times of scarce resources and severe budgetary tensions...

The present article, in particular, deals with such events and several others with the objective of providing for the benefit of all policy makers such as network operators or service providers, the basic trends which characterize telecommunications and space sectors, at the end of this century.

I - TECHNOLOGICAL INNOVATIONS.

Technological choices of industrialists are not only based upon economic or commercial considerations of production costs or market perspectives. They are also governed by an endogenous logic, that of the dynamics of developed and commercialized technologies. The latter is a dynamic of integration which, from rapprochements to convergence, may even lead to merger of several technologies of different origins. As for example, the opto-electronic, the recent and remarkably fertile marriage of optics and electronics. It could revolutionize telecommunications of this end of the century.

The year 1994 has witnessed heads of companies and governments giving more attention to such phenomenom of convergence of technologies in order to integrate it into their strategies or promote it especially in the telecommunications and space sectors. In the United Kingdom, a new strategy for science and technology, based upon significant work of identifying research areas with high potential for development ("technology foresight"), is in the definition phase. In Japan, the MITI has just published a white paper on the promising technologies of the end of the decade. As for the increasing forms of inter-company partnerships which have marked the telecommunications and space sectors in 1994, and to which we shall return in the following developments, it is sometimes not necessary to look for any other logic than that of the dynamics of technological development. There has not been until the initiatives of the American FCC which by authorizing baby bells to penetrate - subject to certain conditions - the televisual cable liaisons market, any consideration given to convergence between audiovisual and telecommunications. Elsewhere, in Europe and especially in France or Germany, we are beginning to manifest a definite interest in an adjustment of training systems by favouring interdisciplinarity, ie., work of research teams of various scientific and technical cultures.

II - PROFESSIONAL FAIRS.

As in previous years, professional fairs have increased, each with its quota of innovations or promises of new technological developments. It is certainly not easy to select one or two over the rest.

The CeBIT of Hanover, set up in the Spring of 1994, has kept its habitual promises by delivering its batch of innovations. Curiously enough, they are found in the field of convergence between technologies in current use. Nokia and Siemens have for the first time presented two cards of reduced format (PCMCIA) adopted by all constructors of microcomputers. They enable transmission of data on the GSM telephony network from a notebook. Toshiba, Nokia and Lotus Notes have together set up a mail service on the GSM network. The efforts of network operators for high transmission speeds, especially in Germany, will oppurtunely support their initiatives in a manner all the more durable since the number of subscribers is regularly increasing (1.2 million in Germany). Success of the radiotelephone which is confirmed elsewhere also supports the efforts of constructors of portables. Nokia, Ericson and Motorola are close to the mythical limit of 200 grammes. Siemens' Notephone combines the potentialities of the traditional telephone handset with Apple's Newton. Alcatel combines sound and image by presenting its visiophone connected to the Numerics network. Cordless data networks continue to benefit from the progress of two technologies : infra red and radio. The development of the former is at present supported by the creation of an international association of approximately 50 companies, the Infrared Data Association. One of the most noteworthy innovations was the experimentation for the first time in Europe of a video conference service between Stockholm and Hanover (1000 kilometers), called IsoEnet, from two MD110-BPN commutators.

ATM technology is as yet experimental. The standard has not completely stabilized and as a result the adoption of a norm will still take several years. The expected development of multimedia will most likely offer it the perspective of a promising market as well as the simplification which it represents for production of terminals. In any event, alliances are already taking place between industrialists of commutation and interconnection. After the operations conducted by CISCO at the end of 1993 (acquisition of Crescendo, agreement with Alcatel), 1994 has seen the conclusion of negotiations between Foresystem and NetEdge Systems, LightStream and Tellalbs and GeneralDatacomm and Lannet.

Simultaneously, on the other side of the Atlantic, the Summer Consumer Electronic Show of Chicago has been distinguished by the coming together of entertainment and electronic industries in the United States for the conquest of the promising market of interactive leisure activities. A new term, a sign of the times, has made its appearance into telecommunications and space vocabulary, that of "Siliwood" which is a product of the contraction of two mythical words "Silicon" (from Silicon Valley) and "Hollywood". In addition to "Siliwood", the Summer Consumer Electronic Show has witnessed a merciless battle develop between producers of consuls and computers, a battle which is all the more fierce as technologies converge and their prices move closer together.

In this sector as in others, the time is right for convergence of the various information and communication technologies.

III - MOBILE TELECOMMUNICATIONS.

Convergence of technologies is at the origin of the success of mobile telecommunications. In this regard, 1994 witnessed the confirmation of considerable growth perspectives in the data transmission sector thanks to mobile equipment. Considering the results on record and the possibility of development of the market, today valued at 100 million dollars, an average annual growth of 72% is expected between now and 1998 which will represent a market share of 1.5 billion dollars on such date, ie., 6% of the global mobile communications market (data transmission networks, analogic or GSM mobile telephone networks, paging or transmission by satellite). The convergence of several factors explains such expected development : the progressive liberalization of markets, particularly in Europe, new possibilities offered by operation of cellular networks, generalization of portable computers, the appearance of new forms of activity (teleworking). Such prospects in Europe contrast with those in Japan where only 2% of Japanese use the radiotelephone. The cost of same still remains relatively high. Real growth in the

Japanese market, which has not taken place in 1994, is expected in 1995, all the more since a deregulation movement is underway which should result in the possibility of a choice between four cellular radiotelephone operators, four personal portable telephone operators, twenty manufacturers of appliances and two transmission frequencies ranging from 800 MHz to 1.5 GHz. On the other hand, the auction sale of several frequencies available in the United States will have marked a world first, and will most likely be copied in Europe in the near future, by enabling provision of a lasting solution to problems of hertzian spectrum management

Interesting development perspectives of mobile telecommunications are at the origin of large scale strategic moves in the United States. The rupture of negotiations between Sprint, Bell Atlantic and Nynex put an end to the project of creating the most important cellular telephone company in the United States. Instead, Sprint seems to have given its preference to pending discussions with TCI, Comcast, Cox Enterprises and Continental Cablevision with a view to launching a national cordless telecommunications service. At the same time, the FCC gave its approval to the merger of ATT and Mac Caw Cellular which resulted in the provision of some 400 cellular telephone networks to the leading American operator of long distance traffic.

In general, the telecommunications market has always been very promising. According to the UIT, it was worth 535 billion dollars in 1994, 78% for services and 22% for equipment and with annual revenue starting from 8 and 9% for services and telecommunications respectively. NTT, the Japanese operator, remains the leading service supplier with revenue in excess of 51 billion dollars. After it, the grading of the top ten is as follows : ATT (39.5 billion dollars), DBP Telekom (34.5 billion dollars), British Telecom (23.3 billion dollars), France Telecom (23.1 billion dollars), SIP (17.5 billion dollars), BellSouth (15.3 billion dollars), Nynex (13.1 billion dollars), GTE (12.6 billion dollars) and Bell Atlantic (12 billion dollars). Furthermore, in 1994 the UIT estimates the number of telephone lines wordwide at 575 million with 28 million new installations.

The year 1994 has finally highlighted the development of enterprise networks (essentially virtual private networks) as opposed to those of the large operators, particularly in the United States. Accordingly, we should note the appearance followed by the confirmed development throughout 1994 of resale services of capacities over leased lines at the initiative of American long distance operators, especially ATT and ACC Corp. This development was notably assisted by the conclusion of a transatlantic agreement between the American and British governments, an agreement which demonstrates the existence of a new market, ie., that of multiform services, from Call Back to more sophisticated services designed for universities or research laboratories.

IV - SOCIAL TRANSFORMATIONS.

Telecommunications technologies like space technologies continue to shape a new society. Their convergence is a factor of social transformation. Although, the labour market's prospects retain everyone's attention in an economic climate in the process of improving after many years of recession, everyone is conscious of the fact that neither the new found economic growth nor the reduction of working hours nor the development of service jobs nor the launching of ambitious civil engineering projects, albeit absolutely necessary, will be sufficient to give rise to full employment. The discussions raised by the connection between employment figures and information technologies has ceased to focus upon the concept of technologies which destroy jobs. It is now clear that development of new information and communication technologies may create new jobs and even promote a new approach to the unemployment problem. In an educated society, all information becomes a source of richness and income for those who create it, just as for those who manipulate it, transport it and store it. Should it not be the same for those who receive it ? In other words, doesn't the training, nowadays indispensable, of qualified employees justify spending money in the area of in-house training ? Permanent training would thus be rehabilitated and with it the status of employee whose competences have been dulled and no longer correspond to the needs of his enterprise. The above implies a review of the initial as well as permanent forms and methods of training.

Teleworking is making progress everywhere, not only in the statistics, but also in everyday life. From an official report completed at the request of the French Minister of the Interior (Breton Report), it notably emerges, that 16,000 people are already using teleworking in France. They were hardly a thousand in number up to 1992, and will most likely increase to between 300,000 and 500,000 between now and the end of this century. Such figures obviously pale in comparison with those presented by other countries, especially the United States where one employee out of five is a teleworker. This study has not only the merit of identifying for the first time in France the statistical importance of teleworking. It also demonstrates the considerable changes which teleworking has effected in the commercial policies of enterprises which, in turn, benefit from an improved penetration of the market, in the forms of organization and production, the costs of which it has noticeably reduced, in the national planning policies in relation to which it favours local economy, and in work practices in relation to which it calls for new regulations.

However, new information technologies continue to give rise to studies with surprising if not contradictory results. It emerges from an inquiry completed by the American Software Association on interactive applications in the United States that more than half of the American population is aware of the existence of interactive services but few amongst them really know how they can benefit from such services. Nonetheless, such services have 6 million subscribers in the United States, a figure which should double between now and 1998. In 1995, according to the same source, 17 million multimedia PC's should be sold in the United States, ie., triple the figure in 1993. Results of such inquiry may a posteriori lend credence to the analysis developed by Howard Rheingold in a recently published book entitled "The Virtual Community : Homesteading on the Electronic Frontier" (Addison Wesley, 1994, p. 325). It effectively demonstrates how the adept of Cyberspace are not attracted by perfecting techniques or in the increase in services offered to them. They are looking for new forms of more convivial relations in a society undergoing dehuamanisation. This is another way of understanding the expression "Global Village" often employed in the world of information technologies.

One thing is sure, the International Labour Organization confirmed it in a report completed from studies conducted in nineteen industrialized countries, that employees are victims of intrusions on their private life as a result of wide scale use of information technologies in their workpl a ces.80% of people in the United States who work in the banking, insurance and telecommunications sectors are complaining about this. Intrusions involve control of their productivity, implementation of a video surveillance system, identification of calls from the workplace, generalization of tapping appliances and the obligatory wearing of magnetic badges. This is how, whilst claiming to liberate employees, information technologies have given rise to new constraints.

V - INTERNATIONAL AGREEMENTS.

Convergence of technologies and, beyond that, of the industrial or commercial operators operating them explains why telecommunications, including space telecommunications, are now at the center of large scale international negotiations. From this point of view, 1994 will remain that of the signature of the final Agreement of the Uruguay Round. This document of over 12,000 pages put an end to almost 10 years of difficult discussions whose outcome was uncertain in that negotiations continued until the final hour, especially in the area of audiovisual communications. It also marks the end of GATT in the sense of an international organization set up in the post-war years to ensure respect for the General Agreement of 1948. A new organization has been called upon to replace it, ie., the World Trade Organization, entrusted with its own powers and an original mechanism for settling disputes between Member States. Telecommunications are not concerned except to the extent that international trade rules are henceforth extended to services within the framework of the agreement on services entitled "GATS". The latter report also contains the necessary provisions to ensure immediate access to telecommunications networks for service suppliers from each signatory State. However, as everyone knows, telecommunications were subject to special treatment which led to the entrusting of a work group created for this purpose with the task of defining its legal international trade rules. These discussions are scheduled to close by 30 April 1996. In the telecommunications field, signature of the Marrakech agreement marks a starting rather than a destination point.

On the other hand, we must surely welcome with relief the conclusion of a telecommunications agreement between the United States and Japan. It averts the risk of a trade war between the two most powerful economies in today's world. After nine years and three fruitless agreements, this new arrangement seems to provide increased access to American technology on the basis of equality of opportunity with its Japanese partners and, in particular, for Motorola which is in partnership with the Japanese company IDO (Nippon Idou Tsushin), for the radiotelephone market of Tokyo-Nagoya.

The prospect of an easing-off in current trade conflicts is manifesting itself ; and one can only rejoice, even if lawyers seem to naively believe that a single agreement would immediately put an end to the trade war and keep it at bay for a period of time.

Although, strictly speaking, it does not concern telecommunications or even space, the United States and Europe have agreed to begin by opening their public markets within the framework of two agreements made alongside the Marrakech negotiations, the agreement on public markets entitled "AGP" and a memorandum of agreement in the form of an exchange of letters between Europe and the United States.

VI - INDUSTRIAL ALLIANCES.

After two years of manoeuvres, the year 1994 has witnessed the consolidation of three transcontinental blocks which should dominate the international communications field over the forthcoming years. The first brings together the British operator, British Telecom, and the American operator MCI. The participation of the first in the capital of the second (20%) and, especially, the creation of a joint subsidiary, Concert, for international users sealed the first of such recent major alliances, presenting a cumulated turnover of 180 billion Francs, representing 4,270 million telephone minutes in 1994. The second group unites Deutsche Telekom and France Télécom whose gradual merger of activities has progressed in 1994, despite postponement of privatization of the French operator to most likely after the Presidential elections in the Spring of 1995, and the American operator Sprint, the first two acquiring a 20% participation in the capital of the third. The joint provision of services destined for international clients and the setting-up by the three operators of a joint subsidiary in international services which is provisionally called "Global Partnership" represents the first stage in the course of a closer rapprochement provided that the American and European authorities give the green light in relation to competition rules in force on both sides of the Atlantic. The group already has a turnover of over 380 billion Francs which corresponds to some 7,480 million minutes of telephonic traffic in 1994. The third and most recent alliance is that of the American giant ATT with the European operator Unisource which already regroups PTT Nederland, Swiss PTT and Telia. KDD and Singapore Telecom can also be counted within the framework of an already very active partnership within the Worldpartners network. This third group represents a cumulated turnover of 300 billion Francs and some 11,250 million telephonic traffic minutes in 1994.

VII - INDUSTRIAL STRATEGIES.

In addition to the above-mentioned alliances, 1994 has witnessed the development of industrial strategies of international implantation by the leading operators in such a way as to enable them make their mark on the world market and occupy the field in opposition to their direct competitors. It is especially true of European operators. During the course of 1994, the British operator, British Telecom, concluded a strategic cooperation agreement with Scandinavian telecommunications operations (Tele Danmark, Televerket and Telecom Finland) pursuant to which the Scandinavian partners undertook to offer their clients the services of Concert, the alliance concluded between BT and the American MCI. The Spanish operator, Telefonica, pursued its South American implantation strategy by consolidating its position in Argentina and affirming its claims to the Bolivian market. After a year of waiting, it also received authorisation from the American Federal Communication Commission (FCC) to participate through its subsidiary Telefonica de Larga Distancia, created in Porto Rico, in the submarine cable systems and to operate between the Bahamas and Holland. TRT, which is a subsidiary of Philips, established itself in the Indian Ocean notably by being engaged by the Malgache government for the supply and installation of a turnkey system of rural telephony. Deutsche Telecom pursued its implantation strategy in Eastern European countries by declaring its interest in a share of the privatization of telecommunications in the Czech Republic. Unisource was authorised to extend its services in France. France Telecom invested in Russia by acquiring a share in the Moscovite consortium MTS, which operates the GSM telephone network in the Russian capital, and, in Sweden by inaugurating new services of its Nordic subsidiary, FT Network Services.

Licences for the second or third mobile telephone network were eventually awarded in France, Spain and Belgium at the end of a long and uncertain decision-making process and of bitterly disputed candidature battles. They were marked by the arrival of new entrants onto the telecommunications market or the confirmation of industrial ambitions in this sector of activity by operators coming from other sectors like the Bouygues group in France. On the other hand, the strategies of traditional operators like Alcatel were considerably limited by a series of legal proceedings taken against their high ranking directors, without, however, the complete disappearance of such strategies as illustrated particularly by the alliance concluded between Alcatel and Pirelli in view of the privatization of STET, the holding of Italian telecommunications, or the purchase of STC Submarine by Cable Alcatel which was made official in February 1994 by the British authorities. The Société Générale of Belgium acquired a share of Sagem's capital through the holding COFICEM. On the sectorial level, we witnessed during the course of 1994, technological convergence obliging, a redeployment of telecommunications operators in the audiovisual communication sector.

VIII - REGIONAL POLICIES.

During the course of 1994, liberalisation of telecommunications and space sectors has progressed despite difficulties and disagreements. It is true in the United States where the Clinton Administration's project of regulatory reform finally gave way, faced with of a coalition of opposing interests, shortly before the Congress elections. However, it is also true in Europe where one witnessed the emergence of a disagreement over doctrines between the Member States of the Corfu Summit in July and during the meeting of the Council of Ministers of Telecommunications of the twelve Member States in October. In Europe, it was confirmed that the date of 1 January 1998 should be that of the complete liberalisation of services and telecommunications infrastructures. However, such undertaking has not succeeded in completely dissimulating the numerous confrontations between States and the Commission and between States themselves. Examination of a draft Directive concerning liberalisation of alternative infrastructures provided an illustration of the seemingly profound divisions between States in favour of rapid deregulation, which would notably permit cable television networks to provide telecommunications services (vocal telephony excepted), and those which remain supporters of liberalisation in stages in accordance with the pre-established calendar and results of current studies on the effects of liberalisation. Publication by the Commission of a new Green Paper containing a programme of actions to be undertaken and a timetable for the forthcoming years has most likely largely contributed to dissipating possible misunderstandings and reassuring any misgivings.

Moreover, the examination at the Corfu Summit of the proposals of an industrial group co-ordinated by the European Commissioner Bangemann, at the request of the Council of Europe, represented a growing momentum in European construction, in that a plan for the construction of a particularly ambitious system of electronic highways was adopted thereat.

Attentive observers of the case law of the Court of Justice of the European Communities took note of, more or less satisfactorily according to their doctrine of adherence, the decision rendered on 27 April 1994 in the Commune of Almelo case. This decision concerned the electrical energy sector and not the telecommunications sector, but it seemed to mark a reversal of opinion in the Court's case law, which was particularly favourable to the Commission during the previous years, especially in the policy of dismantling monopolies, and which appears at present to set limits thereto. In its decision, the Court recalls that the restrictions on competition inherent to monopolies may be justified by the completion of a public service mission. It thus extends the arguments which it presented in a previous decision of 19 May 1993, entitled the Corbeau decision.

IX - ECONOMIC PRIVATIZATIONS.

During the course of 1994, privatization of public telecommunications operators slowed down without however totally ceasing to be news. Convergence of technologies was most likely the cause of it in that it provoked de facto privatization of the telecommunications sector, by the arrival of new entrants from the private sector or from increasing pressures in favour of opening up hitherto monopolistic activities to competition. Privatization of the French operator was definitively postponed as a result of trade union opposition and ... electoral concerns, despite the fact that the French operator accumulated concrete successes in exports to the point of being distinguished by the Swedish Academy and that President Roulet specified the plans for the privatization of the French public operator, upon the request of its supervisory authorities. Agreement was thus reached on a privatization scheme which would lead to the creation of a company whose capital is controlled by the State in place of the existing "établissement public" whilst maintaining the status of its personnel. Privatization of its German homologue has entered into a very active phase but has given rise to serious political opposition in spite of significant preparatory work, which might compromise its smooth development. Privatization of the Italian operator was initiated in good conditions which should quickly lead observers to familiarise themselves with a new name, that of Telecom Italia, the symbol of a long awaited renewal of telecommunications in the European peninsula. If the privatization of OTE, the Greek operator, was provisionally postponed, it is scheduled to be pursued once economic conditions are favourable.

It is true that for all such public enterprises in the course of privatization, the example of the former State-owned Japanese company, NTT, is hardly encouraging. During 1994, NTT had to face, with varying degrees of success, serious competition from its three competitors especially in relation to telephone rates. Its profits before tax were reduced by 62 billion yens by comparision to its forecasts, whilst those of its three competitors increased by 11 to 21%.

Paradoxically, during 1994 the debate on privatization took place on another plane, that of the international cooperatives, INTELSAT, EUTELSAT and INMARSAT.

Privatization of INTELSAT was on the agenda of the last meeting of the Member States of INTELSAT which occurred at the end of October 1994 in Por Lamar in Portugal. The idea, vigorously defended by the Anglo-Saxons, is that the organisation shall evolve into a private company whose capital shall thus become accessible to institutional investors, bankers and industrialists of the sector. Such idea is not shared by all the remaining Member States of INTELSAT. However, it constitutes a major strategic stake.

The issue of INTELSAT's status and its possible evolution is raised in order to better adapt it to its missions and in a general manner to market changes. Need we recall that INTELSAT is today the leading worldwide satellite organisation. Its turnover for 1994 was 3.5 billion Francs and it currently controls a satellite system of more than 70 launched satellites. It is the leading supplier of satellite commercial telecommunications services in the world. It represents on its own a third of the world market as against a third for the specialists of North American domestic needs and a third for the special domestic needs of other countries, such as, the "TELECOM" satellite system for France. INTELSAT represents today over half of the international telephony traffic with 200,000 permanent telephonic voices, the quasi-totality of transoceanic television broadcasting and numerous national services for approximately 40 countries. INTELSAT has 134 members in its organisation. Created in 1964 by only 11 countries, INTELSAT made comfortable profits in 1994, which will most likely enable it to redistribute during the course of 1995 the equivalent of some 18% of the amount of its capital, as provided for in its articles of association.

One may add that the prices offered by INTELSAT are amongst the lowest in the world, in the region of one dollar per day to lease an international telephone line. This demonstrates that the international organisation would be particularly favourable to small States which are not in a position to offer themselves the luxury of a satellite system and could furthermore benefit through INTELSAT from free training or missions, for example, in the context of the UNO.

In such context, the privatization of INTELSAT may appear surprising. However, it is part of the seemingly inexorable changes of the framework of satellite communications and it surely constitutes one of the strategic stakes which international organisations will have to face in the forthcoming years, if not, in the forthcoming months. Has the satellite organisation INMARSAT not recently chosen a more commercial structure for the specific satellite project in low orbit "P 21" ?

X - EMERGING MARKETS [1] .

The rise of Asia in telecommunications and space sectors was confirmed during 1994 to the extent of extending beyond the geographical framework of the Far East. The Korean company, Samsung, confirmed its desire to expand in Europe by revealing new ambitions in the field of mobile telephony, after having heavily invested in consumer electronics, household appliances and microcomputers. Meanwhile, the prospects of growth of the Chinese market, in the region of 59% per year, continued to fuel jealousies in the West, jealousies of which the Chinese government is well aware. In the course of 1994, the Chinese government decided to authorise an operator which is a competitor of the Ministry of Post and Telecommunications to operate infrastructures of national telecommunications (local and long distance communications). By so doing, they probably benefitted from the competition between European and American groups in this area which constitutes a new world Eldorado, by regularly raising their requirements to the limits of what would be acceptable. This behaviour is perhaps not alien to the decision taken at the end of 1994 to refuse China the benefit of the GATT provisions. As confirmed by all the available studies ic., those of CNUCED (Conférences des Nations Unies pour le Commerce et le Développement), BAD (Banque Asiatique de Développement) it remains that with annual growth rates of 5% to 9% depending on the country, Asia escaped from the world recession, that it possesses a marvellous potential for development and that foreign investments are henceforth largely orientated towards Asia to the detriment of Central European countries and especially developing countries.

For the latter countries, the report on development of the International Union of Telecommunications confirmed that a gulf is forming between developed and developing countries. It contains incredible figures. The annual duration of international calls in 1993 was 7 minutes per each inhabitant of the planet. Such average duration conceals considerable disparities from nation to nation. In Luxembourg, for example, it is 7 hours based on the number of inhabitants. The law of proportions does not explain everything. On this basis, Luxembourg is the window of OECD countries which have produced and cashed in 1993 three quarters of the proceeds and income from worldwide telecommunications services, a figure which exceeds 2000 billion Francs. In African countries, the average figure, established in accordance with the same statistical methods, is 7 seconds. 7 seconds only which covers incredible differences between the inhabitants of Djibouti (10 minutes), Lesotho (11 minutes), Swaziland (14 minutes) or Bostwana (17 minutes) and those of

Sierra Leone (3 seconds), Burkina Fasso or Ethiopia (2 seconds) and Tanzania (1 second). A total of 7 seconds for countries which along with several States of the Pacific and States of the former Soviet Union benefit from only 5% of the revenue generated in 1993 by telecommunications activities worldwide.

With almost ten telephone lines per one hundred inhabitants, Saudi Arabia is not, strictly speaking, a country under-equipped with telecommunications networks. In fact, its level of equipment places it at a respectable world average. However, during the course of 1994, ATT concluded there one of the most significant equipment contracts signed over the last number of years. This is because Saudi Arabia represents along with several other countries a market with very strong development potential. Moreover, if one wishes to identify such markets, it is sufficient to follow the international strategy of the American operator. It is present in South-East Asia, notably in China where it has doubled its turnover in 1994. It is also very active in India, Korea, Taiwan, Hong Kong and Singapore. It can also be located in Latin America, in particular, in Argentina where it participates, for example, in a consortium headed by GTE for operation of a cellular telephone license.

XI - MULTIMEDIA SERVICES.

1994 has been that of multimedia services, not because they still occupy centre stage, but rather because we have witnessed such new market affirm itself and manifest particularly interesting development perspectives. Viacom thus appears as the company of the year by completing after the purchase of Block Buster Entertainment, the purchase of Paramount Communications in July 1994. It must be acknowledged that its President Sumner Redstone is probably one of the most combative businessmen in the United States today. However, his success does not allow us to forget other initiatives, especially in Europe, geared towards development via cable networks or hertzian networks, including satellite networks of interactive television services (principally "pay per view") : those of Canal Plus or Lyonnaise des Eaux-Dumez in France, of CLT in Luxembourg, of Philips Electronics in Holland or of BSkyB in Great Britain. The launching of new satellites dedicated to the numeric system (Hotbird 1 and Hotbird 2 for EUTELSAT, Astra 1E and 1F for SES, Telecom 2C and 2D for France Telecom) in 1995, just like the agreement on numerical norms entered in 1994 by the European industrialists of television (Digital Video Broadcasting), should logically favour the development of new initiatives.

In the same spirit, telepurchase services kindled jealousies, especially in the United States with the episodes of the merger between the television channels CBS and QVC. One was hardly surprised to learn that Walt Disney Company launched itself in the adventure of tomorrow's television by associating with three "baby bells" : the same applies to the decision of France Telecom to participate in the capital of General Magic and in so doing join Apple, Sony, Toshiba, Fujitsu, NTT, ATT and Motorola.

Success of the multimedia in Europe more or less depends on the capacity of lawyers to conceive new rules principally for the protection of intellectual property rights. We can probably hope for a contribution from the Commission of the European Union in favour of the adoption of a uniform legal framework for all Member States which would be particularly welcome.

In the multimedia sector as in others, Asia does not trail behind. The launch in 1995 of the satellite Apstar 2 will offer to on-site operators, especially in Hong Kong, Singapore and Taiwan, a market which groups together the equivalent of two-thirds of the planet's inhabitants for their multimedia services. The Turner group has already declared its intentions by reserving 10 million dollars for the launching of its TNT and Cartoon programmes and by so doing is endeavouring to counter the influence of the Murdoch group with StarTV.

XII - PERIPHERAL SECTORS.

The convergence of technologies has similarly touched upon the peripheral sector of computer science. It is no accident that 1994 will be renowned as the year of the biggest acquisition ever completed to date in the software field, that of Intuit by Microsoft (1.5 billion dollars). Bill Gates' firm (with a turnover of more than 6 billion dollars and a profit of 1 billion dollars in 1994) acquired in this way a front line position in one of the most promising sectors of microcomputing, that of management of personal finances. Intuit's software package, Quicken, is today installed in almost 6 million PC's which represents nearly half of the market.

Such acquisition has a dual significance :

- It falls within the framework of ambitious manoeuvres of various operators to master a developing market, that of telepayment. Need we recall that 1994 was also the year of the launching of several direct banking projects capable of providing their services around the clock by telephone, Minitel or computer.

- It corresponds to a wave of large-scale restructuring in the computer sector which affects the creators of competing software Lotus or Borland, but also manufacturers such as IBM or Apple. It should be pointed out here that the rapprochement of IBM and Apple which strenghtened in 1994 notably belongs to the logic of competition generated between Microsoft and the Californian manufacturer by their respective products : how else could Apple counter the all powerful world software leader whose star product Windows controls 80% of the market in 1995 ?

1994 was similarly marked by numerous court battles in the computer sector which made the "giants of yesterday" "bite the dust" or "eat their hats" to adopt some of the expressions heard or read in the international press :

- the unambiguous victory of the manufacturer Compaq, the leading worldwide manufacturer of computers before IBM and Apple, which forced IBM to embark on a suicidal policy of lowering prices of its micro-computers by up to 27%,
- the change in strategy of the Cuppertino firm (Apple), with its decision to open its Power PC technology to other manufacturers and, consequently, to accept the cloning of its new models,
- the difficulties of Intel whose Pentium microprocessor showed weaknesses.

Even if such events, which are essentially American, interest world economics, they should not lead us to overlook others which might well dominate tomorrow's news. During the course of 1994, the European micro-computers market re-established a growth rate close to that of the eighties with 2.58 million computers delivered and an increase of 17.8% in the second quarter which was twice that recorded during the first quarter. The Japanese micro-computer market which was for a long time protected, is in the course of opening up under pressure from foreign manufacturers. It is important to note that Apple or IBM have seen their respective share of the Japanese market regularly increase over the last three years, as have sales of the 486 Intel microprocessor (over 27% in the course of 1994). However, it is not necessarily bad news for Japan. A more open and consequently more competitive internal market will incite Japanese manufacturers to leave their reserve and engage in more agressive strategies on the world market. They have the means therefor. Must we recall that the manufacturers NEC and Toshiba are listed amongst the top ten worldwide and that Sega and Nintendo today dominate the world market in software games.

In such conditions what do other events represent : the launching by Aiwa of TV-PC, at the same time a television set and IBM compatible microcomputer fitted with a CD-ROM reader, an operation system Window, a stereo tuner and a central memory of 8 megabits ; the setting in motion of the super computer CS2 which henceforth equips the CERN (European Laboratory for the physics of particles) ; or the decision of the Taiwainese manufacturer ACER to open up the capital of all its subsidiaries ? Nothing in appearance, but a lot in reality. The merger of computer and television set is the technology of tomorrow which will most likely witness the combination of computer, telecommunications and audiovisual communication techniques. The super calculator CS2 was created within the framework of the ESPRIT programme of the European Union, which demonstrates what Europeans are capable of once they join their resources. The decision of the founder and President of the ACER, Stan Shih, probably marks the end of an essentially family style concept of capitalism in Eastern and Southern Asia and its entry into a new era which is closer to Western management methods.

XIII - SPACE ACTIVITIES [2].

Finally, in the space sector, 1994 confirmed the hopes which were placed in it. By announcing the forthcoming launching of the television satellite HotBird 1, EUTELSAT clearly demonstrated its desire to propose a credible alternative to the Astra satellite system of the private company SES. HotBird 1 is equipped with 16 canals and will be placed at 13 degrees in Eastern orbit. It will thus join the EUTELSAT II-F1 satellite which already broadcasts 16 television channels and 12 radio stations, representing more than 47 million households serviced. However, the announcement of a tender for a third generation of telecommunications satellites "EUTELSAT III", whose selected supplier should be made known in the course of the first semester of 1995 and the expected launching of the Spot 5 programme (fifth generation of French teledetection satellites) brought to world industry development perspectives which it imperatively requires after many years of recession.

During the course of 1994, the French company Aérospatiale achieved a dual revolution, both internal (by organising its activities in operational centres at the initiative of its new MD) and external by accepting the principle of a rapprochement and probable merger with DASA, which the American Space Systems Loral views favourably. However, beyond such operation, we shall certainly recall the announcement of "stock swap" of Lockheed and Martin Marietta which provoked a spectacular increase in the share price of both American firms and will certainly prove to be one of the most magnificent operations of the year. The gulf thus opens up between two different concepts of restructuring, which are imperatively necessary in the world space sector, that of the industrial alliance which is in vogue in Europe and that of the merger which is preferred across the Atlantic.

Finally, the year 1994 has been that of count down. If all the projects announced in the United States are completed in less than ten years, more than a thousand satellites will gravitate in low orbit around the earth. The deceptive and promising aspects of such figure are immediately perceptible. Deceptive in the current state of known launching capacities, but promising for new States entering into such a considerable market : India, Russia or China. China moreover has confirmed its competences by placing an Australian satellite (Optus B3) in orbit by means of a Longue Marche 2E launcher. Should we infer therefrom that we are living through the end of an era, that of absolute world domination by Western operators, in particular Arianespace ? Such would probably be a premature conclusion. Arianespace has continued in the path of success despite a number of disappointments, in particular, the failure of Flight 70 transporting the PanAmSat 3 satellite, and, one can only look forward to the setting-up of the Ariane 5 launcher.

The space financing and insurance industries will most likely be in high demand during 1995 as they were in 1994. Their future is thus assured - but did we doubt it ? - , as long as they keep developing new legal techniques required to deal with the launching of more and more numerous and costly satellites.

XIV - FINANCIAL SERVICES.

In a period of scarce resources and very high budgetary tensions, financing entities were probably delighted that 1994 was the year which marked the successful launching of several specialised investment funds in telecommunications : Emerging Market Telecommunications Fund (BEA Associates, 182 million dollars), Gabelli Global Telecommunications Fund (Gabelli Funds, 50 million dollars), Gartmore Global Utilities Fund (Gartmore Investment, 20 million dollars), GT Global Telecommunications Fund (GT Capital Management, 2.1 billion dollars), GT Telecommunications Fund (GT Capital Management, 70 million dollars), Kleinwort European Privatization Investment Trust (Kleinwort Benson International Investment, 500 million pounds), Montgomery Global Communications Fund (Montgomery Asset Management, 230 million dollars), Telecom Invest (BNP, 140 million francs). Market development, privatization measures, deregulation policies, infrastructure modernisation programmes : all the conditions for success of such funds are thus present to meet the financial requirements which, albeit considerable, are accompanied by particularly interesting perspectives of return on investment. According to Salomon Brothers, 60 billion shares should enter the market over the next three years. Privatization of the Dutch operator should yield 3 billion dollars for its public shareholders and that of the Danish operator 2 billion dollars. They should reinvest in the modernisation of telecommunications networks.

Such was the year 1994 with its procession of conspicuous events and hidden trends. "Sous la paille des mots, retrouve le grain des choses" ... in order to grind it, as it has just been done. The dough of information and communication technologies shall thus rise to form bread (the famous French baguette !) to be consumed in the months or years to come. Will 1995 be that of sampling (in the good as in the bad sense of the term) ? It is probably too early to say. For the time being, we must allow the leaven of the logic of convergence which has traversed 1994, produce its effects.

REFERENCES

1. See : the excellent "Telecommunications Laws in Latin America", Baker & Mc Kenzie, 1994
2. See : Stewart White, Stephen Bate & Timothy Johnson, "Satellite Communications in Europe. Law and Regulation", Longman, London, 1994, a reference book on all the important aspects of the law and regulation of satellite communications in Europe.

INTERNATIONAL TELECOMMUNICATIONS SOCIETY

CALL FOR PAPERS

For the ITS European Regional Conference in Juan Les Pins near Nice, France, September 1-3, 1995 (in conjunction with the 22nd annual EARIE Conference)

ITS is soliciting research papers for presentation in the following areas

Telecommunications in the Mediterranean
Regulation / Privatization / Liberalization
Changes in the Structure of the Telecommunications Industry
Industrial Policy Issues
Changes in the Structure of the Telecommunications Industry
Telecommunication Services
Convergence among Telecommunications, Broadcasting and Computing
Other Sessions

Papers should be based on current research on theoretical and/or empirical issues relevant to the making of telecommunications policy. Submission should consist of **3 copies** of **a one-page abstract** and **the paper of no more than 25 pages** (typewritten, double-spaced). Submissions must be received **no later than June 30, 1995. Please address all contributions to** the **ITS European Office, Dr. Jürgen Müller c/o DIW, Königin-Luise-Str.5, D-14195 Berlin, tel. +49 30 89789-662, fax +49 30 89789 200, email: jmueller @ fhw.fu-berlin.de.**

- *II* -

HOT TOPICS

AN OVERVIEW OF THE APPLICATION OF EUROPEAN AND NATIONAL COMPETITION LAW IN THE FIELD OF TELECOMMUNICATIONS

MICHAEL SCHMITTMANN*, INGE DE VRIES**

* Michael Schmittmann, Rechtsanwalt, Heuking Kühn Kunz Wojtek, Elisabethstrasse, D-40217 Düsseldorf Germany ; ** Inge de Vries, Advocaat in The Hague

INTRODUCTION

In all Member States of the European Union, the telecommunications sector has traditionally been the exclusive province of state or public organisations [1]. Member States acknowledged that the inflexible state administration - in contrast to the developments in the USA and in Asia - considerably harmed this future-orientated and economically important sector. Independent of the parallel developments in the Member States, the policy of the European Union resulting from the Green Paper on the Development of the Common Market for Telecommunications Services and Equipment of 30th June 1987 acted as a stimulus [2]. Since then, markets for services and terminal equipment have undergone a fundamental change. European standardisation and networks - also including European states outside the Union - are required and are being created. Although exclusive rights for national telecommunications organisations were originally recognised, such rights are being progressively cut back [3].

Today, the legal environment of the telecommunications sector looks completely different from 1987, when the Green Paper first appeared. The dominance of public regulations is being replaced by a growing importance of the general competition law, partly filling the gaps resulting from European and national deregulation, privatisation and liberalisation. The purpose of this article is to introduce the major legal changes and to give an overview of the application of competition law as practised by the authorities and courts within the last few years. The main decisions and the literature referred to in the article are listed in the appendix. The emphasis will be placed on the European level rather than on the national level. However, examining the parallel application of the national German competition rules will show how EU Member States react to the challenges mainly originating from Brussels and how the dualism of competition law systems in the area of telecommunications works at the moment.

The telecommunications sector was, until recently, dominated by government-controlled telecommunications organisations (TOs) [4]. The path to free competition is characterised by three progressive steps which were taken on a European level with a view to increasing the efficiency of the sector. These three steps can be characterised as liberalisation, harmonisation and, finally, safeguarding competition in the market created by the first two steps.

Liberalisation was a move directed against the postal monopolies in order to free the sector from their crippling stronghold. Harmonisation was needed to create a level playing field in terms of technical standards in order to facilitate market entry and effective competition between enterprises offering comparable equipment and services. In the market created on the basis of these two steps, liberalisation and harmonisation, competition must be safeguarded by the application of Community rules on dominant positions (Art. 86 EC-Treaty) and anti-competitive agreements (Art. 85 EC-Treaty).

I - LIBERALISATION

1.1. The basis of Art. 90 EC-Treaty

Liberalisation in the telecommunications sector was based on the application of a competition provision, namely, Art. 90 EC-Treaty. This article permits Member States to grant special rights to undertakings providing services of public interest. On the other hand, it ensures that these undertakings are subject to the competition rules.

Art. 90 EC-Treaty and its interpretation by the European Court of Justice (ECJ or the Court) gained particular significance in the field of telecommunications in which old TOs had been around for a long time and tolerated as static monopolies by Member States' legislation. This situation stood in sharp contrast to other economic sectors in which European liberalisation of trade in goods and services only addressed regulatory obstacles relating to specific products or services. Free trade between Member States was obtained there by enforcing the basic Community freedoms. In the telecommunications sector, however, monopolies were the principle obstacles to cross-border trade. The justification of these Member State monopolies has been examined by the Commission on the basis of competition rules.

Economic activities in the telecommunications sector have been opened to competition by Community law. Some areas can still be reserved to the TOs by Member States. This remaining reservation against competition is based on the general public interest provided for in Art. 90 (2) of the EC-Treaty.

a). Special or exclusive rights under Art. 90 (1) EC-Treaty

Art. 90 (1) EC-Treaty permits Member States to grant special or exclusive rights to undertakings [5]. If such an undertaking enters into an anti-competitive agreement or abuses its dominant position as a result of a state measure, this restriction on competition can be prohibited under Art. 90 (1) EC-Treaty in connection with Art. 85 or 86 EC-Treaty. In the telecommunications field, the European Court of Justice held, for example, that Art. 90 and 86 EC-Treaty are infringed if a Member State grants to an undertaking operating the public telecommunications network the power to set standards for telephone equipment and to check that other equipment providers meet those standards when it is itself competing with those providers on the market for that equipment [6]. The Court considered the extension of the network monopoly to the market for terminal equipment without objective necessity as another infringement of Art. 90 (1) and 86 EC-Treaty [7].

b). Exemption from competition rules under Art. 90 (2) EC-Treaty

Although the competition rules of the EC-Treaty apply to public or private enterprises, including those active in the telecommunications sector Member States cannot avoid the effect of the competition rules by adopting measures to the contrary. The only exception is contained in Art. 90 (2) EC-Treaty which provides that

"undertakings entrusted with the operation of services of general economic interest or having the character of revenue producing monopoly shall be subject

to the rules contained in this Treaty, in particular to the rules on competition, in so far as the application of such rules does not obstruct the performance, in law or in fact, of the particular tasks assigned to them." (boldface added)

Although the wording of this article is not very clear, the general purpose is to provide for a deviation from the competition rules for undertakings with the specific tasks mentioned in Art. 90 (2) EC-Treaty if compliance with these rules makes the performance of their tasks impossible. As an exemption, Art. 90 (2) EC-Treaty must be narrowly interpreted [8]. In the British Telecommunications case of 1982, the Commission implicitly accepted the provision of telecommunications systems throughout the U.K. as "the particular task assigned to them" [9]. Now, the Commission considers that the task of telecommunications undertakings covered by Art. 90 (2) EC-Treaty consists of providing and exploiting a universal network, i.e., a network having general geographical coverage, to any service provider or user upon request within a reasonable period of time [10]. In the RTT v. GB-Inno-BM case, the Court also held that a monoply for the operation of such a network would, for the time being, fall under Art. 90 (2) EC-Treaty. However, an exclusion from or restriction of competition on the market for terminal equipment was not justified on the basis of Art. 90 (2) for the operation of a network [11].

c). Measures of the Commission under Art. 90 (3) EC-Treaty

If governmental measures based on Art. 90 (1) or (2) EC-Treaty eradicate the beneficial effect of the EC rules, in particular the competition rules [12], the Commission can promulgate appropriate directives or decisions addressed to the Member States on the basis of Art. 90 (3) EC-Treaty. Member States may only intervene in the functioning of the free market mechanism if this is necessary for purposes of general interest. Only the Commission - without intervention by the Council [13]- has the right and the obligation to ensure the proper application of Art. 90 (1) and (2) EC-Treaty by measures based on Art. 90 (3) EC-Treaty. The Commission assumes from the case law of the European Court of Justice [14] that it has the exclusive competence, under the control of the Court, to decide when the exception of Art. 90 (2) EC-Treaty applies.

The Commission is also of the opinion that national authorities can determine that this exception does not apply, i.e., the competition rules are applicable, if they find that these rules do not obstruct the performance of the task of general economic interest assigned to the undertakings. Otherwise, they should suspend their decision so that the Commission can assess the case [15]. Which of the ECJ's considerations the Commission bases its opinion on is unclear though. For instance, the Court held in the Ahmed Saeed case [16] that Art. 90 (3) EC-Treaty "does not preclude the application of paragraphs (1) and (2) of that article where the Commission fails to act." This ruling and the subsequent development of the case law on Art. 90 EC-Treaty has, in the opinion of the authors, shown that the national judicial authorities can also determine that the exception applies [17].

In the field of telecommunications, the Commission has adopted two directives based on Art. 90 (3) EC-Treaty, which are to open up the telecommunications market for terminal equipment and value added services to competition. Since the financial resources for the development of networks still derived mainly from the operation of the telephone service, the Commission abstained for the time being from also opening up basic voice telephony and infrastructure operations to competition in the 1990 telecommunications services directive [18]. In the meantime the Council agreed that monopolies in these areas will end by 1 January 1998 [19]. Furthermore, in October 1994, the Commission amended the two Directives mentioned above so as to include satellite communications [20] and in December 1994 it proposed a further amendment to the 1990 Directive in order to give liberalised (especially multi-media) telecommunications services access to cable infrastructure [21].

Mobile and personal communications were also exempted from the 1990 Directive. On the basis of its Green Paper of April 1994 [22], on 23 November 1994 the Commission adopted a proposal for a Council Resolution setting priorities for future legislative action. This would include the abolishment of the remaining special and exclusive rights by 1 January 1996 [23].

1.2. Developments in Germany

a). The Postal Restructuring Act of 1989 - Postal Reform Package I

In Germany, these European measures basically affected the German postal and telecommunications organisation in existence at that time, the Deutsche Bundespost ("DBP"). Historically, all postal and telecommunications matters were exclusively dealt with by the DBP. The constitution of the DBP and the regulatory framework of telecommunications sector were substantially restructured by the Postal

Restructuring Act ("Poststrukturgesetz") of 8th June 1989 [24]. The Post Office was split into three completely separate public enterprises: postal services, banking, and telecommunications. The German public telecommunications provider, the Deutsche Bundespost TELEKOM ("DBP TELEKOM") is subject to the supervision of the Minister for the Post and Telecommunications. Its only monopolies are a network and a telephone service monopoly. In other services and terminal equipment, the DBP TELEKOM must compete with private suppliers. The same is true in the mobile radio communication and individual satellite communication sectors. Commercial relations are now governed by the private law of contract, not by public administrative law.

In the meantime, DBP TELEKOM and private competitors are fiercly competing to the benefit of customers in various areas. At least in mobile communication, Germany can be seen as one of the pacemakers in Europe.

b). The Law for the Reorganisation of the Post and Telecommunications 1994 - Postal Reform Package II

In September 1994, a second reform package has been adopted. The main feature was that the DBP TELEKOM has been transformed into a stock corporation company, called the Deutsche Telekom AG. With regard to the liberalisation of the network and voice telephony monopoly, the law does not anticipate the European-wide liberalisation currently contemplated before 1998.

1.3. Secondary Community Law

a). The Terminal Equipment Directive

In 1988, the Commission adopted a directive with the objective of introducing competition in the market for telecommunications terminal equipment (telephone sets, modems, telex terminals, data terminals, mobile telephones, etc.) [25]. The so-called Terminal Equipment Directive was based on Art. 90 (3) of the Treaty. France, supported by Italy, Belgium, Germany and Greece challenged the legal basis for this before the European Court of Justice by claiming that the Commission did not have the competence to issue the directive. On 19th march 1991, the Court in principle upheld the Commission's competence to issue a directive under Art. 90 (3) EC-Treaty in this case, but annulled certain provisions contained in the directive [26].

i) Content of the Directive in light of the court judgment of 19th March 1991

To reach effective competition in the terminal equipment market, the directive basically imposed three obligations on the Member States : Firstly, Member States had to withdraw all special or exclusive rights granted to TOs for the import, marketing, connection or bringing into service of telecommunications terminal equipment and the maintenance of such equipment [27]. The ECJ found that the Commission did not specify the term "special rights" and therefore the part of the Directive relating to special rights was annulled [28]. The Member States may only refuse telecommunications equipment not satisfying certain "essential requirements" relating to the safety of users and workers, network security, and inter-operability [29]. Moreover, the Member States may require that economic operators fulfill certain technical qualifications. These qualifications have to be established on the basis of objective, non-discriminatory and publicly available criteria [30]
Secondly, the Directive required the Member States to separate the operational and regulatory functions of the TOs. The responsibility for drawing up the technical specifications for terminal equipment, monitoring their application, and granting type approval must be entrusted to a body independent from the TO. The technical specifications and type-approval procedures used for terminal equipment, moreover, must be published [31].
Thirdly and finally, the Directive allowed customers, who had concluded a long-term leasing or maintenance contract for terminal equipment during the prior period when exclusive or special rights in this respect still existed, to terminate these contracts with a maximum notice of one year [32]. The ECJ also annulled this provision arguing that Art. 90 EC-Treaty was not the appropriate basis to address the autonomous acts of undertakings. Art. 90 EC-Treaty could only be used in relation to state measures. Therefore, the Commission will have to make its decisions pursuant to Art. 85 or 86 EC-Treaty [33].
As a result of this Directive, the biggest obstacle for competition in the terminal equipment market - a legal monopoly - was removed.

ii) Implementation in Germany

In Germany, the Directive was implemented by the

Telecommunications Installation Act of 1989. This Act provided that everyone may set up and operate type-approved terminal equipment if the conditions laid down for the proper handling of telecommunications traffic are ensured [34]. The maintenance monopoly for telex services and the exclusive distribution right in the terminal equipment market were automatically withdrawn under this act [35]. The separation of regulatory and operational tasks was introduced by the Postal Reform Package I and further defined by the Postal Reform Package II (see above, under II).
The terminal equipment must implement the essential safety requirements and qualifications requirements mentioned in the Directive [36]. The Minister for Post and Telecommunications (Bundesminister für Post und Telekommunikation) may to stipulate the procedure for the type approval of terminal equipment. Already in 1982, type approval was established in the Telecommunications Type Approval Regulation (as newly drafted in the Telekommunikationszulassungsverordnung of 22 March 1991).

b). The Services Directive

The equivalent of the Terminal Equipment Directive for telecommunications services is the Commission Directive on competition in the markets for telecommunications services [37] ("the Services Directive").

i) Content of the Directive in light of the court judgment of 17th November 1992

This Directive has also been challenged before the European Court of Justice because of the Commission's use of Art. 90 (3) EC-Treaty, this time by Spain, supported by France, Belgium, Italy, Germany, and Greece [38]. The outcome was similar to the judgment in the terminal equipment case [39]. The Commission obliged the Member States to withdraw all special and exclusive rights [40] for the supply of telecommunications services [41], other than voice telephony, and to ensure that any operator would be entitled to supply such services [42] The Member States are allowed to subject the supply of competitive services to licensing or declaration procedures which are aimed at ensuring compliance with the essential safety requirements. The procedures must be based on specific objective criteria and applied without discrimination.

Secondly, the separation of regulatory and commercial functions requires that the grant of operating licenses, the control of the type approval and mandatory specifications, the allocation of frequencies, and surveillance of usage conditions are carried out by an entity independent of the TOs [43].
The third obligation has, like in the Terminal Equipment case, been annulled by the Court. It concerned the right of users to terminate long-term contracts imposed by the TOs [44]. The Court reiterated that Art. 90 EC-Treaty conferred powers on the Commission only as regards state measures.
Although the basic voice telephony monopoly and the monopoly over telecommunications infrastructure have so far been exempted from these liberalisation measures, these will end in most countries of the EU in 1998 (see above). These liberalisation measures were highly urged by the Bangemann group. In June 1994, it submitted a report on Europe and the Global Information Society to the European Council in Corfu in which it called for full liberalisation in the telecommunications market and for a unified regulatory framework.

ii) Implementation in Germany

In Germany, the Telecommunications Installation Act in its version of 1989 and as amended by the Postal Reform Package II of 1994 is the essential basis for the regulation of the telecommunications sector. It provides for the exclusive right of the Deutsche Telekom AG to set up and operate networks and to provide voice telephony services [45]. The details of monopoly services, including the obligation to render such services without unfair discrimination, are specified by the Federal Minister in administrative rules. The telecommunications ordinance (Telekommunikationsverordnung - TKV) of June 1991 contains further conditions.
Furthermore, the law divides the services in two categories: so-called "mandatory services" and "optional services". Mandatory services are those which Deutsche Telekom AG is obliged to render in the public interest outside the monopoly which private competitors may also render. An ordinance describes the services, e.g., the publication of telephone books, etc. [46] Optional services are those which Deutsche Telekom AG may freely provide within its constitutionally defined area of competence.
The Telecommunications Installation Act provides for the private service providers that are entitled to provide telecommunications

services for third parties over fixed or switched connections to be made available by the Deutsche Telekom AG. The only obligation is to give notice in writing of the start-up of operations as well as any modifications in or the cessation of this operation to the Telecommunications Minister within one month [47].

II - HARMONISATION

In the wake of the liberalisation of access to the markets for certain telecommunications equipment manufacturers and service providers, competition was further facilitated by a number of measures aiming at the harmoniation of market conditions.

2.1. Equipment

With regard to terminal equipment, harmonisation measures occurred in two phases. Under the Council Directive on the initial stage concerning type approval for mass-produced terminal equipment [48], the Member States were required to mutually recognize conformity tests carried out in another Member State in relation to common technological specifications.

The second phase was based on a new approach to technical harmonisation and standardisation. The Council Directive on the approximation of the laws of the Member States concerning telecommunications terminal equipment, including the mutual recognition of their conformity [49], 1 defines only the essential requirements to be satisfied by all equipment if it is to be placed on the market. These requirements cover the safety of users and network personnel; electromagnetic compatibility; protection of network integrity; efficient use of the radio frequency spectrum; interworking for establishing, modifying, charging for, holding and clearing connections; and interworking of terminal equipment via the public telecommunications network [50]. The task of drawing up detailed technical standards for practical application falls upon the standardisation bodies CEN, Cenelec and ETSI [51]. The difference with the first Directive is that not only the type approval test will be recognised in another Member State, but equipment will also not be subjected to any new test or procedure.

In Germany, the essential requirements mentioned in the Council Directive are implemented by Section 4 of the Telecommunications Type Approval Regulation ("Telekommunikationszulassungsverordnung") of

22nd March 1991. If the appropriate conditions are fulfilled, the Federal Approval Office for Telecommunications shall grant approval [52]. Conformity tests carried out in another Member State in relation to common technological specifications are recognised in Germany [53]. Technical tests are generally not carried out again [54].

2.2. Services

With regard to competitive services, the Council adopted a Directive on the establishment of the internal market for telecommunications services through the implementation of an open network provision (so-called "ONP Directive") [55].

The Directive concerns the harmonisation of conditions for open and efficient access to and use of public telecommunications networks and, if applicable, public telecommunications services. It is a framework directive identifying areas for which ONP conditions now have been defined [56] (for example, leased lines [57]), public packet switched data services [58], and ISDN [59]) or still are to be defined. The basic principles are that the conditions of access must be based on objective criteria, that they are transparent and published in an appropriate manner, and that they guarantee equality of access and are non-discriminatory [60].

In Germany, the conditions for use of the network are established in the amended Telecommunications Regulation ("Verordnung über die Rahmenvorschriften für die Inanspruchnahme von Dienstleistungen des Unternehmens DBP TELEKOM") of 16th September 1992 and expressly refer to the European ONP principles [61].

III - SAFEGUARDING COMPETITION IN A COMPLEX MARKET

3.1. European Community law - Art. 85 and 86 EC-Treaty and the Merger Control Regulation

Following the basic steps of liberalisation and harmonisation, the telecommunications market can be roughly split into three major business segments:

- reserved areas still under exclusive control of TOs, such as networks and basic telephony;
- competitive services open for competition between private enterprises or TOs or both;

- equipment open for competition between private enterprises or TOs or both.

A peculiarity of the telecommunications sector is the existence of an area in which competition is excluded. This area is the exlusive domain of the TOs. The justification of such an exclusive right should be the performance of a task of general public interest. In order to guarantee the performance of this task, the Member States financially support or regulate these tasks of the TOs. At the same time, there is a field in which the TOs are active and in which they face competition by other companies. The telecommunications sector is furthermore characterised by the high level of technology involved. This can require cooperation to reach inter-connection and inter-operability, but also to share the high investments, both in operations and in the development of new products.

These aspects make the telecommunications sector particularly prone to encountering difficult questions of competition law, both in relation to Art. 85 EC-Treaty (anti-competitive agreements or concerted practices between undertakings) and Art. 86 EC-Treaty (abuse of dominant positions). Naturally, the TOs will be the major targets of scrutiny under Art. 86 EC-Treaty. They will be dominant within the fields under their exclusive control, but often also in the fields open to competition, because they might still have a de facto dominant position in these fields.

With regard to Art. 85 (1) EC-Treaty (prohibition of anti-competitive agreements), particular attention will paid to agreements between TOs and between TOs and non-TOs. Two specific restrictions on competition are cross-subsidisation from the area reserved for the TO in favour of the other party (or eventually of a joint undertaking), and discrimination of third parties.

Characteristics of the telecommunications market might justify coordination of activities between actual or potential competitors under Art. 85 (3) EC-Treaty if the following are met: such a coordination should produce economical or technical benefits which are passed on to the consumer to a fair extent; the restrictions on competition must be indispensable to obtain these benefits and in any event, competition may not be eliminated for a substantial part of the products in question.

To the extent that the TO holds a dominant position in facilities, services, and equipment markets, its conduct leading to anti-competitive agreements could also infringe Art. 86 of the Treaty.

Due to the complexity of the various issues, in 1991 the Commission published guidelines on the application of EC competition rules to the telecommunications sector (hereinafter the Guidelines [62]). The

Guidelines must be read in connection with the major European political aim of developing efficient Europe-wide networks [63] and services, at the lowest cost and of the highest quality, to provide the European user with a basic infrastructure for efficient operation. In these Guidelines, the Commission announced how it intended to apply articles 85 and 86 to the telecommunications sector. Since then, a number of cases have been and are being investigated by the Commission. The actual application of the competition rules by the Commission, including its policy to send comfort letters if no action is intended, will be described hereinafter.

a).Art. 85 EC-Treaty

The structure of the market as outlined above basically raises concerns relating to:

- agreements concerning networks and reserved services;
- agreements concerning the provision of non-reserved services and terminal equipment; and
- agreements on research and development.

i) Agreements concerning networks and reserved services

Agreements concerning networks and reserved services will usually involve TOs on both sides. The most potentially harmful form such agreements can take are price agreements restricting competition between TOs mainly for telecommunication "hubs" to major customers. The Guidelines mention the Commission's examination of charges for the use of international leased lines. The TOs, united in the European Conference of Postal and Telecommunications administrations (CEPT), decided in a recommendation to raise their charges for leased lines by the imposition of a 30% surcharge or an access charge when third-party traffic was carried on a leased circuit or if such a circuit was inter-connected to the public telecommunications network. In the same recommendation, the CEPT also decided for the application of uniform tariff coefficients in order to determine the relative price level of international leased circuits. The Commission was of the opinion that the recommendation constituted a price agreement falling under Art. 85 EC-Treaty [64]. Without opening a formal procedure, it intervened successfully at the CEPT and competition between the TOs in the supply of leased international lines was restored to the benefit of users and particularly to the

benefit of suppliers of value added services [65].
Other agreements which seem to be typical between TOs and which may be violative of by Art. 85 (1) EC-Treaty prohibiting anti-competitive agreements are, according to the Guidelines [66] :

- agreements on other conditions for the provision of facilities;
- agreements on the choice of telecommunications routes;
- agreements foreseeing special treatment for TOs' terminal equipment or other companies' equipment for the inter-connection or inter-operation of terminal equipment with reserved services and facilities;
- agreements on the exchange of information;
- agreements on the imposition of technical and quality standards for services provided via the public network.

Especially in case of the very last agreement mentioned, the potentially substantial economic benefits of the agreement can be relevant for the granting of an exemption under Art. 85 (3) EC-Treaty. Balancing economic benefits, on the one hand, and the negative impact of restrictions on competition, on the other hand, is a complex exercise. With regard to standardisation agreements, exemptions could in principle be granted if an agreement brings more openness to the market and facilitates access to the market, and these benefits outweigh the restrictions caused by the agreement [67]. The TOs of France, Germany, Italy, and Spain and British Telecom recently notified of such a technical cooperation agreement that sets common rules for a "Global European Network" (GEN) for data transmission via fibre optic capacities. After preliminary scrutiny, the Commission decided that the agreement will be further examined [68].

ii) Agreements concerning the provision of non-reserved services and terminal equipment

As to agreements concerning the provision of non-reserved services and terminal equipment, three constellations are possible, namely

- Agreements between TOs;
- Agreements between TOs and providers of competitive services or equipment;
- Agreements between providers of competitive services or equipment.

Agreements between TOs could restrict competition between TOs, but also bring about economic benefits that outweigh the restrictions and thus render them compatible with the competition rules. An example of such a benefit is the introduction of "one stop shopping" (one single sales point for cross-border services) to the advantage of the consumer. Agreements between TOs should not strengthen dominant positions nor should they have restrictive effects on third party undertakings [69]. An example of the latter can be the refusal to provide facilities to third party suppliers of services [70]. The danger of discrimination and cross-subsidisation by Community TOs in favour of a joint undertaking providing global value added network services existed in the case of Infonet [71]. Infonet is owned by five Community TOs [72] (including Deutsche Telekom AG) as well as public and private telecommunications operators from outside the Community. The global value added services included data communications services (e.g., electronic mail) and voice communications services (e.g., in private networks). The Commission intervened during the examination of the agreements relating to the organisation of Infonet. The dangers mentioned were removed to its satisfaction. The Community TOs and Infonet agreed that Infonet will not be granted special terms or conditions for the use of reserved services nor will it be provided with technical or commercial information concerning the reserved area and unavailable to its competitors. Moreover, Infonet would be charged for the costs of the reserved services on an arm's length basis. Obligations for keeping records should enable the Commission to verify these assurances.

The alliance between the British TO, British Telecom, and MCI could also be mentioned here since MCI can be regarded as a non-Community TO. MCI provides basic telecom services in the US. The transaction consisted of the purchase by BT of a 20% stake in MCI and the creation of a joint venture, Newco, linking international voice and data services. The transaction was originally notified under the Merger Control Regulation [73]. The Commission, however, decided that the transaction was not a concentrative [74] but a cooperative operation [75] since the parties retained their autonomy. For one part of the agreement (the purchase of the shares) the Commission decided that it did not fall under Art. 85 (1). This is in conformity with the general rule that Art. 85 (1) EC Treaty does not apply to agreements for the purchase of shares, provided that there is no coordination of competitive behaviour possible. For the other part (the creation of the joint venture) the Commission granted an exemption given its positive effects for consumers (the offer of new services and the existence of strong competition [76]) This decision has been considered as an indication how the Commission looked at the growing number of strategic alliances in the telecommunications sector [77]. The balance between cooperation and (no foreclosure of) competition in any case has to be ensured. An important element seems to be whether there is already competition in the national markets concerned (BT and MCI operated in open markets).

Another notification in this area concerns the joint venture "Eunetcom", formed by France Telecom and Deutsche Telekom AG. Eunetcom will offer international corporate networks and relevant value added services to private users as an outsourcing partner [78]. The same TOs recently also notified to the Commission the formation of a global strategic alliance. A joint venture formed under the name "Atlas" will provide global business communications services to cooperate customers worldwide. The services include data network services, national and international virtual private network, customized networks, outsourcing and VSAT services [79].

On 17 May 1994, the Commission published the notification of a joint venture agreement between BT Jersey (a subsidiary of British Telecom) and France Cables et Radio (a subsidiary of France Telecom) under which the parties are to develop, through a joint venture (Jetphone) telecommunications services operating on board aircraft and relayed by terrestrial means (TFTS: terrestrial flight telephone system [80]).

Agreements between TOs and providers of competitive services or equipment may also restrict competition between the parties to the agreement or restrict competition by third parties. Such agreements will often take the form of a joint venture. The Commission will examine each individual case in order to see whether the economic benefits outweigh the harmful effect on competition [81].

The practice of the Commission was first illustrated in the Eirpage decision [82]. In this case, the Irish TO, Bord Telecom Eireann, and Motorola Ireland Ltd., a subsidiary of the U.S. Motorola Group, a world market leader in mobile telecommunications, jointly set up a company, Eirpage, for the purpose of providing a paging service throughout the Member State inter-connected to the network. Since the parent companies could have set up the business individually (as potential competitors) and they are prohibited from competing with the joint venture, the formation of one company infringes Art. 85 (1) EC-Treaty. Competition is even further restricted since the joint venture is the only provider of inter-connected paging services in Ireland. The Commission granted an individual exemption for a period of ten years because the joint venture contributed to the development of telecommunications services in Ireland. The Commission made sure it could verify the assurance that no cross-subsidisation would take place from the reserved services to Eirpage's activities and that Eirpage would not enjoy preferential tariffs for the use of Bord Telecom Eireann's services.

A similar approach was taken by the Commission in the FNA case [83]. Twelve Community and non-Community TOs and other companies [84] set up a joint venture "FNA" (Financial Network Association). FNA was set up to provide a wide range of specialised telecommunications services for banks and insurance companies. The possible infringements of Art. 85 (1) EC-Treaty related to cross-subsidisation by revenue from monopoly operations, bundling of reserved and non-reserved services and discrimination. The parties represented that they would avoid these restrictions on competition. The Commission accepted that the agreements offered benefits to users and consumers (like "one-stop shopping"). Moreover, the agreements improved the efficiency of telecommunications and global competitiveness in the outsourcing market. Therefore, the Commission closed the procedure by sending a comfort letter.

If a TO sets up a joint venture with an actual or potential competitor, and this joint venture does not bring about any improvements and benefits on the market in question nor is it indispensable in order to ensure the entry into the market for example because of heavy costs, the Commission will declare the joint venture agreement void. It accordingly did in the Astra-case [85]. The British TO BT is also one of the two operators in the U.K. licensed to uplink signals to satellites and the only provider of uplink services for international television distribution. Before the launch of the Astra satellite, all satellites were individually or collectively operated by the national TOs. Since 1983, BT had been offering space segment capacity on Eutelsat (and Intelsat) satellites to programme providers. Another satellite would provide an alternative source of supply.

BT agreed with Sociéte Européenne des Satellites SA ("SES") to set up a joint venture company, BT Astra SA. SES is a Luxembourg corporation established in 1985 for the purpose of operating satellites. Its first satellite, the Astra I A satellite, was launched in December 1988. SES was licensed to provide uplink services in Luxembourg. The joint venture would offer operators of TV programmes originating from the U.K. a packaged service consisting of a BT uplink in the U.K. and transponder space on SES's satellite. The Commission found that the joint venture restricted competition both for the provision of space segment capacity for the distribution for television channels and for uplinking services. Because of the lack of benefits mentioned above, the arrangements were not eligible for exemption. Consequently, the arrangements were (retroactively) declared void.

It is an increasingly common phenomenon that TOs join together with companies not operating in the telecommunications area in order to venture into new, not strictly telecommunications-related business activities. An example is the joint venture between the Dutch PTT Telecom and NOB, the main television facilities group in the Netherlands. The new company, Intrax, would provide satellite news-gathering services both within and outside of the Netherlands.

In the Netherlands, the uplinking of signals to satellites was liberalised in 1991. Only a license is needed. Capacity on Eutelsat satellites can only be obtained via several Eutelsat Signatories. The Dutch PTT Telecom is such Signatory. This situation could place Intrax in an unjustifiably favourable position vis-à-vis its competitors. The PTT, however, assured the Commission that it will deal with Intrax on the same footing as competing companies and that the allotment of capacity will be carried out in a non-discriminatory fashion. Since the Commission did not receive further comments on this joint venture, it closed the file by sending a comfort letter [86].

Another example is the creation of International Private Satelltie Partners LP. Nine partners formed a limited partnership, International Private Satellite Partners LP to provide international telecommunications services via satellites to businesses in Europe and North America and, to the extent that capacity is not fully utilized by it and its partners, also to offer transmission capacity to third parties. Most of the parties were private companies active in the telecommunications and aerospace areas but among them was also the Italian TO STET. As none of the partners were considered competitors in the relevant markets, the Commission decided that the venture fell outside the scope of Art. 85 (1) EC-Treaty. The Commission even considered that the creation of IPSP would increase the level of competition in both the market of advanced telecommunications services and that of satellite transmission capacity since it would be a new competitor in these markets mainly dominated by the TOs. A number of provisions (like a non-competition obligation imposed on the general partner of IPSP, Orion Satellite Co) were considered as restraints ancillary to the creation and successful operation of IPSP. Exclusive rigths for STET, like to promote the sale of the IPSP services in Italy once the Italian telecommunications market has been deregulated, were considered as non-appreciable restrictions to competition [87].

A further example is the Encompass Europe case. For this project, the participation of a TO was required in order to share the substantial risk associated with huge investments. Encompass Europe is a joint venture between a wholly-owned subsidiary of the Dutch PTT Telecom and two American companies in the transport industry. The joint venture would introduce an integrated logistic management system. This system provides for comprehensive information on the movements of cargo across transport modes. Apart from the financial contribution, PTT Telecom is to bring its expertise in the European market to the joint venture. The parties maintained that the cooperation between them does not restrict competition since they are introducing a new advanced product to the market through substantial investment and technological development, and thus the competitiveness of existing users of logistics systems can be improved. The agreements would contribute to technical progress and also serve the interests of the consumers. The Commission intended to take a favourable decision on the agreements on the basis of certain representations. In particular, the parties agreed to respect the principles contained in the block exemption for air transport computerised reservation systems [88].

On 29 April 1994, an agreement between Groupement de Cartes Bancaires and France Télécom was notified to the Commission. The purpose of the agreement is the use of CB cards as a means of accessing the public telephone network, alongside those cards issued by France Télécom (prepaid cards and subscription cards [89]).

In 1992, agreements with regard to equipment have been notified to the Commission by STET, Italtel, AT&T and AT&T-NSI. These agreements provided for technological and commercial cooperation in the field of telecommunications equipment, notably public switching transmission systems and some terminal equipment. AT&T sufficiently showed to the Commission that it was not in a position, through cross-subsidisation from one of its activities (network operator) to another (equipment manufacturer), to sell components and sub-assemblies for telecommunications equipment to its European subsidiaries at prices that did not reflect costs. STET and Italtel gave similar representations to the Commission. On this basis, the Commission published a notice proposing to take a favourable view of the agreements [90].

With regard to agreements between providers of competitive services or equipment, the Commission will emphasise the avoidance of cooperations which could strengthen a dominant position.

The Commission did not see a reason to prohibit an agreement notified by Aérospatiale and Alcatel Espace. The agreement regards telecommunications satellites and satellites having related applications, in particular military ones. The agreement provides for business cooperation coupled with technical and industrial specialisation. It aimed at rationalising the activities of the parties but does not provide for integration of their production lines. The Commission stated that it intended to take a favourable view of the agreement [91].

It also intended to take a favourable view towards the agreement, whereby Communications de Mobile Cellulaire (France), the company jointly set up by Matra Communications SA and Cellcom Ltd. (United Kingdom), was authorized to issue capital to two more shareholders Talkline PS PhoneService GmbH and Norauto (France [92]).

iii) Agreements on research and development

Research and Development (R&D) agreements - if ensured by Art. 85 (1) EC-Treaty - are, in general, likely to be exempted because R&D is essential for technical progress and for the competitiveness of the industry to the benefit of users. They could often be covered by the block exemption found in the Regulation on R&D [93] or be granted individual exemption. The block exemption Regulation concerns joint R&D and joint exploitation. If such agreements are vertically extended to joint marketing and distribution, the Commission could be asked for a negative clearance or an exemption.

In the telecommunications sector, new products or services are a major competitive advantage, not only for individual companies but for the whole Community. Moreover, any company wishing to remain competitive will have to supply the most technologically advanced equipment and services available. The Community's industrial policy [94] will play an important role in the Commission's competition decisions. The controversy between international competitiveness, which could eliminate competition on the common market, and European competition policy becomes particularly clear in this area. Examples mentioned in the Guidelines are the Optical Fibres' case [95] and the "Canon-Olivetti" case [96].

The considerable expenditure on R&D, which the nature of the market for telecommunications products require, has been particular noted by the Commission in the GEC-Siemens/Plessey case. General Electric Company plc and Siemens AG formed a jointly-owned company, GEC-Siemens, to acquire Plessey plc. A part of Plessey's business would thereafter be allocated between the parent companies. Another part of Plessey would constitute the joint venture, among others for joint R&D. Plessey made an application to the Commission alleging that these arrangements infringed Art. 85 and 86 EC-Treaty. The application was rejected on the grounds that the agreements did not result in an appreciable restriction of competition or that the conditions for an exemption were met, in particular because of the promotion of technical progress which involved high costs [97].

A similar approach was taken in the Alcatel Espace/ANT Nachrichtentechnik decision. R&D cooperation without joint marketing was in this case not considered practicable, given the particular nature of satellites. An exemption was therefore granted [98].

The Commission took a more flexible position towards competition rules in the Konsortium ECR 900 case [99]. The companies AEG AG, Alcatel NV and Oy Nokia AB formed a consortium, ECR 900, for the joint development, manufacture, and distribution of a pan-European digital cellular mobile telephone system known as "GSM" (Groupe spécial mobile). The only potential buyers for the GSM system at the time were the national network operators. The Commission decided that this cooperation agreement did not even constitute a restriction of competition in part because the facts showed that the development and

manufacture by individual companies would not take place because of the high costs involved.
The Commission seems to stress the effect of the joint venture on the market. In the Konsortium ECR 900 case, the Commission was of the opinion that no individual company was able to entry the market. In the cases of Alcatel Espace/ANT Nachrichtentechnik and GEC-Siemens/Plessey, the cooperation did not appreciably restrict competition because of the competitive environment.

b). Art. 86 EC-Treaty

Art. 86 of the Treaty prohibits the abuse by one or more undertakings of a dominant position within the common market or in a substantial part of it that may affect trade between Member States. Art. 86 EC-Treaty gives examples of such an abuse:

"a) the direct or indirect imposition of unfair purchase or selling prices or other unfair trading conditions;

b) the limitation of production, markets, or technical developments to the prejudice of consumers;

c) the application of dissimilar conditions to equivalent transactions with other trading parties, thereby placing them at a competitive disadvantage;

d) the conclusion of contracts subject to acceptance by other parties of supplementary obligations which, by their nature or according to commercial usage, have no connection with the subject of such contracts."

A dominant position is defined as

"a position of economic strength enjoyed by an undertaking which enables it to hinder the maintenance of effective competition on the relevant market by allowing it to behave to an appreciable extent independently of its competitors and customers and ultimately of its customers" [100].

TOs will in general hold a de jure dominant position over the network and with regard to voice telephony services. Before true competition is established in the non-reserved market, they often also hold a de facto dominant position in that market. In case of a monopoly, competition does not exist. An abuse is committed

"where, without any objective necessity, an undertaking holding a dominant position on a particular market reserves to itself or to an undertaking belonging to the same group an ancillary activity which might be carried out by another undertaking as part of its activities on a neighbouring but separate market, with the possibility of eliminating all competition from such undertaking" [101].

In its Guidelines, the Commission broadly divides abuses according to the undertaking involved into:

a) TOs' abuses
b) abuses by an undertaking other than TOs
c) abuses of a dominant purchasing position.

i) Telecommunications Organisations' abuse

The particular abuses by TOs are related to their legal or de facto monopoly in one market and their competitive activities in another related market. The Guidelines mentions: a) the refusal or restriction of the usage of the reserved service so as to limit the provision of non-reserved services by third parties; and b) predatory behaviour as a result of cross-subsidisation.
An example of a refusal to supply reserved services is found in the Belgian leased lines case [102]. In Belgium, the TO at the time, RTT, imposed various restrictions on the lease of lines. Even before the adoption of the Services Directive, the Commission's approach was that such restrictions are not necessary for the task allowed under Art. 90 (2) EC-Treaty. After intervention by the Commission, the RTT eliminated the restrictions.
The Commission is also alert to the risk of discrimination and cross-subsidisation in assessing notified cooperation agreements between a TO and a non-TO. In the Eirpage decision [103], the Commission exempted the joint venture agreement on the special condition that the Commission would receive the joint venture's annual accounts. The Commission made sure it could verify the assurance that no cross-subsidistion would take place from the reserved services to Eirpage's activities and that Eirpage would not enjoy preferential tariffs for the use of TO's services. Also in the Infonet case [104], the Commission intended to take a favourable view of a cooperation agreement due to representations on the avoidance of cross-subsidisation and discrimination in favour of the TOs joint undertaking.

ii) Abuses by an undertaking other than Telecommunications Organisations

With regard to other telecommunications undertakings having a dominant position, the Commission is paying particular attention to the exercise of industrial property rights in relation to standards and to the availability of often crucial technical information.
The example mentioned in the Guidelines is the IBM case. IBM

agreed to provide other manufacturers with the technical interface information needed to permit products competitive with IBM computers [105].

In the Decca case of 1989, the Commission found an abuse by Decca because it concluded market dividing agreements and it changed transmission signals. Decca exploited a navigation system consisting of transmission of signals by land-based stations to devices placed on bord a ship in this case. The system and receivers were originally covered by a patent. After the patent expired, other undertakings started to sell receivers. In order to hinder competition, Decca concluded license agreements with these competitors, resulting in a division of the market, and made changes in signals for the purpose of impeding the functioning of competing receivers. These actions constituted an abuse because Decca intended to exclude those competitors with which it entered into license agreements from the market for commercial receivers. By changing the signals, it obstructed and coerced those competitors who were not willing to enter into such agreements [106].

iii) Abuses of a dominant purchasing position

Art. 86 EC-Treaty also applies to the conduct of an undertaking holding a dominant purchasing position. European public procurement directives require TOs to follow transparent procurement procedures in the reserved market [107]. Both in reserved and non-reserved markets, abusive practices might violate Art. 86 EC-Treaty. The Commission will subtantively examine whether the same criteria for awarding the contract has been followed by the dominant undertaking for all suppliers.

c). The Merger Control Regulation of 1989

Last but not least, of course the Regulation on the control of concentrations between undertakings [108] is also applied to the telecommunications sector. A detailed discussion, including the question of whether a dominant position in the European market in some areas of telecommunications services and satellite communications is not necessary to compete on a global market, must be left for a later occasion. It is noted here that the second decision delcaring a concentration incompatible with the common market (Art. 8 (3) of the Regulation [109]) concerned a case involving also a TO (The MSG case [110]). One of the main considerations was that MSG (a German pay-television venture between Bertelsmann AG, Leo Kirch and Deutsche Telekom AG) would strengthen the dominant position of Deutsche Telekom AG in the market for cable infrastructure.

d). Conclusion

The Commission practice relating to the enforcement of Art. 85 and 86 EC-Treaty in the telecommunications sector is based on two pillars, namely, the preventive force of the Guidelines, and selective decisions in order to demonstrate the willingness of the Commission to enforce a policy as outlined in the general guidelines.

Since the issuance of the Guidelines, the Commission has made four decisions. The exemption decision in the Eirpage case, the prohibition decision in the Astra case, the decision concerning a strategic alliance in the BT-MCI case which was partly considered not to fall under Art. 85 (1) EC-Treaty, partly provided an exemption, and the International Private Satellite Partners case which also did not fall under Art. 85 (1) EC-Treaty. On the other hand, many agreements have been notified to the Commission on which the Commission did not find it necessary to make a decision. Almost all of these involved TOs. The two main elements of applying the competition rules to TOs are apparently the avoidance of discrimination and cross-subsidisation. These issues will certainly remain the focus of attention, especially for those undertakings which compete with TOs, at least in a segment of the market, while probably relying at the same time on services and/or networks. With the growing importance of global services another important element seems to be whether the telecommunications operations operate in open markets. This could force Member States to open up their markets in still reserved areas even before 1998.

Since the issuance of the Guidelines, the Commission did not make any decision related to Art. 86 EC-Treaty. The possible abuse of TOs is, as mentioned above, taken into account in assessing notified agreements. The Commission's general practice of the last years shows that the Commission is more often applying Art. 85 and 86 EC-Treaty in the same decision [111]. As long as the telecommunications sector is characterised by a high number of legal or factual monopolies, this practice is expected to continue.

While the Commission's Guidelines demonstrate that this institution has acquired a considerable expertise in the field of supervising competition in the field of telecommunications, the practice of settling cases informally might over time reduce the preventive force of the Guidelines since it deprives parts of the market of knowledge about the development of the Commission's policy implementation.

Although the considerable advantage of settling a case informally for the undertaking concerned shall not be negated, the Commission ought to consider how the Guidelines can be kept up-to-date, if not by published "case law". The law must reflect technological and economical changes, while, on the other hand, the leading players need to stay fully abreast with these legal developments.

3.2.. German competition law

a). Competition guarantees under telecommunications law

Before the parallel application of the German rules are described, it is recalled that the Deutsche Telekom AG has a network monopoly and a voice telephony monopoly.

In the monopoly area, a first regulatory restriction on competition is contained in Section 7 of the Telecommunications and Posts Regulation Act. Financial compensation between the Deutsche Telekom AG, Deutsche Postdienst AG and Deutsche Postbank AG) is expressly allowed. Also, financial compensation from the services reserved to Deutsche Telekom AG in favour of its services in the non-reserved area has been declared permissible. If in the non-reserved area, persistent financial deficit impairs the competitive possibilities of other enterprises on the market without any justifiable reason, the Minister, in consultation with the Minister of Economics, may take the necessary measures to eliminate this impairment. The Minister of Economics, in consultation with the Telecommunications Minister, will decide whether there is a case of inadmissible impairment. If the performance of this task so requires, the Minister of Economics will call in the Federal Cartel Office which, according to 48 of the Act against Restraints of Competition (ARC), is authorised to decide on such matters. Third parties cannot derive any rights from the provisions of section 7 of the Telecommunications and Posts Regulation Act. The application of the GWB is not affected by this provision.

The network and telephone monopoly of the Deutsche Telekom AG are not absolute. The Minister may grant licenses to set up and operate telecommunications installations insofar as the Deutsche Telekom AG has an exclusive right [112]. Licenses have, for example, been issued for the operation of mobile radio communications systems, radio paging systems (trunk call and bundle networks), and satellite networks.

In the field of mandatory services, the Telecommunications Installation Act provides for another possible far-reaching restriction on competition. In this area, the Minister may impose certain obligations on competitors of the Deutsche Telekom AG which are suitable to remove the competitive disadvantages of the Deutsche Telekom AG. Such obligation may relate to the conditions of supply as far as coverage and quality are concerned and to the pricing factors. The current level of business activities achieved by these competitors may not be impaired [113].

It follows that the competition particularly in the non-reserved area can still be subject to regulatory restrictions.

One of the essential features of the reform of 1989 was that the relations of the Deutsche Telekom AG with its customers were henceforth governed by private law, regardless of whether these relations concern its services in the monopoly area or in a competitive area. For the use of the Deutsche Telekom AG services, the federal government can issue framework acts [114]. In particular, it may lay down regulations concerning the conclusion, the subject and termination of contracts, and the rights and duties of the contracting parties and of any other parties involved in telecommunications traffic, including liability regulations as well as the conditions for the connection of terminal equipment. These conditions have been laid down in the Telecommunications Regulation ("Verordnung über die Rahmenvorschriften für die Inanspruchnahme von Dienstleistungen des Unternehmens DBP TELEKOM") of 16th September 1992 [115]. This regulation applies to the use of monopoly services [116] and basically provides for equal competitive conditions. These conditions include the unbundling of its monopoly services in accordance with market demand as far as reasonably possible [117] and the obligation to provide its monopoly services purusant to EC law, in particular the Open Network Provision Directive [118].

b). Applicability of the Act against Restraints on Competition (ARC) to the German national TO

Without a doubt, the competition rules do not apply to state measures such as laws, regulations, and other administrative acts based on the respective provisions. The competition rules only apply insofar as a competitive relation exists between competitors. The competition rules do not apply in areas in which the federation excludes competition. This is especially true in the case of a legal monopoly [119]. Before the first postal reform of 1989 took place, the competition rules were rarely applied, e.g., to public procurement by the Deutsche Telekom AG [120].

According to 98 (1) ARC, the law also applies to enterprises which are wholly- or partly- owned by the state or which are governed or operated by the state insofar as 99 to 103 do not provide otherwise. Deutsche Telekom AG can clearly be defined as such an enterprise. The competition laws came up for amendment for the fifth time on 1st January 1990. Until then, the ARC, pursuant to the (then) 99 (1), was not applicable to contracts made by the Deutsche Bundespost (Deutsche Telekom AG predecessor) to the extent that the tariffs or conditions arising from those contracts, decisions, or recommendations were established or allowed by law or implementing regulations. This exception for tariffs and conditions established by the state in practice meant that the exception from the ARC in favour of the Deutsche Bundespost actually covered all commercial activities of the Bundespost since all dealings with customers were established by regulation [121]. Even after the postal restructuring, a broad power to issue regulations still exists, in particular concerning the mandatory services of the Deutsche Telekom AG.

In the fifth renewal of the competition laws, 99 (1) ARC was deleted. One might have thought that this deletion would result in the public enterprise Deutsche Telekom AG being subject to competition law. However, the Explanatory Notes to the new wording of 99 ARC assume that the exception regarding the Bundespost mentioned before is only of an editorial nature, i.e., that 99 (1) ARC had only affirmed the law in this area as it otherwise existed and an explicit provision was not required. If the state were to impose certain prices or conditions, as it does for the telephone service monopoly or for the minimal services of the Deutsche Telekom AG, no prohibited agreements or concerted practices could exist, with the result that the old provision of 99 (1) ARC would have be superfluous [122].

Consequently, one should assume an important exception from the ARC for activities of the Deutsche Telekom AG. This uncertainty could only be clarified by a judgment of a German court. It seems inappropriate that German competition law is more lenient in its treatment of public telecommunications enterprises than is EC law.

c). German case law/substantative issues

Compared to the EC case law, the guidance in Germany is poor. Apart from a single bold decision of the Higher Regional Court in Düsseldorf in 1980 that fully applied the ARC to the market demand conduct by the Deutsche Bundespost in public procurement procedures [123], there is no important competition law judgment from a German court. One intervention by the Federal Cartel Office in a case comparable to the IBM representation to the Commission is recorded [124]. The Deutsche Bundespost was required to make available at no expense support documents on technical transfer installations and interface conditions to those producers which had not participated in the standardisation work of the working groups of the Bundespost. The basis for this intervention was 26 (2) ARC, which is comparable to Art. 86 of the EC-Treaty [125].

In both the monopoly and competitive area, the Federal Cartel Office may one day use its instruments under 1 ARC (parallel provision to Art. 85 EC-Treaty) and 22, 26 ARC (parallel provisions to Art. 86 EC-Treaty). The relatively strict German merger control under 23 et seq. ARC may also restrict joint ventures within or outside Germany which fall under the German law due to the "effects doctrine" under 98 (2) ARC as long as the EC Merger Control Regulation allows national control [126]. New alliances of Deutsche Telekom AG with private companies give reason to expect new case law.

However, for the time being, it is obvious that the EC law and its application has priority over national control for various reasons. The main reason is the globalisation and Europeanisation of telecommunications markets as a result of business activities across national borders. The modern EC policy, the creation of secondary Community law, the know-how of the European Commission, and the broad applicability of the EC cartel law also play a decisive role for the stewed comparative view of EC and national competition law at work.

IV - FINAL REMARKS

The brief overview of the liberalisation and harmonisation concepts behind the European telecommunications policy and law, the more intensive look at the case law applying European competition law, and finally a quick look at the parallel developments in Germany show that competition law is the coordinate axis for the future of this market. Liberalised and partly-deregulated national or European telecommunications markets will not be an open playground for TOs or private major players, but have to be seen within the framework of the competition law. The guidelines on the application of EC competition rules in the telecommunications sector of 1991 help to delineate the decisive borders. Nevertheless, it cannot be ignored that no other area of law is more in a state of continual change than telecommunications: technical inventions of tomorrow cannot be foreseen today, and it is not a disadvantage but rather an inherent characteristic of the law that it is not proactive but reactive. The competition law must follow and accompany the technical and economic developments instead of preventing or trying to harnass them into a precautionary set of rules which proves to be inappropriate when it comes to its application. Any business activities in the telecommunications markets have to consider this reactive approach of competition law and requires a careful look at the necessarily constantly updated legal practice of the courts and authorities.

Appendix

Selected decisions and publications

Selection of some relevant decisions and publications

I. COURT OF JUSTICE/COURT OF FIRST INSTANCE

Italian State v. Sacchi, Case 155/73, Judgement of the ECJ of 30th April 1974, (1974) ECR 409, (1974) 2 CMLR 177

Italy v. EC Commission, Case 41/83, Judgement of the ECJ of 20th March 1985, (1985) 2 CMLR 368

Centre Belge d'Études de Marché-Telemarketing SA v. Compagnie Luxembourgeoise de Telediffusion SA, Case 311/84, Judgement of the ECJ of 3rd October 1985, (1986) 2 CMLR 556

Bond van Adverteerders v. Netherlands, Case 352/85, Judgement of the ECJ of 26th April 1988, (1988) ECR 2085

Alsatel v. SA Novasam, Case C 247/87, Judgement of the ECJ of 5th October 1988, CMLR June 1990

French Republic v. Commission of the European Communities, case C 202/88, Judgement of the ECJ of 19th March 1991, (1991) ECR 1223.

Elliniki Radiophonia Tiléorassi - Anonimi Etairia v. Dimotiki Etairia Pliroforissis and Sotirios Kouvelas, Case C 260/89, Judgement of the EC of 18th June 1991, not yet published,OJ C 201 of 31st July 1991, p. 5

Régie des Télégraphes et des Téléphones v. GB-Inno-BM SA, Case C 18/88, Judgement of the ECJ of 13th December 1991, not yet published,OJ C 15 of 21st January 1992, p. 13

Kindgom of Spain and others v. Commission of the European Communities, Cases C-271, C 281 and C 289/90, Judgement of the ECJ of 17th November 1992, not yet published,OJ C 326 of 11th December 1992, p. 8

Ministère Public v. F. Decoster and Others, Case C-69/91, Judgement of the ECJ of 27th October 1993, not yet published

Ministère Public v. Taillandier, Case C-92/91, Judgement of the ECJ of 27th October 1993, not yet published

Procureur du Roi v. J.M. Lagauche and Others, Joint cases C-46/90 and C-93/91, Judgement of the ECJ of 27th October 1993, not yet published,OJ C 316 of 23rd November 1993, p. 4.

II. COMMISSION DECISIONS

British Telecommunications, Commission decision of 10th December 1982, L 360 of 21st December 1982, p. 31

Optical Fibres, OJ L 236 of 22nd August 1986, p. 30

Olivetti/Canon, Commission decision of 22nd December 1987,OJ L 52 of 26th February 1988, p. 51

Decca Navigator System, Commission decision of 21st December 1989,L 43 of 15th February 1988, p. 27

Alcatel Espace/ANT Nachrichtentechnik, Commission decision of 12th January 1990,L 32 of 3rd February 1990, p. 19

Konsortium ECR 900, Commission decision of 27th July 1990,L 228 of 22nd August 1990, p. 31

Eirpage, Commission decision of 18th October 1991,OJ L 306 of 7th November 1991, p. 22

Astra, Commission decision of 23rd December 1992, OJ L 20 of 28th January 1993, p. 23

III. INFORMAL POLICY

Cordless telephone in Germany, Press Release of 26th March 1985

GEC-Siemens/Plessey, letter of 1st September 1989 OJ C 239 of 25th September 1990, p. 2

CCITT Competition Report 1990, p. 52

CEPT Competition Report 1990, p. 52

MDMS Competition Report 1990, p. 53

RTT Competition Report 1990, p. 52

Infonet, notice pursuant to Art. 19 (3) of Regulation 17 OJ C 7 of 11th January 1992, p. 3

Encompass Europe (GLV/ELTS BV), notice pursuant to Art. 19 (3) of Regulation 17OJ C 233 of 11th September 1992, p. 2

STET, Italtel, AT&T and AT&T-NSI, notice pursuant to Art. 19 (3) of Regulation 17OJ C 333 of 17th December 1992, p. 3

Intrax, notice pursuant to Art. 19 (3) of Regulation 17, comfort letterPress release of 25th October 1993OJ C 117 of 28th April 1993, p. 3

Eunetcom, notification of a joint ventureOJ C 175 of 26th June 1993, p. 11

International Private Satellite Partners, notification of a joint ventureOJ C 193 of 17th July 1993, p. 5

FNA, comfort letterPress rlease of 15th November 1993

Aérospatial/Alcatel Espace, notice pursuant to Art. 19 (3) of

Regulation 17OJ C 47 of 15th February 1994, p. 6

GEN, notice pursuant to Art. 19 (3) of Regulation 17OJ C 55 of 23rd February 1994, p. 3

BT-MCI notice pursuant to Art. 19 (3) of Regulation 17OJ C 93 of 30th April 1994, p. 3

IV. LITERATURE

Books:

Denton Hall : "EC Telecommunications Law", 1993, Chancery Law Publishing

R. Ellger, T-S. Kluth : "Das Wirtschaftsrecht der Internationalen Telekommunikation in der Bundesrepublik Deutschland", 1. Aufl., Baden-Baden: Nomos Verl.-Ges., 1992.

Elizabeth Hiester : "Telecommunications" in: Butterworths Competition Law, Butterworths & Co. Ltd. 1991-1994

Articles :

B. Armory : "Telecommunications in the European Communities; The new regulatory framework"in: EuZW-Heft 3/1992, p. 75

M. Coleman : "European Competition Law in the Telecommunications and Broadcasting Sectors" in: (1990) 5 ECLR, p. 204

C-D. Ehlermann : "Telekommunikation und Europäischisches Wettbewerbsrecht" in: EuR-Heft 2-1993, p. 134

Christoph Engel : "Telekommunikationsrecht - Grundstrukturen eines neuen Rechtsgebietes am Gegenstand des Europarechts, ZUM 1992, 279

M. Hoskins : "A Review of EEC Telecommunications Policy: Too Much of a Good Thing?": European Business Law Review, January 1992, p. 8

J.M. Naftel : "The Natural Death of a Natural Monopoly: Commission in EC Telecommunications after the Telecommunications Terminals Judgement" in: (1993) 3 ECLR , p. 105

C. Scott : "Liberalising and re-regulating European Community telecommunications: the Commission's proposals" in: Summer 1993, Utilities Law Review, p. 98

P-A. Trepte, in: Autumn 1991, Utilities Law Review, p. 120

REFERENCES

1. See the survey in the Green Paper of the EC Commission on the "Development of the Common Market for Telecommunications Services and Telecommunications Equipment", COM (87), 290 fin., p. 58 et seq.
2. See footnote 1.
3. For the state of the European telecommunications policy and its application see Bernhard Amory, EuZW (1992) p. 75-85; communication of the Commission on the consultation on the review of the situation in the telecommunications services sector, COM (93) 159 fin. of 28th April 1993.
4. Telecommunications organisations means public or private bodies to which a Member State grants special or exclusive rights for the provision of a public telecommunications network and, if applicable, public telecommunications services; compare Art. 2 of Council Directive 90/387/EEC, OJ L 192 of 24th July 1990, p. 1 and Article 1 of Commission Directive 90/388/EEC, OJ L 192 of 24th July 1990, p. 10; In general, these are the old PTTs.
5. Compare consideration 14 of the European Court of Justice's judgment in the Sacchi case, 155/73 (1974) ECR 409.
6. See judgment of the ECJ of 19th March 1991 in case C-202/88 (France v. Commission) (consideration 51) (1991), ECR 1223; judgment of the ECJ of 13th December 1991 in case C-18/88 (RTT v. GB-Inno-BM SA) (Consideration 25), not yet reported; judgment of the ECJ of 27th October 1993 in case C-69/91 (Ministère Public v. F. Decoster and Others), not yet reported.
7. See judgment of the ECJ of 13th December 1991 in case C-18/88 (RTT v. GB-Inno-BM SA) (consideration 24), not yet reported.
8. The definition of the undertaking entitled to benefit from the exemption must be narrowly construed and requires that the undertakings involved be entrusted with a service of general economic interest by a sovereign act of the public authority (see Decca, 218; BRT/Sabam ECJ 23 March 1973, 127/73, (1974) ECR p. 318).
9. OJ L 366 of 21st December 1982, p. 31.
10. Compare consideration 18 of Commission Directive 90/338/EEC of 28th June 1990 on competition in the markets for telecommunications services, OJ L 192 of 24th July 1990, p. 10.
11. See judgment of the ECJ of 13th December 1991 in case C-18/88 (RTT v. GB-Inno-BM SA) (consideration 22), not yet reported.
12. Compare, e.g., consideration 11 of the judgment of the European Court of Justice of 19th May 1993 in case C-320/91 (Corbeau), not yet reported, and judgment of the ECJ of 27th October 1993 in joint cases C-46/90 and C-93/91 (Procureur du Roi v. J.M. Lagauche and Others.
13. The Commission has direct executive and rule-making powers in this area, the latter is especially unusual if compared to other chapters of the Treaty.
14. The case law referred to in the Guidelines on the Application of the EEC Competition Rules in the Telecommunications Sector (91/233/02), OJ C 233 of 6th September 1991, pp. 2-26, no. 23, are the Case 10/71, Müller-Hein, (1971) ECR 723 and the judgment in Case 66/86, Ahmed Saeed (1989), ECR 803.
15. Guidelines 23.
16. Consideration 53.
17. See judgment in Corbeau case, consideration 20 and 21, footnote 12. Compare the Commission's power related to Art. 85 (3) exemptions. Art. 9 (1) of Regulation 17 states that "Subject to review of its decision by the Court of Justice, the Commission shall have sole power to declare Art. 85 (1) inapplicable pursuant to Art. 85 (3)." Only notified agreements can be exempted at all and,

secondly, a comparable exemption does not exist for Art. 86. Art. 90 (2) can, in the authors' view, be considered as a general exemption for certain state measures, the effect of which is contrary, in particular to Art. 85 (1) or Art. 86. Member States are not obliged to notify state measures under the meaning of Art. 90. Apart from the absence of a procedure comparable to the one under to Art. 85 (3), the ability of the Commission to exempt such acts is therefore limited. Similar to Art. 85 (3), it could, however, be argued that a "positive" decision or directive on the basis of Art. 90 (3), limited in time, could prevent the direct applicability of Art. 90 by the national authorities and a "negative" decision or directive could not be transformed by more flexible national rules. In the absence of a procedure in which an exemption could be obtained by the Commission, in the authors' opinion, the Commission's inaction cannot prevent national courts from declaring a state measure either contrary or not contrary to Art. 90.

18. See Art. 2 and 4 of Commission Directive 90/388/EEC of 28th June 1990 on competition in the markets for telecommunications services, OJ L 192 of 24th July 1990, p. 10.
19. Council Resolution of 22 July 1993 on the review of the situation in the telecommunications sector and the need for further development in that market, OJ C 213 of 6 August 1993, p. 1; Council Resolution of 22 December 1994 on the principles and timetables for the liberalisation of telecommunications infrastructures, OJ C 79 of 31 December 1994, p. 4.
20. OJ L 268 of 19 October 1994, p. 15.
21. IP/94/1262 of 21 December 1994.
22. COM (94) 145 of 27 April 1994
23. COM (94) 492 of 23 November 1994.
24. Bundesgesetzblatt I 1, p. 1026.
25. Commission Directive 88/301/EEC of 16th May 1988 on competition in the markets in the telecommunications terminal equipment; OJ L 131 of 27th May 1988, p. 73.
26. Judgment of the ECJ of 19th March 1991 in Case 202/88: French Republic v. Commission of the European Communities; ECJ Reports 1991-3, p. 1223 et seq.
27. Art. 2 of the Terminal Equipment Directive.
28. See considerations 45-47 of the Judgment of the European Court of Justice of 19th March 1991 in Case 202/88. By Directive 94/46/EC of 13 October 1994 (OJ L 268 of 19 October 1994, p. 15), by which the Commission amended Directive 88/301/EEC for the first time, a definition of special rights has then been added (See Art. 1 (b)).
29. Art. 3 of the Terminal Equipment Directive.
30. Art. 3 of the Terminal Equipment Directive.
31. Articles 6 in connection with Art. 5 of the Terminal Equipment Directive; Art. 6 has been the subject of the judgment of the ECJ of 27th October 1993 in joint cases C-46/90 and C-93/91 (Procureur du Roi v. J.-M. Lagauche and Others, not yet reported; judgment of 27th October 1993 in case C-92/91 (Ministère Public v. Taillandier), not yet reported; judgment of 27th October 1993 in case C-69/91 (Ministère Public v. F. Decoster and Others), not yet reported. In the latter two cases, the Court ruled that different departments of a single administration cannot be considered to be independent from each other for the purpose of Art. 6.
32. Art. 7 of the Terminal Equipment Directive.
33. Consideration 55-57 of the Judgment.
34. Section 1 (3) of the Telecommunications Installation Act.
35. With regard to ordinary telephone terminal equipment, subject to the transitional period until 1st July 1990 of Section 25 of the Telecommunications Installation Act.
36. Section 2a (1) and (2) of the Telecommunications Installation Act.
37. Commission Directive 90/388/EEC of 28th June 1990, OJ L 192 of 24th July 1990, p. 10.
38. Judgment of the ECJ of 17th November 1992 in Joined Cases 271, 281 and 289/90: Kingdom of Spain and Others v. Commission of the European Communities; not yet published.
39. See for an analysis of both judgments "European Court provides key to open public utilities to competition" by Martin Coleman, in: Autumn 1991, Utilities Law Review; note to the judgments by H.M. Gilliams, in: SEW 4 (1993), p. 368.
40. Again, the Court annulled the provisions relating to "special rights" in the absence of any specification: compare terminal equipment judgment. By Directive 94/46/EC of 13 October 1994 (OJ L 268 of 19 October 1994, p. 15) the Commission amended Directive 90/388/EEC by adding a definition of special rights (See Art. 1 (a) (ii)).
41. The Services Directive does not apply to telex services and terrestrial mobile communications (Art. 1 (2)) and telecommunications services do not include broadcasting and television (Art. 1 (1)).
42. Art. 2 of the Services Directive.
43. Art. 7 of the Services Directive.

44. Art. 8 of the Services Directive.
45. Section 1 (2) and (4) of the Telecommunications Installation Act.
46. Verordnung zur Regelung der Pflichtleistungen der Deutschen Bundespost TELEKOM - TPflLV of 16th September 1992, Amtsblatt BMPT no. 19/92 of 7th October 1992.
47. Section 1a (1) of the Telecommunications Installation Act.
48. Council Directive 86/361/EEC of 24th July 1986, OJ L 217 of 5th August 1986, p. 21.
49. Council Directive 91/263/EEC of 29th April 1991, OJ L 128 of 23rd May 1991, p. 1.
50. Art. 4 of Council Directive 91/263/EEC.
51. CEN is the European Committee for Standardisation; Cenelec is the European Committee for Electromagnetical Standardisation; ETSI is the European Telecommunications Standards Institute.
52. Section 2a (1) and (3) of the Telecommunications Installation Act.
53. Section 15 (2) of the Telecommunications Type Approval Regulation of 22nd March 1991.
54. Section 15 of the Telecommunications Type Approval Regulation of 22nd March 1191.
55. Council Directive 90/387/EEC, OJ L 192 of 24th July 1990, p.1.
56. See Annex I to the ONP Directive.
57. Council Directive 92/44/EEC of 5th June 1992 on the application of open network provision to leased lines, OJ L 165 of 19th June 1992, p. 27.
58. See Council Recommendation 92/382/EEC of 5th June 1992 on the application of open network provision to public packet switched data services, OJ L 200 of 18th July 1992, p. 1.
59. See Council Recommendation 92/383/EEC of 5th June 1992 on the application of open network provision to ISDN, OJ L 200 of 18th July 1992, p. 10.6)
60. Art. 3 of Council Directive 90/387/EEC.
61. Section 9 of the Telecommunications Regulation.
62. OJ C 233 of 6th September 1991, p. 2.
63. See on this, the new chapter on trans-European network (Article 129b et seq. EC-Treaty) which has been introduced by the Maastricht Treaty.
64. And a collective infringement of Art. 86 EC-Treaty.
65. Guidelines, no. 46; Competition Report EC 1990, Point 56: in case the Society of Worldwide Interbank Financial Telecommunications (Swift), a major carrier for interbank electronic funds transfer.
66. Guidelines, no. 47-53.
67. Guidelines, no. 49.
68. OJ C 55 of 23rd February 1994, p. 3.
69. Guidelines 56-63.
70. See Guidelines no. 59.
71. OJ C 7 of 11th January 1992, p. 3.
72. Transpac (subsidiary of France Telecom); DBP TELEKOM, Telefonica International Holding BV (subsidiary of Telefonica, the Spanish TO); RTT (Belgian TO); PTT Telecom BV (Dutch TO).
73. Council Regulation No. 4064/89/EEC of 21st December 1989 on the control of concentrations between undertakings, OJ L 395 of 30th December 1989, p. 1; corrected by OJ L 257 of 21st September 1990, p. 13.
74. OJ C 259 of 23rd September 1993, p. 3; European Commission Press Release IP/93/757 of 14th September 1993.
75. See for the difference between concentrative and cooperative operations: Commission Notice regarding concentrative and cooperative operations of 23 December 1994, not yet published, and Commission Notice concerning the assessment of cooperative joint ventures pursuant to Art. 85 of the Treaty, OJ C 43 of 16th February 1993, p. 2.
76. OJ L 223 of 27 August 1994, p. 36.
77. The strategy concerns the future expected, global provision of telecommunications and multi-media services. The press has reported the following strategic alliances concerning Europe:
- one between AT&T and Unisource (the Dutch-based joint venture between TOs from Sweden, the Netherlands, Switzerland and Spain);
- one between Sprint Corp, France Telekom and DBP Telekom AG;
- one between German energy group Veba and UK's Cable & Wireless plc;
- a joint venture between steel and engineering group Thyssen AG and the US biggest regional phone operator Bellsouth International;
- a joint venture between the German industrial group Viag AG and BT;
- CNI Communications Network International, a joint venture between Mannesmann AG, utility RWE Energie AG and Deutsche Bank AG (allegedly looking for an international partner). Most of these alliances seem to target the German market first because of its size, its growing expectations

and expectations about its liberalisation measures before 1998.

78. OJ C 175 of 26th June 1993, p. 11.
79. OJ C 377 of 31 December 1994, p. 9
80. OJ C 134 of 17 May 1994, p. 5.
81. Guidelines no. 64-70.
82. OJ L 306 of 7th November 1991, p. 22 (notification: OJ C 294 of 7th November 1991).
83. European Commission Press Release IP/93/988 of 15th November 1993.
84. France Telecom; Belgacom; Italcable SpA; Mercury Communications Ltd; Telefonica de Espana SA; DBP TELEKOM; Telstra; HongKong Telecom International Ltd.; Kokusai Denshin Denwa Co, Ltd. (KDD), MCI; Singapore Telecom; Stentor.
85. OJ L 20 of 28th January 1993, p. 1; see for an analysis of the case: "The European Commission's Astra decision" by Helen Charlton, in (1993) 3 ECLR; an action for anullment of the Commission's decision was brought before the Court of First Instance of the European Communities on 17th March 1993 (Case T-22/93), OJ C 113 of 23rd April 1993, p. 4, and, only possible upon request of the applicant British Telecom, removed by order of the President of Court of First Instance of 22nd September 1993, OJ C 277 of 15th October 1993, p. 18.
86. European Commission Press Release IP/93/907 of 25th October 1993; OJ C 117 of 28th April 1993, p. 3.
87. OJ C 159 of 10 June 1994, p. 2; IP/94/1236 of 16 December 1994, PJ L 354 of 31 December 1994, p. 75.
88. OJ C 233 of 11th September 1992, p. 2.
89. OJ C 264 of 21 September 1994, p. 3.
90. OJ C 333 of 17th December 1992, p. 3.
91. OJ C 47 of 15th February 1994, p. 6.
92. OJ C 221 of 9 August 1994, p. 9.
93. Commission Regulation 418/85/EEC, OJ L 53 of 22nd February 1985, p. 5 as amended by Regulation 151/93/EEC, OJ L 21 of 29th January 1993, p. 8.
94. See Art. 130 as introduced by the Maastricht-Treaty.
95. OJ L 236 of 22nd August 1986, p. 30; Guidelines no. 138.
96. OJ L 52 of 26th February 1988, p. 51, Guidelines no. 138.
97. OJ C 239 of 25th September 1990, p. 2.
98. OJ L 32 of 3rd February 1990, p. 19.
99. OJ L 228 of 22nd August 1990, p. 31.
100. Case 322/81 Michelin v. Commission (1983) ECR 3461, 3503 para. 30.
101. Guidelines, no. 87.
102. Commission Press Release of 29th January 1990.
103. OJ L 306 of 7th November 1991, p. 22 (notification: OJ C 294 of 7th November 1991).
104. OJ C 7 of 11th January 1992, p. 3.
105. See Guidelines no. 114.
106. OJ L 43 of 15th February 1989, p. 27.
107. See in particular Council Directive 93/98 EEC of 14th June 1993 coordinating the procurement procedures of entities operating in the water, energy, transport, and telecommunications sectors, OJ L 199/84 of 9th August 1993.
108. Council Regulation no. 4064/89/EEC of 21st December 1989 on the control of concentrations between undertakings, OJ L 385 of 30th December 1989, p. 1; corrected by OJ L 257 of 21st September 1990, p. 13.
109. See Christopher Jones/Enrique González-Díaz, The EC Merger Regulation, London 1992 with a list of all cases including a number of companies from the telecommunications industry, p. XIV seq.
110. OJ L 364 of 31 December 1994, p. 1
111. See "Kartellrechtelijke beschikkingen EG-Commissie, 1988-1989" by Mr. M.H. van der Woude, in: SEW 12 (1990) with references (e.g. to the Vlakglas decision).
112. Section 2 (1) of the Telecommunications Installation Act.
113. Section 1a of the Telecommunications Installation Act.
114. 9 of the Telecommunications and Posts Regulation Act.
115. Although this Regulation was based on the DNP Constitution Act and the DBP Constitution act has been repealed by the Postal Reform Package II of 1994, by virtue of section 16 of the Law concerning the Transformation of the Deutsche Bundespost Enterprises into the Legal Structure of Stock Corporation - Stock Corporation Act, this regulation remains in force.
116. Compare Sections 26 and 27 of the Telecommunications Regulation.
117. Section 6 of the Telecommunications Regulation.
118. Section 9 (2) of the Telecommunications Regulation.

119. See, e.g., OLG München, WuW/E OLG 2925, WuW, 9/83, 751.
120. Higher Regional Court of Düsseldorf (Oberlandesgericht Düsseldorf), decision of 12th February 1980, NJW 1980, p. 585-587.
121. See Ellger/Kluth, Wirtschaftsrecht der Internationalen Telekommunikation, p. 155 et. seq.
122. Explanatory Notes to the fifth renewal of the competition law, WuW 4/1990, pp. 283, 354.
123. Higher Regional Court of Düsseldorf, decision of 12th February 1980, NJW, pp. 585-587.
124. The IBM commitment is one of the earliest decisions on Article 86 concerning the System Network Architecture (SNA), a standard issued by IBM for inter-connecting computer systems and networks. The Commission accepted representation from IBM to provide manufacturers with the technical interface information needed to permit competitive products to be used with IBM's computers; see Bull. EC 10-1984, p. 96, press release of the Commission IP (88), 814 and here footnote 99.
125. FCO, Tätigkeitsbericht 1978, 88, special report 9, 35, no. 72.
126. See Art. 21 (1, 2) of the Regulation prohibiting Member States to apply national competition legislation to community-scale concentrations falling within the scope of the Regulation.

TELECOMMUNICATIONS AND SPACE JOURNAL, Volume 2, 1995 ©

UNIVERSAL SERVICE : AN INSTRUMENT FOR REGULATORY CAPTURE? *

DR TIM KELLY,

Head of Operations Analysis, International Telecommunications Union, Place des Nations, 1211 Genève 20 Suisse

I - REGULATORY CAPTURE?

The deliberately provocative title of this paper questions whether Universal Service might sometimes be used as an instrument for regulatory capture. While the concept of Universal Service may well be as pure as the driven snow, the problem lies not so much with how the term is defined as with the way it is used, or rather manipulated, in the cause of certain aims. There are many different players in the modern telecommunications drama — incumbents, new market entrants, resellers, cable TV operators, user groups, consumer groups — and it would not be surprising if they each adapted the script to suit their own interests. This paper asserts that the current debate over Universal Service has more to do with politics than economics and that it is insufficient to provide an economic answer to the question of estimating the cost of providing Universal Service obligations without first providing a political answer.

So what exactly is regulatory capture? A working definition would be that:

> *Regulatory capture describes the situation in which a particular market player uses its influence or superior resources to extract a regulatory decision (or lack of decision) which brings greater benefits to that particular player, or its client interests, than tosociety as a whole.*

It should be borne in mind that the boundary between legitimate lobbying and undue influence is difficult to draw, and of course regulatory capture strategies are not necessarily malign. In many instance, both society and individual players will benefit in relative terms from regulatory decisions. But where the resources available to one player are much greater than those available to other market players, then regulatory capture can be a real problem. The telecommunications industry in its current phase of transition from a monopoly to a competitive market structure is characterised, in many countries, by a "David and Goliath" type market structure. Furthermore, the mindset of many new regulatory bodies is still more concerned with protecting the industry (against foreign rivals) rather than protecting the consumer. Thus the telecommunications sector is particularly vulnerable to regulatory capture strategiesor the activities of special interest groups.

There are many examples of regulatory capture strategies:

1. The most obvious strategy is where the player with superior resources (either deeper pockets or better access to information) attempts to "buy" a particular regulatory outcome through means which are legitimate (e.g., employing consultants or lobbying firms, sponsoring research, organising seminars) or illegitimate means (e.g., bribes, hospitality, poaching etc). The use of due judicial process and the establishment of appeals procedures can reduce (but not eliminate) the likelihood of regulatory capture, but it isinvariably possible for lobbying groups to go above the heads of regulators and attempt to influence politicians directly.

2. A second approach is where the player which owns the main network, and therefore controls much of the network information (concerning traffic flows, client lists, usage patterns, interconnection agreements technical standards etc), discloses information to regulators or firms seeking interconnection rights on a selective basis or, worse still, provides information which is deliberatelymisleading or false. Technical information can be disclosed in such a way as to confuse, obfuscate or overload regulators. There are certain legitimate reasons why a company may wish to withhold information (e.g., customer privacy, commercial confidentiality, national security) but these criteria are often applied in a blanket manner to excuse the selective disclosure of information;

3. A third regulatory capture strategy is the use of the "xenophobia" argument which is designed to create the impression that "letting foreigners in" will somehow lead to widescale loss of jobs, capital flight, loss of national sovereignty or some other terrible disaster. A variant on this xenophobia argument is the "miss the boat" argument which states that other countries are subsidising the development of some strategic new technology and that if the national government does not do the same, then local firms will miss the boat and the "foreigners" will have gained a competitive advantage;

4. Finally, the most insidious form of regulatory capture strategy is the "special pleading" argument which says that infrastructure provision, or some particular service such as voice telephony, is somehow different from all other goods and services and needs to be treated with kid gloves. This is the most insidious argument because it is true, up to a point. No two sectors are the same and there are certain aspects of telecommunications, privacy for example, which may need special regulatory treatment. But these limited cases are often extended to support unjustified arguments for special treatment for the sector as a whole.

So, to what extent does Universal Service fit into these categories of regulatory capture?

II - UNIVERSAL SERVICE: MAKE UP YOUR OWN DEFINITION

Universal Service issues can be used in any of the above strategies for regulatory capture. For instance, most studies of the definition and costs of providing Universal Service are financed by the industry (1). Most studies are based on incomplete and potentially misleading information (2). Many studies conclude that other countries do it better (3), and Universal Service is the most common type of special pleading argument (4). Thus Universal Service can be a powerful weapon to divert, distract or disrupt regulatory actions.

The concept of Universal Service has changed over time. Originally, it meant connecting all subscribers to the same network with the emphasis on interconnection. Later the emphasis changed to a social justice slant of reaching all households and providing a similar level of service. Now the emphasis has changed again towards evaluating the level of compensation payable to the principal network operator (the incumbent) by new market entrants in terms of interconnection fees and a contribution to meeting the costs of uneconomic service obligations (in UK parlance, Access Deficit Contributions or ADCs). The changing meaning of the word reflects the evolution of the telecommunications industry from unconnected island networks, to a monopoly-provided national network, to competing interconnected, privately-provided networks.

Given this shift in the meaning of the term, it is not surprising that there are at least two quite distinct meanings attached to the term Universal Service; the first essentially social/political and the second economic/legal:

- To some, Universal Service means access to telecommunication services as a basic right to all citizens (the right to communicate) which is essential for full participation in the community and as a basic element of the right to freedom of expression.
- To others, Universal Service is a regulatory obligation which can be broken down into a number of subsidiary concepts which include availability, accessibility and affordability:
  — universal geographical availability, or the provision of service on demand irrespective of whoever, wherever and wheneverthe user requests it. This provision is often extended to cover the provision of emergency service calls on a free-to-subscriber basis;
  — non-discriminatory access, which means that all users are treated alike in terms of price and provision of service quality. This provision may be extended to cover access for people with a disability that need a specially adapted service (e.g., videophones for the deaf) or translation and interpretation services for non-native speakers ;
  — reasonable costs or affordability, which is a more subjective concept relating to the comparative cost of the service andthe relative purchasing power of the user. In some countries, the concept of affordability is extended to the provision of untimed or free calls in the subscriber's local area.

The common tendency, especially in international fora in which developing country interests are represented alongside those of developed countries, is to mix and match elements of the two definitions. This is unfortunate and leads to confusion. In the first concept, the implicit responsibility lies with society, or the state, to provide the service directly (possibly implying state ownership) or at least to make provision for users that are cannot reached by normal means by imposing licence obligations on operators. In the second definition, the implicit responsibility lies with the public telecommunication operator(s) to adhere as closely as possible to the principles of Universal Service, without necessarily requiring to cross-subsidise one group of users or one type of service with income from another.

III - UNIVERSAL SERVICE AND INFRASTRUCTURE COMPETITION

In most countries, Universal Service was largely achieved under monopoly conditions and competition was introduced at a fairly advanced stage of network development. However, this historical fact does not necessarily mean that Universal Service can only be achieved under monopoly. Indeed, there is mounting evidence to suggest it might be achieved quicker, and more efficiently and at lower cost under competitive conditions:

- Under the social/political definition, the basic measurement of success is the number of households which have access to a telephone or, if this statistic is not available, the relative level of teledensity in a country. Teledensity has increased in all countries that have introduced competition, and competition has actually speeded up the level of investment in countries where teledensity is low. For instance, in the Philippines, more new lines (around 200'000) were added in 1993 following the award of new licences for infrastructure competition than in all the previous ten years added together (Figure 1). In the UK, the Cable Communications Association is fond of quoting the statistic that some 14 per cent of all subscribers taking joint cable TV and telephony service are first time telephone subscribers. These two examples from different parts of the development scale suggest that competition can assist the achievement service coverage, for the first five per cent as much as for the last five per cent;

- Under the economic definition, the basic measurement is cost of service. Recent research by the OECD appears to confirm that competition has had beneficial effects in reducing prices. Over the last five years, prices for a basket of business telecommunication services have fallen much faster in competitive markets than in non-competitive markets. For residential telecommunication services, prices in competitive markets have fallen while those in non-competitive markets continue to rise (see Figure 2).

Furthermore there is evidence that service availability has grown faster in competitive markets than in non-competitive markets. Digitisation of networkswitches is proceeding faster (and for a cost per line which is roughly half as much) in competitive OECD markets than in non-competitive markets (Table 1). But the clearest evidence comes in cellular markets where both the level of growth (new subscribers added) and the geographical roll-out, has been faster in competitive than in non-competitive markets (Figure 3). Penetration rates in competitive markets averaged more than 3 per 100 inhabitants compared with just 0.22 in monopoly markets.

Figure 1: Teledensity, 1983-93, and new licensees awarded in the Philippines

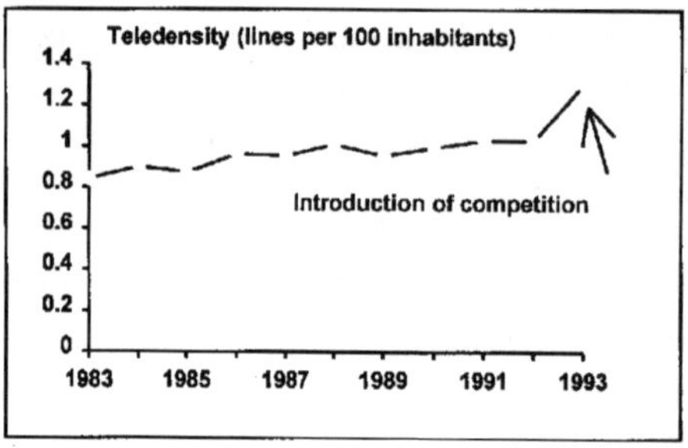

Company	*Licenses*	*Lines to install*
Capwire	IGF	300'000
Digitel	IGF	300'000
ETPI	IGF	300'000
Globe	IGF, CMTS	700'000
ICC	IGF	300'000
Isla	IGF, CMTS	700'000
Philcom	IGF	300'000
Piltel	CMTS	400'000
Smart	IGF, CMTS	700'000
TOTAL		**4'000'000**

Note: IGF = International Gateway Facility. CMTS = Cellular Mobile Telephone System.
Source: ITU World Telecommunications Indicators. Philippines National Telecommunications Council.

Figure 2: Tariff changes, 1990-94, in competitive and non-competitive OECD markets
Index 1990 = 100

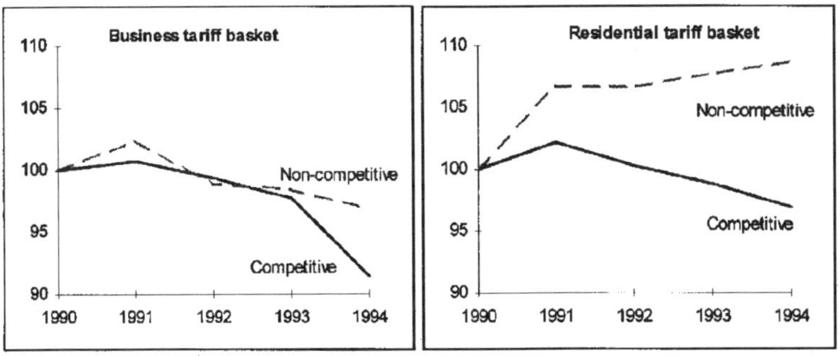

Note: Competitive market defined as Japan, NZ, UK, UK (1990-94), Canada, Australia (1992-94), Sweden (1993-94). Calculations based on OECD tariff comparison methodology, using PPP exchange rates.
Source: OECD Communications Outlook, 1995 (forthcoming, November 1994).

Figure 3: Competition stimulates growth
Subscribers and penetration rates for competitive and monopoly cellular radio markets, 1987-93

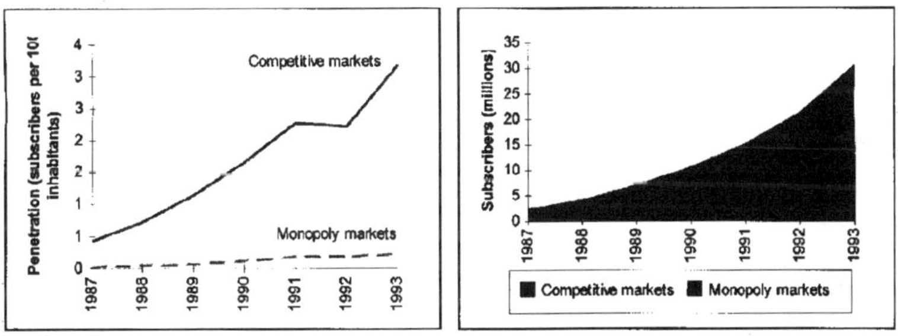

Note: Analysis includes 26 countries which had initiated a cellular radio service by 1987 and had more than 20'000 subscribers by the end of 1993. These markets cover around 90 per cent of the total world market. Competitive markets are defined as those with more than one firm competing for business in the same geographical area. At the start of the period, there were six competitive markets. By the end of 1994 some 16 out of the 26 countries were competitive.

Source: ITU World Telecommunication Indicators Database.

So, at a simple level, there are fairly convincing arguments to show that competition can assist the development of Universal Service both directly, in promoting service growth, and indirectly, by making the cost of service provision cheaper. If this is the case, then perhaps the task of the regulator should be simply to permit competition in the marketplace wherever possible, and to let market forces do the rest. One of the great ironies of the debate is that some of the strongest proponents of maintaining Universal Service are monopoly service providers who have singularly failed to provide Universal Service themselves and have long waiting lists. And yet they claim that the people without a telephone will have their rights endangered if another service provider is allowed to offer service! This is a clear example of a regulatory capture strategy whereby monopolists claim they are protecting Universal Service but are actually protecting their monopoly privileges.

Table 1: Digitisation statistics and investment efficiency in competitive and non-competitive OECD markets

Country	*Level of digitisation (%)*		*New digital lines added (k)*		*Average investment per digital line In US$*
	1990	*1992*	*1990-92*	*% replacement*	
Netherlands	32	83	3'917	88%	712
Portugal	30	54	994	21%	1'158
United Kingdom	**47**	**64**	**5'408**	**71%**	**1'254**
United States	**44**	**60**	**26'286**	**71%**	**1'328**
Luxembourg	31	70	87	74%	1'417
New Zealand	**72**	**95**	**400**	**84%**	**1'504**
Greece (1)	1	11	455	-20%	1'554
Canada	**60**	**80**	**3'785**	**75%**	**1'571**
Ireland	55	68	216	40%	1'834
Australia	**15**	**40**	**2'135**	**78%**	**1'910**
Belgium	37	48	591	41%	2'120
Denmark	28	39	363	74%	2'165
Iceland	39	49	18	46%	2'247
Sweden	**38**	**54**	**968**	**93%**	**2'372**
Japan	**39**	**60**	**13'325**	**77%**	**2'577**
France	75	83	3'817	52%	2'623
Finland	35	51	448	84%	2'669
Norway	38	50	324	58%	2'818
Italy	33	48	4'100	67%	3'665
Austria	11	27	581	58%	5'356
Germany	12	25	5'029	30%	5'783
Spain	28	36	1'492	20%	5'990
Switzerland	29	42	614	61%	6'819
OECD	40%	57%	75'353	67%	2'238
OECD competitive	43%	61%	52'307	73%	1'700
OECD non-competitive	35%	49%	23'046	52%	3'459

Note: "% replacement" = Percentage of digital lines which are replacement lines rather than net new additions to the network.
"Average investment per digital line" calculated by dividing total investment (1991 and 1992) minus spending on net new lines (estimated to be US$1'000 per line) by new digital lines installed in 1991 and 1992.
(1) "Percentage of digital lines which are replacement lines" in Greece is negative because pace of digitisation is slower than the pace of installing new lines (in other words, some new lines being installed are analogue).
Countries highlighted in bold type are those which permit some degree of infrastructure competition. Table includes all OECD countries except Turkey and Mexico.

Source: ITU World Telecommunication Indicators Database.

IV - COSTING OUT UNIVERSAL SERVICE OBLIGATIONS

But the task of defining Universal Service Obligations (USOs) is more complex than this simplistic analysis would allow. The problems start when elements of the social/political and the economic/legal definitions become mixed up. For instance, policy-makers often impose obligations to provide services to uneconomic users on incumbent operators rather than on the state. This mixes social obligation (definition 1) with non-discriminatory access (part of definition 2). But what does non-discriminatory access mean?

- Geographically averaged prices (i.e. applying the same tariff structure to all consumers irrespective of where they live). This does not necessarily exclude tariff options such as discounts for volume users, but it does imply the availability of such options on a uniform national basis;
- Specific and limited criteria for refusal of service (e.g., evidence of criminal use of telephone, consistent non-payment of bills) which are universally applied;
- Uniform service quality, at least for a standard rate subscription. Users requiring a higher service quality or faster maintenance times could be given the option of paying a higher tariff;
- Assistance for disadvantaged members of the community who need special help in gaining access to the basic telephone service (e.g., people with disabilities);

These obligations could be applied equally well to new market entrants as to incumbents. However, not all new market entrants will have a network which is extensive enough to reach all potential subscribers. Indeed many service providers may choose not to construct their own networks but will lease facilities instead. Furthermore, at least in the early stages of the transition to a competitive telecommunications environment, the incumbent operator is much better known by the majority of low income subscribers and will therefore be the first point of call when requesting service. So, during this transition phase, it is logical that the universal service burden will fall unevenly, though not exclusively, on the incumbent service provider. To what extent should the incumbent be compensated and how is the burden to be measured?

Some of the classic research on costing out the Universal Service obligation was carried out in Australia. The Bureau of Transport and Communications Economics (BTCE) in 1989 published a study aimed at estimating the cost of the costs and subsidies associated with meeting specific community service obligations. The study was carried out in the context of setting financial targets for Telecom Australia (now Telstra). The company also produced its own study. Unsurprisingly there was a wide gap between the two estimates. The BTCE adopted an approach based on an "avoidable cost" methodology whereas Telecom Australia took a "fully distributed cost" methodology. The two approaches each produced a range of cost estimates according to different assumptions on the cost of capital and other input variables. The BTCE study produced a basic cost estimate of A$240 million for the company plus a further A$50 million for the Government (the Australian Departments of Social Security and Veterans Affairs provide concessions to pensioners towards the cost of telephone line rental). Using the same costs of capital, the company's study produced an estimated cost of A$800 million using the fully distributed cost model. The BTCE study concluded that the cost of providing Universal Service obligations was almost exactly matched by the profits generated from other customers, for instance from calls made to the uneconomic subscribers, and therefore that no external subsidy was necessary. Needless to say, the company disagreed.

The logical conclusion of the Australian experience is that everything depends on the methodology adopted and the purpose for which a study is undertaken. It is equally possible to show that the cost of providing USOs is high or low depending on the particular research objective. As noted earlier, Universal Service is especially prone to regulatory capture in that it most of the studies are funded by a player in the market with a particular interest to defend (e.g., the incumbent, the new market entrant, the regulator) and there can be little objectivity, even on the part of the regulator.

V - DECONSTRUCTING UNIVERSAL SERVICE

It is difficult to argue against Universal Service. It stands for all that is best in human nature — social justice, equality, freedom of speech, the right to communicate and other wholesome qualities. But it is this very goodness that makes the concept particularly vulnerable to regulatory capture. It also makes it difficult to separate out the myth from the reality. For instance:

- It is often claimed that monopoly is the natural bastion of Universal Service. In practice, the evidence presented in this paper shows that Universal Service is often more efficiently provided under competitive market conditions;
- The concept of "uneconomic customers" is often poorly defined.

There are actually two types: those that cost a lot to serve (for instance, because they live in remote or rural areas) and those which cannot afford to pay for the service or use it so sparingly as to make it not worthwhile for the supplier to carry out billing, debt collection and maintenance. Both groups are diminishing in size.Technological improvements are making low density service provision much cheaper (e.g., fixed cellular and other radio-based solutions). Cost-oriented tariff rebalancing and the introduction of low-usage tariffs is reducing the number of cross subsidised customers. The uneconomic customers which remain are more likely to be in inner city housing estates than in remote rural areas.

- It is often claimed that geographic averaging of pricing subsidises uneconomic customers. In practice, geographical de-averaging may well lead to lower prices in cities and other highly densely populated areas which is where the majority of low income users live;
- It is often claimed that the cost of providing universal service falls solely on the incumbent. In practice the burden is more evenly shared, especially after the first few years of transition to a competitive market environment and when market share is more equally distributed;
- It is often claimed that the cost of providing universal service is high. In practice, the cost is quite mo dest and is often offset by the network externality gains of having a larger network (e.g., extra calls generated);
- It is often claimed that serving the local loop is a fundamentally loss-making activity which needs to be subsidised by profitable long-distance and international services. In practice, local loop operators have more often than not proved more profitable than long-distance operators (for instance in the United States, or in isolated cities such as Hull or Edmonton) and, because of their direct access to customers, are often in a stronger position to develop new services;
- It is often assumed that all people would install a telephone if they could afford it. In practice many people decide not to install one for reasons of privacy, a desire for tranquillity or due to technophobia, as well as for reasons of financial prudence.

Nevertheless, even after taking a deconstructionist approach and setting these false assumptions aside, there still remains the problem of how to finance uneconomic users. There is no single solution which is likely to answer all needs, but a few principles are proposed here for discussion:

- Subsidies should be targeted to specific needs wherever possible. For instance, an unemployed job-seeker needs to be able to make and receive calls but not necessarily to own a telephone. This could be arranged through the provision of free phonecards and/or a voice-mailbox to provide a "virtual telephone" for a specific period of time. Similarly, an elderly person living alone needs to be able to make calls to a limited range of numbers (to relatives, for security etc). A subsidised phone rental scheme, such as that operated in Australia, could provide this service.
- The regulatory and operational aspects of providing service should be separated wherever possible. For instance, the task of defining users who are likely to require support should be left to the relevant professional experts (social services, the police, unemployment agencies) not to the operator. Where possible, the subsidies should be administered directly by these agencies, not by the operator.
- Universal Service Obligations should be opened to competitive tendering wherever possible. For instance, the obligation to provide service to a particular group of uneconomic users could be "auctioned" using a type of Dutch auction whereby the company proposing the lowest subsidy gains the right to provide the service. This should facilitate market entry and should make it easier for regulators to judge the real cost of the Universal Service burden.
- The residual costs of providing Universal Service should be provided by the telecommunications industry as a whole rather than by individual players or by the state. This can be done either by hypothecation, whereby a proportion of overall tax revenues from telecommunications operators (from sales tax and/or corporation tax) is devoted to the purpose, or by establishing a common fund to which all players contribute, such as the Universal Service High Cost Funds (USF) in the USA.

CONCLUSION : A CAPTURED CONCEPT?

For too long, Universal Service has been a captured concept; a bit like a fairground curiosity that is wheeled out to frighten and to impress. In reality, the beast is quite harmless when reduced to its true dimensions. Universal Service Obligations will not go away — as Jesus said, "the poor are always with us" — but the problem can be managed. Three steps are necessary:

- Recognising how and why the concept of Universal Service is manipulated by different players to achieve specific aims and taking steps to avoid regulatory capture;
- Separating the social/political and the economic definitions of Universal Service and breaking each down into its component parts;
- Applying principles of social justice and equality of access in order to define specific, measurable objectives to be met under a Universal Service programme, and developing policies which utilise market mechanisms to achieve them.

Hopefully following these guidelines it may be possible to establish a fresh approach to the definition and management of Universal Service Obligations.

* The opinions expressed in this paper are those of the author and do not necessarily reffect the opinions of the ITU or its members . A version of this paper was given at the Analysys/Mercury symposium on Universal Service Obligations, held in Cambridge, December 1994.

TELECOMMUNICATIONS AND SPACE JOURNAL, Volume 2, 1995 ©

DEVELOPMENTS IN ACCESS TO SPACE SEGMENT CAPACITY OF INTERNATIONAL SATELLITE ORGANISATIONS

TIMOTHY JOHNSON AND SEONG SIN HAN

S J Berwin & Co, Solicitors, 222 Grays Inn Road, London, WC1X 8HB, United Kingdom Tel: (44) 0171 837 2222 ; Fax: (44) 0171 833 2860.

INTRODUCTION

The International Telecommunications Satellite Organisation ("INTELSAT") Agreement entered into force on 12 February 1973. The European Telecommunications Satellite Organisation ("EUTELSAT") Convention came into force on 1 September 1985 with its Convention and Operating Agreement largely based on the INTELSAT Agreements. Since their inception, due to the provisions of their constitutive documents, examined in more detail below, access to the satellite capacity (often referred to as "space segment capacity") of these international satellite organisations ("ISOs") has generally only been available through the local Signatory (normally the national telecommunications operator) of the ISO.

In recent years the rapid developments in technology and the establishment and increase in competition in this sector has meant that the issue of access to the ISOs' space segment capacity has become keenly debated.

Responding to the increasing pressures for change, much emphasis and attention has been placed on this issue, in particular on the part of the ISOs and the European Union. Steps are finally being taken to liberalise the position in respect of access to the ISOs' space segment capacity. The satellite communications equipment and services industry has clearly underlined the crucial importance of resolving the issues of access to the space segment in achieving further progress. It has been argued within industry circles that an uncertain or unstable policy would have a significant negative impact on investment and development for space segment operators, satellite service providers and the space industry within Europe [1]. This article reviews the current state of affairs with a view to considering the merits and likely impact of the proposals for change that have been put forward.

I - GENERAL STRUCTURE OF THE ISOs

Both INTELSAT and EUTELSAT have an essentially similar organisational framework. The constitutive agreements are in two parts, the first (the Agreement and the Convention for INTELSAT and EUTELSAT respectively) establishes the organisation itself along with their structure and scope, and the Operating Agreement defines the procedures and details of the normal running of the business of the organisations.

The framework of these organisations is such that individual states are Party to the INTELSAT Agreement/EUTELSAT Convention and constitute the Assembly of Parties [2], while a telecommunications operator in each country is designated by the respective Parties as Signatory to the Operating Agreement, and these Signatories compose the Meeting of Signatories/Board of Signatories (for EUTELSAT) [3]. INTELSAT has a further Board of Governors [4] composed of certain Signatories who meet specified minimum investment requirements. Both organisations then have an executive organ headed by a Director-General [5].

Taking a broad overall view, the Assembly of Parties is responsible for general policy and long-term objectives of the organisations [6]. The EUTELSAT Board of Signatories and the INTELSAT Board of Governors has responsibility for the operation and maintenance of (the organisation's) Space Segment and for any other activities which (the organisation) is authorised to undertake [7], including the approval of earth-stations and the adoption of terms and conditions governing the allotment of Space Segment capacity.

II - ACCESS TO SPACE SEGMENT UNDER THE CONVENTIONS AND OPERATING AGREEMENTS

There are three main problems arising from the provisions of the EUTELSAT and INTELSAT constitutive documents which have been seen as hindering direct access by non-Signatories to the space segment of INTELSAT and EUTELSAT.

2.1. Earth Station Approval and Allotment of Space Segment Capacity

All applications for approval of an earth station to utilise the ISOs' space segment [8] and for access to space segment capacity [9] are made through the Signatory in whose territory the earth station is, or will be, located. This has been perceived as giving rise to an inherent conflict of interest between the Signatory's role as a provider of satellite and telecommunications services and as monopoly supplier of space segment capacity.

Such applications may also be made by a duly authorised telecommunications entity ("DATE") [10], authorised by its national regulatory body, with respect to earth stations located in Territories not under any Party's jurisdiction. This has the odd consequence that access to ISO space segment capacity may be easier to obtain by an entity from a country that is not a Party to the ISO than by a non-Signatory entity based in a country that is a Party to the ISO.

2.2. Liability

The responsibility for ensuring compliance by the earth station with the rules and standards set out in the approval document and the terms and conditions for allotment of space segment capacity lies with the Signatory or DATE making the application. With respect to EUTELSAT, where a Signatory has made the application for either earth station approval or access, its designating Party can assume the Signatory's responsibility [11]. INTELSAT, however, as originally constituted, allows a Signatory's Party to assume responsibility only with respect to some or all of the earth stations not owned or operated by the Signatory itself [12].

2.3. Investment Share/Utilisation Charges

Another difficulty that faces the ISOs is that the investment shares of Signatories varies according to the volume of use by each respective Signatory (which includes use by any entity obtaining access to space segment capacity through that Signatory) [13] which is calculated as the percentage of the total utilisation of the Space Segment by all Signatories.

There are also utilisation charges which the Signatory or DATE is also liable to pay. These charges are intended to cover the operating, maintenance and administrative costs, the provision of operating funds, the amortisation of Signatories' investment and compensation for the use of Signatories' capital [14] (essentially a return on their investment). In practice this charge is not as significant as the investment share which usually equates roughly to two to three times the yearly utilisation charge. Further, the utilisation charges are not considered a major issue as they are ordinarily passed on to the user under the basic arrangements for the use of the capacity. It is therefore seen as less of a burden on the Signatory but nevertheless might influence the impartiality of a Signatory's handling of an application since in the event of default in payments the liable Signatory would have to pay the utilisation charge out of its own funds or otherwise be liable to suffer any sanctions which may be imposed.

2.4. Summary

These arrangements have led to difficulties because, in the majority of cases, the national Telecommunications Operator ("TO"), eg. BT in the UK, is the designated Signatory. This of course means that while providing and promoting their own satellite services, the national TO also fulfils the role of providing its potential competitors with space segment capacity and/or as (partially) substitute regulator. The likelihood of conflicts of interest arising is therefore very high as is the temptation for the Signatory to impose discriminatory or anti-competitive terms, including monopoly pricing.

III - LOCAL ATTEMPTS TO IMPROVE ACCESS

In the absence of any concerted attempt to reform the ISOs to allow easier access, a number of steps have been taken to try and address the problems in some jurisdictions.

For example, in the UK, following a review in 1989 of BT's position as Signatory to the ISOs, the UK telecommunications regulatory body, OFTEL [15], required BT to set up an independent arm to deal with its function as Signatory, known as the Signatory Affairs Office ("SAO") [16]. The SAO's objective is to segregate the administrative role of providing access arrangements for other service providers from the commercial business of BT [17]. It should be noted that in order to carry out its role objectively and fairly, the SAO guarantees that it will not greatly mark up the charges payable by the applicant. The SAO agreed at the time of its formation to make no more than a 7% mark up, which is intended purely to cover the administrative costs [18]. OFTEL made it clear at the time that this was only an interim solution, with the long term objective being the reform of the ISOs to allow direct access. This has been viewed by the European Commission [19] as a possibly acceptable short-term measure, pending changes in the ISOs' treaties.

More recently, in September 1992 arrangements were put in place between a select few EUTELSAT Parties and Signatories to make access more freely available. Four EU States; France, Germany, the Netherlands, and the UK, have an arrangement of multiple access whereby an entity from one of these countries wishing to obtain access to EUTELSAT space segment may make an application to the Signatory of any of these countries. In addition, Switzerland, while not an EU Member, joined this arrangement at the end of 1994. This example has not however been followed by the other Parties, although they are encouraged by EUTELSAT to do so. Within this arrangement, the allocation of the financial liability and responsibility is determined by bilateral agreements entered into by and between the Signatories with the Parties informing EUTELSAT of their consent. These agreements basically provide that all responsibility relating to the earth station and financial liability is assumed by the Signatory through whom access was granted and the use by the applicant is considered part of that Signatory's total space segment capacity use.

IV - EUROPEAN UNION

As the ISOs are the largest providers of space segment capacity in Europe, the attention of the European Commission has been focused on changing their constituent instruments, or as a minimum measure, the achievement of a flexible interpretation of those instruments [20]. Aside from the need to meet the demands of the evolving market, the European Commission considers change to be a necessity particularly in light of the EU's competition rules and telecommunications policy. With the Treaty on European Union identifying the establishment of trans-European networks in telecommunications as a key EU policy, the free availability of the required space segment capacity is thought to be fundamental by the Commission. The institutions of the EU have recognised the importance of the field of satellite communications not only in terms of industry but also as a factor in achieving the complete integration of Europe as a single community [21].

The EC has been very active in recent times and in the past few years in particular. As can be seen from the Green Paper on a common approach in the field of satellite communications in the European Community [22], adopted by the Commission in November 1990, it was recognised that major changes were required in the regulatory environment in order to fully exploit the potential of satellites. This Green Paper proposed four major changes in extending the application of the general agreed principles of Community telecommunications policy to satellite communications:

i) Full liberalisation of the earth segment.
ii) Free (unrestricted) access to space segment capacity, subject to licensing procedures, such access being on an equitable, non-discriminatory and cost-oriented basis.
iii) Full commercial freedom for space segment providers, including direct marketing of satellite capacity to service providers and users.
iv) Harmonisation measures as far as required to facilitate the provision of Europe-wide services.

In implementing these principles, the attention of Member States was initially focused on earth segment issues and a certain amount of progress has been made in this area in terms of both liberalisation and harmonisation. However with the rapid changes occurring in the market, the Commission turned its attention to access to the space segment resulting in a Communication to the Council and the European Parliament concerning the provision of, and access to, space segment capacity, on 10 June 1994 [23]. The Commission stated that it intended to use all the available powers under the EU competition rules to eliminate restrictions relating to access and called for steps to be taken to liberalise fully the arrangements for access to space segment capacity of the ISOs. After due consideration of this Communication, the Council adopted a Resolution [24] in November last year agreeing with the Commission and calling upon Member States to provide non-discriminatory access to space segment capacity at a national level and for Member States to collaborate closely in the reforms being considered by the ISOs and in managing orbital frequency resources.

In addition, an EC Directive on the liberalisation of satellite services and equipment was adopted on 13 October 1994 [25], making amendments to previous Directives with particular regard to satellite communications. This Directive requires Member States to abolish all "special" and exclusive rights which have the effect of restricting the ability of other undertakings to enter the market. Member States are required to ensure that any fees imposed on providers of services as part of authorisation procedures be based on objective, non-discriminatory criteria; "any regulatory prohibitions or restrictions on the offer of space-segment capacity to any authorised satellite earth station network operator" be abolished and to authorise "any space-segment supplier within their territory to verify that the satellite earth station network for use in connection with the space-segment of the supplier in question is in conformity with the published conditions for access to his space segment" [26].

V - PROPOSED CHANGES/CHANGES THAT HAVE BEEN IMPLEMENTED

The evolution of the satellite telecommunications market has prompted the ISOs and the EU to re-evaluate the present arrangements for access to space segment capacity in order to meet the realisation of the need for easier access as a prerequisite in the maintenance of a commercially viable Pan-European and global satellite communications network. The changes proposed by EUTELSAT and implemented by INTELSAT appear to take different approaches, due in part to the actual wording of the constitutive Agreements and the commercial differences between the two organisations.

5.1. EUTELSAT

Considering the nature of the EUTELSAT organisation, the attentions of the European Commission and the developments in the market, clearly long-term measures are needed which can only be brought about by changes in the constitutive agreements. The provisions for access and earth station approval state that applications may only be made by the Signatory [27]. In order to have any long-term direct-access arrangements, this will obviously need to be amended.

In this context, EUTELSAT's Parties and Signatories are continuing to study how best this can be achieved. While it has been agreed by the Parties and Signatories to EUTELSAT that amendments to the Convention and Operating Agreement are required in several significant areas, the reality of tackling this is being approached with a great deal of caution.

EUTELSAT is considering the commercialisation of the organisation with the possibility of transforming into a corporate entity, either under the umbrella of an international treaty, or as a completely private company.

It is projected that this process will be completed by the end of 1999, although EUTELSAT has conceded that to achieve even a "quasi-company" will be difficult and in order for this to be achieved, the details involved must be clear and finalised by the end of 1998. If changes are desired sooner, in EUTELSAT's view, an acceptable compromise of some sort would have to be reached, retaining some of the features and objectives of the present EUTELSAT with only a partial redefinition of its aims and philosophy.

EUTELSAT is also addressing the provision of unrestricted access to space segment capacity. In December 1994, the Assembly of Parties decided that Non-Signatory Entities ("NSEs") will be allowed direct access to EUTELSAT Space Segment, provided they invest in the Organisation in proportion to their utilisation in the same way as the Signatories at present, although without any voting powers. It has further been decided that such NSEs must invest to the same minimum level as the Signatories, namely 0.05% of the total investment share. The NSEs will in effect have "quasi-Signatory" status. This right will be obtainable by NSEs within a framework of national arrangements between the regulatory authorities, the Signatory and the NSEs in the country in question. Any such arrangements are able to be entered into immediately subject only to a requirement of prior notification to EUTELSAT. At present no arrangements have been notified to EUTELSAT under this scheme.

This is stated by EUTELSAT to be a short term solution which can be introduced without amending the Convention and the Operating Agreement.

One proposed long-term option is that of allowing multiple Signatories to be designated by each Party. While this would lessen the likelihood of conflicts of interest by allowing more channels through which access can be gained, it leaves the issues of investment and compliance unchanged. The effect of such a multiple Signatory structure is uncertain as it will depend on the view taken by individual Parties and their Signatories. In countries such as the UK where full liberalisation is advocated, the Party would likely consider the approval and designation of all significant telecommunications operators within their territory who wanted such status. This would, in effect, provide direct access completely while also solving the issue of liability in that, as Signatories, each user will be under the same obligations but only with respect to their own use. Nevertheless this option would create other problems outwith the issues of direct access. At present all 44 Signatories are members of the Board and any future increase would compound the existing problems of an unwieldy governing structure. If a multiple Signatory system was to be implemented there would need to be corresponding amendments to the Convention to accommodate this. This is nevertheless likely to be agreed by the Assembly of Parties at their Autumn meeting this year.

While the temporary direct access arrangement detailed above may change the nature of the procedure and resolve some of the issues of conflicts of interest, the direct nature of the arrangement does not extend fully to other areas such as the liability for investment share and responsibility for compliance with EUTELSAT's terms and conditions. The NSE seeking direct access will have primary responsibility for meeting its investment requirement, and although EUTELSAT will make strenuous efforts where necessary to recover such sums from the NSE directly, ultimate liability will remain with the Signatory itself. Further, the responsibility for compliance with the required terms and conditions will also remain with the Signatory. While these issues will no doubt be addressed to some extent by the arrangement between the regulatory body, the Signatory and the NSE, it still remains that the perhaps slightly undesirable scenario will remain whereby the Signatory will be responsible for a body over which it exercises no direct control.

With regard to compliance with EUTELSAT's earth station standards, the Board of Signatories decided that the liability of a Signatory be limited to just the initial compliance upon receiving approval. The operation of the earth station is now the sole liability of the actual operator. EUTELSAT envisage that with the EC Directive of 13 October 1994 [28], further changes will be implemented to remove the Signatories from this process completely.

A further facet of this arrangement, at present, is that, unlike INTELSAT's developments (see below), there is no provision for "investment trading" by the NSEs, whereby an entity may request that its investment share be decreased, on the condition that the reduced amount be taken up by other investors. This is allowed by the Signatories under Article 6(h) of the Operating Agreement. This is however stated by EUTELSAT to be able to be easily implemented and is being considered at the moment.

The proposals for amendments to the existing texts are expected to be considered at the next meeting of the Assembly of Parties in May although the only decision that will be made at this meeting will be an amendment to Article XIX of the Convention which governs the procedure for amendments. The proposals to be considered are likely to take into account the likelihood of a changed role for the Signatories, especially considering the recent EC Directive regarding satellite communications [29]. One possibility is that the Signatory's role will be merely regulatory with access being available to all those telecommunications operators that have been duly authorised on a national level.

5.2. Intelsat

Recent changes appear to address all the problem areas highlighted above concerning access, investments shares and charges, as well as responsibility for compliance with the terms and conditions for access and use.

While among the INTELSAT Parties there is no equivalent to the EUTELSAT multiple access arrangement, the Board of Governors of INTELSAT has decided to implement changes in order to provide direct access to the Space Segment by the actual or prospective service supplier. A series of Board of Governors decisions were taken from March 1993 to September 1994 in order to implement changes in access arrangements and signatory liability and so achieve a position where non-Signatory entities could deal directly with and obtain access to space segment directly from INTELSAT itself, if authorised by their Signatory to do so.

a). Utilisation Charges and Sanctions

On the request of a Signatory, it will not be held liable for payment of utilisation charges attributable to those entities authorised by the Signatory to access the INTELSAT space segment ("authorised entities"). Instead the INTELSAT Management is authorised to carry out credit checks on such authorised entities and impose any requirements to ensure timely payment by them. Further, such authorised entities will themselves be directly liable to the imposition of sanctions denying additional use if in default of their utilisation charges. If such default continues beyond 150 days all service to that user may be suspended.

b). Liability for Earth Stations

While the Agreements allow a Party to assume the Signatory's own liability, they may now agree to assume the liability of its designated Signatory with respect to damage resulting from the mis-operation of authorised entities' earth stations, although such liability will not extend to financial liability.

d). Investment Share

In the final meeting to date of the Board of Governors concerning this issue, it was decided that INTELSAT, on the request of a Signatory, would refrain from holding the designating Signatory liable for payment of investment share financial adjustments and capital contributions attributable to its authorised entities.

The Board of Governors noted that the INTELSAT Management, if requested by the Signatory, will deal directly with an authorised entity for the billing and payment of capital contributions, repayments of capital, compensation for use of capital, and financial adjustments related to investment share changes. In these instances the authorised users will be primarily liable for such payments and appropriate steps may be taken to obtain payment, including using the authorised entity's investment share as collateral. However, any payments that might become due under Article 18 of the Operating Agreement, relating to liability for claims for loss or damage, will remain the Signatory's liability.

As regards investment by authorised entities, it was decided that only actual users of the INTELSAT system may be authorised by the Signatory to invest and may only withdraw its investment, while still a user, with the Signatory's agreement.

Investment Trading and Minimum Investment

It was decided by the Board of Governors that a Signatory may permit an entity authorised to invest, to request a greater or lesser investment share under Article 6(d) of the Operating Agreement which allows "investment trading". Year end investment share trading between Signatories has long been a feature of INTELSAT and there has traditionally been an excess of demand over supply. The system now in place provides for a first round of investment share trading restricted to Signatories followed by a second round open to all investors. The only restriction is that an investor must retain a minimum investment.

The calculation of the minimum investment requirements set out in Article 6(h) of the Operating Agreement, namely 0.05% of the total investment shares, have also been redefined to take into account the independent investment by authorised entities, in that it is the combined investment shares of the Signatory and the authorised entities within its country. In addition, at a Signatory's request, INTELSAT Management will apportion the national investment share between the Signatory and its authorised entities. The INTELSAT Management will however play no part in any financial settlements or adjustments between a

Signatory and its authorised entities, except where such adjustment arises solely from the annual adjustment of investment shares comprising part of the general INTELSAT-wide settlement process.

Types of Access and Services

An entity designated to have direct access to INTELSAT is termed an "Appointed Customer", with four ascending and cumulative levels of access available to a non-Signatory entity:

i) Operational/Technical Access - participation in global Traffic and Operations Representatives meetings; access to published capacity availability and assignment information and INTELSAT's Information Systems; receipt of operational and technical material.

ii) Commercial and Service Matter Access - eg. Meetings with INTELSAT Management staff on commercial and capacity availability; receipt of the INTELSAT Service Guide and Tariff Manual; Access to Occasional Use Video capacity availability information through the INTELSAT ESTA/TV-MAX.

iii) Contractual Access - direct ordering of all, or specified, INTELSAT Services, including the declaration of Long Term Commitments; Submission of Earth Station antenna applications; and devolution of billing liability for satellite utilisation charges to the service ordering entity.

iv) Investment.

A Signatory is able to authorise either some or all of the services available under the first three of the above levels, although Signatories are encouraged to authorise all the services in the particular level of access to be granted.

In contrast to the likely outcome of EUTELSAT's review, these arrangements have and will require no rewording of the constitutive Agreements, and are already in place and functioning. However, their effectiveness in providing direct access will be dependent on the particular Party and Signatory as the appointment of an authorised entity is still through the Signatory and the devolution of liability for those entities will be at the discretion of the Party. The nature of the circumstances is such that different Parties are at varying stages of economic development and as such will have different priorities and policies. For example, the UK, which is strongly in favour of full liberalisation, and has been for many years, has submitted a blanket approval for all UK telecommunications operators to have full service and investment access to INTELSAT, whereas many other Parties consider the development of a single national TO of greater importance.

INTELSAT's Assembly of Parties set up a Working Party in order to explore the possibilities for changing INTELSAT's structure and long-term future. The role that has been assumed by this group is that of an advisory body to the Assembly of Parties. The efforts of the Working Party are aimed at enhancing INTELSAT's competitiveness and ability to adapt to the rapidly changing telecommunications market. In a similar vein to EUTELSAT's review, options are being studied to increase the commercial character of the organisation, possibly even the transformation of INTELSAT into an international corporation. This possibility does not, at present, appear likely to be implemented due to the desires of the Assembly of Parties.

Amendments to the current Agreements are also being mooted to accommodate greater direct access and greater flexibility in investment, possibly by means of allowing multiple Signatories. Potential revisions in investment arrangements may eventually allow those directly accessing users making sufficient use of INTELSAT's space segment capacity, who at present have no voting rights, other than those rights included in the national investment share, to achieve parity with the present Signatories in terms of both responsibilities and rights. For example, in the UK, both Cable & Wireless and Mercury Telecommunications plc are both significant users of INTELSAT services and if they were to be granted individual full Signatory status, would likely qualify, under the present rules, for a position on the Board of Governors.

The Working Party's attentions are also turned to the broader range of issues inherent in any structural change, including governance, regulatory policies, finance and personnel. Possible changes being considered by the Working Party include the concept of a subsidiary entity, which would allow for the expansion of services offered by INTELSAT to take into account the developments taking place in the telecommunications market. Presently, INTELSAT's main commercial activities are public voice telephony, business services, broadcasting, and domestic/regional services.

In addition, other short to medium term measures are being considered to enhance INTELSAT's ability to discharge its over-riding mandate, which will remain inviolate throughout the course of any discussions, of providing universal access to global telecommunications and non-discriminatory charging.

CONCLUSIONS

The developments that have been outlined in respect of access to the space segment capacity of both EUTELSAT and INTELSAT are to be welcomed and represent a positive step towards achieving a degree of freedom. However, their effectiveness will need to be judged on the basis of real commercial facts, not simply a review of the legal steps. For

EUTELSAT, it is clear that the proposals for change are at an embryonic stage; there has been no real movement in the commercial position for prospective users. By contrast, the steps taken by INTELSAT, driven by the desire of its members, are more palpable and have constituted a real attempt to address the problems of access. It remains to be seen how many users will actually benefit from the new scheme during its first year of operation and also whether, later this year, there will be any difficulty for users who have acquired investment shares to off-load unwanted capital contribution obligations during the forthcoming rounds of investment share trading.

A more detailed review of developments in access to ISO space segment capacity will be found in the second edition of "Satellite Communications in Europe" by T. Johnson, S. White and S. Bate to be published later this year by FT Law and Tax.

REFERENCES

1. Communication from the Commission to the Council and the European Parliament on Satellite Communications: The Provision of - and Access to - Space Segment Capacity [COM(94)210 Final 10/6/94].
2. INTELSAT Agreement Article VII/ EUTELSAT Convention Article VII.
3. INTELSAT Agreement Article X/ EUTELSAT Convention Article VIII.
4. INTELSAT Agreement Article IX.
5. INTELSAT Agreement Article XI/EUTELSAT Convention Article XIII.
6. INTELSAT Agreement Article VII/EUTELSAT Convention Article IX.
7. INTELSAT Agreement Article X/EUTELSAT Convention Article XII.
8. INTELSAT Operating Agreement Article 14/EUTELSAT Operating Agreement Article 15.
9. INTELSAT Operating Agreement Article 15/EUTELSAT Operating Agreement Article 16.
10. paragraphs (a) of the Articles referred to in notes 8 and 9 above.
11. EUTELSAT Operating Agreement Articles 15(c) and 16(c).
12. INTELSAT Operating Agreement Articles 14(c) and 15(c).
13. INTELSAT Operating Agreement Article 6/EUTELSAT Operating Agreement Article 6.
14. INTELSAT Operating Agreement Article 8/EUTELSAT Operating Agreement Article 8.
15. Office of Telecommunications
16. The SAO was established by an agreement entered into between OFTEL and the SAO in 1989 following an OFTEL statement. This was followed by OFTEL's Review of British Telecom's Arrangements for the Provision of Space Segment (28 November 1989) and comments by Sir Bryan Carsberg, then Director-General of Telecommunications in the UK.
17. Following INTELSAT reforms (see below for details), after 1 April 1994, the SAO effectively only deals with access and earth station arrangements for EUTELSAT. The SAO's contractual arrangements with operators consist of an "Enabling Agreement" which governs the relationship between BT, as Signatory to EUTELSAT, and the Operator, and also contains confidentiality arrangements.
18. This is provided for in the SAO's agreement with OFTEL. The mark up is often referred to as the Management or Signatory fee.
19. see note 1 above.
20. see note 1 above.
21. see note 1 above; see also Communication from the Commission "Green Paper on a common approach in the field of satellite communications in the European Community" [COM(90)490 Final 20/11/90].
22. COM(90)490 Final 20/11/90.
23. see note 1.

24. Council Resolution of 22 December 1994 on further development of the Community's satellite communications policy, especially with regard to the provision of, and access to, space segment capacity - (94/C 379/04)
25. Commission Directive 94/46/EC 13/10/94.
26. Article 2(6)(b) of Commission Directive 94/46/EC 13/10/94.
27. EUTELSAT Operating Agreement Articles 15 and 16.
28. see note 21 above.
29. see note 22 above.

- *III* -

SPACE ISSUES

TELECOMMUNICATIONS ON CELESTIAL BODIES - SOME LEGAL PROBLEMS.

ERNST FASAN

Rechtsanwälte, A 2620 Neunkirchen, Natschbach, Kneissgasse 10, Austria.

INTRODUCTION

When Sputnik I in 1957 transmitted its "beep, beep, beep" back towards a surprised Earth, no legal regulations existed regarding telecommunications from Space to Earth. In the meantime, this shortcoming has been corrected extensivly.

But although it is more than 25 years since Armstrong and Aldrin landed on the Moon and since the former transmitted his famous "Great Leap for Mankind" - message back to our planet, telecommunications on and from the Moon and other Celestial Bodies is still a field only sparsely regulated.

This paper investigates that particular legal situation.

I. A SHORT ACCOUNT OF SOME SPACE FLIGHTS.

1.) On Oct. 4, 1957 the age of Space Travel began with the start of Sputnik I. At this date, for the first time in mankind's history, a space object broadcast a signal by way of telecommunication from Outer Space (the orbit around Earth in an altitude of about 251 to 900 kilometers (km)) to Earth.

The frequency used was 20.005 megacycles (Mc, now MHz). At this time, there were not yet in force the ITU radio regulations of Dec. 21, 1959. Up to then regulations did not provide for allocations regarding Space - to Earth or Earth - to Space services.

Although this frequency of 20.005 mc was the exact frequency assigned to the station PEN in Kootwijk, The Netherlands, it seems that telecommunication from spacecraft such as satellite Sputnik I at that time was not regulated and especially not prohibited by International Law [1].

When ITU was recognized as a specialized agency of the United Nations on Jan. 1, 1949, there was already in force the 1947 (Atlantic City) ITU convention, Art. 6 of which which formulates as tasks of the IFRB among others:

"(a) To effect an orderly recording of frequency assignments made by the different countries so as to establish, in accordance with the procedure provided for in the Radio Regulations, the date, purpose and characteristics of each of these assignments, with a view to insuring formal international recognition thereof.......

(b) To furnish advice to Members and Associate Members with a view to the operation of the maximum practicable number of radio channels in those portions of the spectrum where harmful interference may occur....."

And Art. 4 had endorsed (and still does endorse) international cooperation.

However, with this now it became obvious that astronautical telecommunication services needed regulation [2].

As Leonard Jaffe reported [3], ITU and its consultative committees at once started deliberating regulations and especially allocations of frequencies for space services.

Regulations started with the ITU Radio Regulations of Dec. 21, 1959 and then the Xth plenary assembly of CCIR. The following Extraordinary Administrative Radio Confererence on space communications (both in 1963) made several important allocations for space services.

However, even then ITU foresaw clearly the predominant importance of telecommunications via satellites, especially in geostationary orbit and to work on this was the main task of ITU in these and the following years [4].

In the meantime, space exploration was not limited to sending unmanned (and later manned) salltites into Earth orbit.

2.) Already at the beginning of the year 1959 first Luna 1 of USSR and then Pioneer 4 of USA achieved escape velocity. And in the same year, on Sept. 12, 1959, the Soviet spaceprobe Luna 2 reached the Moon and made hard impact. About a month later, Luna 3 sent back the first pictures from the hidden side of the Moon [5].

Then came a short hiatus in deep space flights and there was the preparation and then successful implementation of the first manned space flight around the Earth by Gagarin (April 12, 1961) and of Glenn (Feb. 20, 1962).

But in 1961 Venera 1 of USSR already had started towards Venus ob Feb. 12.

1962 not only brought the hard impact of Ranger 4 on the far side of the Moon, but the first Mariner flight to Venus as well. On Dec. 14, 1992, Mariner 2 sent signals toward Earth from an altitude of about 35.000 km above our neighbour planet.

Towards Mars, US-Mariner 4 then started on Nov. 28, 1964 and broadcast its arrival in about 10.000 km altitude on July 14, 1965. And Venera 3, having started on Nov. 16, 1965, dropped a Soviet flag towards Venus in March, 1966.

American Spacecraft Surveyer (started on May 30, 1966 and transmitted pictures to Earth and Pioneer 6 and 7 respectivly reagarding their deep space missions) used frequencies of 2.300 MHz. Pioneer 10, which started towars Jupiter on March 2, 1972, and reached an altitude of 137.000 km above that planet on Dec. 3, 1973 and Pioneer 11 (start April 5, 1973, arrival 42.000 km above Jupiter on Dec. 2, 1974), used frequency bands of 2.292 MHz (Pioneer 11 proceeded then towards the planet Saturn, which was reached by it in 1979) [6].

On Jan. 31,1966, the Soviet Spacecraft Lunar 9 successfully made a soft landing on the Moon and radiated picutres back to Earth. Three months later, Luna 10 succeeded in making orbits around the Moon.

Several other successful (and some unsuccessful) starts towards the Moon and other Celestial Bodies followed. However, on July 16, 1969, Apollo 11 with a crew of three (Neil A. Armstrong, Commander; Lt. Col. Michael Collins, Command Module Pilot; and Col. Edwin F. Aldrin, Jr., Lunar Module Pilot) was launched and made its world famous successful Moon landing.

Telecommunication Journal [7] as well as many other professional magazines brought a detailed report and of course described extensively the "Lunar Orbit Communications" and "Lunar Surface Communications". In this article NASA reported that both the Command Service Module (CSM) and the Lunar Module (LM) were prepared to communicate over the S-band with approximately 2300 MHz. And this S-band links allowed communications both to Earth and between the two modules themselves. The LM served as a communications relay "by passing on VHF voice and data from the astronauts walking on the Moon towards Earth." There was even the possibility to send TV pictures to Earth via a TV camera with a "standard 525-line, 30 frame per second TV signal."

There followed several manned Lunar missions from USA,and the flights of Mariner 6 (Feb. 24, 1969) and Mariner 7 (March 27, 1969) towards Mars. Both flights succeeded and in May (and November respecticely) 1971, photographs were relayed back to Earth. Viking 1 (start Aug. 20, 1975) and Viking 2 (start Sept. 9, 1975) made soft landings on Mars. The (then) USSR started spacecraft towards Venus, for instance Venera 7 on Aug. 7, 1970, which examined the Venus atmosphere, Venera 4 (June 12, 1967, with the same task) and consecutive Venera flights; the Soviet Zond 3 flew by the Moon on July 18, 1995 and proceeded towards the Sun and the USA sent Mariner 10 towards Venus on Nov. 3, 1973. The spacecraft flew by Venus on Feb. 5, 1974 and reached the vicinity of Mercury on March 29, 1974.

On Dec. 11, 1972, the US American Apollo 17 made the 6th (and last) manned landing on the Moon and on Aug. 22, 1976, Soviet Luna 24 returned after a soft landing on the Moon with soilprobes from our natural satellite.

After a hiatus in the eighties, Muses A, as the frist Japanese satellite, was launched into a Lunar oribit successfully, and unmanned spaceprobes Galileo and Clementine followed in the next years.

II. THE DEVELOPMENT OF SPACE LAW.

At the time, when Goddard, Oberth und Ziolkowsky published their scientific deliberations regarding the possibility of space flight, already some authors thought about Space Law. In 1932 the first book on Space Law was published by a Czech in German Language "Das Weltraumrecht, ein Problem der Raumfahrt [8]. The author, Vladimir Mandl, dealt with several problems quite farsightedly. But Telecommunication from Space was not among those.

With the beginning of the fifties, space legal literature started in earnest, and one of these early authors, namely Andrew G. Haley was the first to discuss legal problems of Space Telecummunication [9].

The international legal community, however, was not as fast as these scientists. Only one month after the start of Sputnik 1, on Nov. 13, 1957, USA and other states proposed at the United Nations a resolution for controlling the use of Outer Space for peaceful purposes only [10]. There followed (on majority vote) on Dec. 13, 1958, a new resolution of the UN General Assembly (UNGA) for founding an ad hoc Committee on the peaceful uses of Outer Space [11]. In the first report of this Committee already several organisations, including ITU, were mentioned within the framework of the United Nations.

But still, unanimity could not be reached at once.

But finally, UNGA passed unanimously a Resolution [12] on Dec. 20, 1961. This Resolution recommanded to States the following principles:

a) International Law including the United Nation Charter should be applicable for Outer Space and the Celestial Bodies:
b) Outer Space and the Celestial Bodies should be free for exploration and use by all States according to Internatioanl Law and should not be subject to national appropriation. Furthermore, the General Assembly endorsed the plan of ITU for inviting a special conference in order to allocate frequency bands for activities in Outer Space.

There followed the more detailed UN Resolution 1963 (XVIII), unanimously adpoted on Dec. 13, 1963. In the same year, the frequency allocations, made at the 1959 Geneva Conference (for instance the spectrum between 1.000 and 10.000 MHz) was greatly extended.

Finally, on Jan. 27, 1967, there was signed the Treaty on Principles Governing the Activities of States in the exploration and use of Outer space, including the Moon and other Celestial Bodies (the Space Treaty) which entered into force on Oct. 10, 1967 [13].

This Treaty is the basic Document, one can say: the Constitution, of Space Law. It endorsed the principles of UN Resolution 1721 (XVI) and the following Resolutions regarding the freedom of exploration and use of Outer Space, the notion that Outer Space, including the Moon and other Celestial Bodies are not subject to national appropriation, and the applicability of international law.

Regarding Celestial Bodies, its Art. IX. reads as follows:

"In the exploration and use of outer space, including the moon and other celestial bodies, States Parties to the Treaty shall be guided by the principle of the co-operation and mutual assistance and shall conduct all their other activities in outer space, including the moon and other celestial bodies, with due regard to the corresponding interests of all other States Parties to the Treaty...."

There followed the "Rescue Argeement" on April 22, 1968, the "Liability Convention" of Oct. 9, 1973, the "Registration Convention" on Sept. 15, 1976 and the (though not ratified by the space powers) "Moon Agreement "of Dec., 1979.

UNGA Resolution 27/92 of Dec. 10, 1982 adopted the principles for direct TV broadcasting, and UNGA Resolution 41/65 of Dec. 3, 1986, principles related to remote sensing, finally UNGA Resolution 47/68 of Dec. 14, 1992, about principles of the Use of Nuclear Power Sources in Outer Sapce

In its conventions and final acts the International Telecommunication Union (ITU) adhered to these principles and a steady and peaceful development of regulations and frequency allocations for telecommunication in Outer Space followed.

Thus International Law, and the International Law of Telecommunication in particular, became firmly based upon the notions of cooperation and peace. The so called "cold war" was not reflected upon in the relevant norms. On the contrary: Space Law has been one of the main fields of legal cooperation even among nations with different social systems. The way "toward the information age" [14] was not too much affected by international political tensions [15].

III. DEEP SPACE AND CELESTIAL BODIES:

ITU realised that special contingencies might arise at distances far beyond our planet. And therefore deep space was defined as follows:

"Space at distances from the Earth approximately equal to, or greater than, the distance between Earth and the Moon" [16].

With this it became obvious that a special legal regime might someaday might be necessary for those space telecommunications that would take place on or from the Moon or from Earth's neighbouring planets.

This provision was altered in the final acts 1988. Now, it reads

"*Deep Space* : Space at distances from Earth equal to, or greater than, 2x106 kilometers." [17].

With this, the Moon is no longer deemed deep space and there might be intended a different legal system for telecommunication regarding the Moon and telecommunication on (and from) the Planets.

Regarding deep space research, the CCIR Recommandation 576/1 [18] reads as follows:

"PREFERRED FREQUENCIES AND BANDWIDTHS FOR DEEP SPACE RESEARCH.

(Question 22/2, Study Programme 22A/2)

The CCIR,

Considering

a) that frequencies most suited for telecommunications between Earth and spacecraft in deep space are determined partly by atmospheric and interplanetary propagation phenomena;

b) that technology also influences the selection of preferred frequencies;

c) that requirements for telecommunication must be satisfied during periods of adverse atmospheric effects;

d) that the same frequency my be used for spacecraft at different celestial coordinates, but that different spacecraft in the vicinity of the same coordinates and wihtin the beamwidth of an earth station antenna will usually require different frequencies;

e) that is is practical and desriable to effect telemetring and tracking functions on t he same space-to Earth link, and telecommand and tracking functions on the same Earth-to-space link;

f) that to effect precision tracking, a pair of coherently -related Earth-to-space frequencies is desirable;

g) that for more accurate calibration of the effects of charged particles on the velocity of propagation, simultaneous use of links with coherent frequencies in two more widely separated band is required;

h) that voice and video links associated with manned spacecraft in deep space could use frequency bands allocated for telemetring, telecommand and tracking functions;

j) that Report 683 considers the selection of prefrerred frequencies for deep space research in the 1 to 40 GHz range;

k) that Report 849 considers the selection of preferred frequencies for deep space reserach in the 40 to 120 GHz range,

UNANIMOUSLY RECOMMENDS

1. that frequency bands for deep space research in the 1 to 40 GHz range be allocated, with due regard to the feasibility of sharing, in the preferred frequency ranges listed in Table I of Report 683, taking into account the bandwidth requirements and equipment characteristics described in Reports 536 and 683;

2. that frequency bands for deep space research in the 40 to 120 GHz range be allocated, with due regard to the feasibility of sharing, in the preferred frequency ranges listed in Table I of Report 849, taking into account the bandwidth requirements and equipment characteristics described in Report 536 and 849.

Report 683 - 3 gives a detailed examination of the frequency bands in the 1 to 40 GHz range for deep space research. It states that the objective of identifying preferred frequency bands is to provide the basis for allocations from which the Designer can select operating frequencies best suited to mission requirements. It lists the current allocations and the corresponding ranges of preferred frequencies. It is pointed out that the 7 and 8 GHz band pair provide the principle telecommunication links used for current deep space research.

Report 849 - 2 deals with the frequency bands in the 40 to 120 GHz range. The advantages for frequencies above 40 GHz are pointed out as higher link performance, pontential for wider band width, reduces errors in measurement that depends on the velocity of propagation, and the possibilty of shielding from terrestrial interference.

In its Thirty-third Report of the ITU on Telecommunication and the

Peaceful Uses of Outer Space [19], it is consequently pointed out that the Provisional Draft Agenda for WRC 97 includes the following items:

"- review of the propagation information in Appendix 28 used for the determination of coordination areas in the frequency bands between 1 and 40 GHz shared by space and terrestrial services."

Furthermore, "allocation issues for other unplanned space services" are recommended.

On this basis, then, we can expect regulations for the future deep space missions and especially for mission towards the planets Venus, Jupiter (and ist Moons) and especially Mars. If missions towards Mars by manned spacecraft for somewhen in the next century are contemplated, all contingencies, which could arise during such a long term mission, will still have carefully to be investigated in order to protect such an important endeavour. And this would of course be necessary for telecommunications as well.

REFERENCES

1. The freedom to explore and use Outer Space for peaceful purposes is an universally recognized notion of International Law. See (10), (11), (12), (13).
2. Haley , Space Law and Goverenment, Meredith, 1963, 169 ff, and Fasan Weltraumrecht, Krausskopf Flugwelt, 1965 138 ff.
3. Telecommunication Journal, Vol. 32 No. 5, May 1965, 194.
4. A very good overview is given by Wolfrum, Telekommunikation in Boeckstiegel, ed. Handbuch des Weltraumrechts, Heymann, Koeln, 1991, 367 ff.
5. A thorough overview of Lunar flights is given by the Subcommittee in the Lunar Development of the International Astronautical Academy (IAA), Chairman H. Koelle, see Doc. - L02, July, 1994.
6. Koelle (5), Reports in "Weltraumfahrt", Umschauverlag, Frankfurt, p.e. No. 5, Sept./Oct. 1964, ed by Buedeler etc.
7. "Apollo - 11 Communications", NASA Telecommunication Journal, Vol. 36 VIII/1969, 369.
8. Bensheimer, Berlin, Mannheim, 1932
9. Basic Concepts of Space Law, Jet Propulson XXVI, 1956, 951 et al.
10. UN - Doc. A/C 1/PV. 821, 6
11. UNGA Doc. A/3818,
12. UNGA Doc. 1721 (XVI)
13. See "United Nations Treaties and Principles on Outer Space", published by the UN Office for Outer Space Affairs under the directorship of Jasentuliyana; furthermore the Reports on Space Law and Related Documents, Reports for the US Senate, ed. by Galloway.
14. Rapp, Foreword Vol. 1 of Telecommunications & Space Journal, Serdi, 1994, 11
15. For more details see Diederiks - Verschoor "Legal aspects effecting telecommunications activities in Space" and the literature, quoted there Vol. 1 of Telecommunications & Space Journal, Serdi, 1994, 81.
16. WARC 1979, Section VIII, 47, earlier Final Acts, Geneva, 1963, BA 84.
17. Orb 88, Final Acts, Section 8, 22.
18. CCIR Duesseldorf, 1990, Section 2 E, Recommandation 576/1.
19. ITU-Booklet No. 42, Geneva, 1994.

EARTH AND SPACE DIALECTICS, SOVEREIGNTY AND THE RIGHT TO PRIVACY

ADOLFO SERGIO SPADONI

University of Milan, Institute of Philosophy and Sociology of Law, Via Festa del Perdono, 7, 20100 Milan ; Private Address: Via Battistotti Sassi 13, 20133 Milan.

Abstract *: A dialectic approach to problems is a betrayal if it abdicates the critical task of the "unhappy conscience" of knowledge which underlines the contradictions of "reality". Remote Sensing cannot escape immanent contradiciton. As an advanced form of technology, it is entrapped in the ambiguity of the oppositional couples: freedom/bond, progress/regression. This is the paradox of an amazing combination of intellect (science - technology - productive capacity) working freely: outer space is removed from the régime of State sovereignty and has power of commanding and controlling people and things; at the same time it brings with itself bonds which go beyond national State borders, thus cancelling the little which remains of individual personal freedom which can be expressed in the juridical discipline of privacy protection.*

Key Words *: Dialectics, Outer Space, Sovereignty, Remote Sensing, Right to Privacy.*

INTRODUCTION

In an era characterized by the unchallenged advance of "total capital" and the inexorable forward march of science and technology, as the main factors of growth and the widespread reproduction of wealth, the aim of this work is to introduce a note of dissonance in the harmonious scenario underlying the progressive results of human society.

In the history of thought, from Eraclito to Adorno [1], the holistic or negative dialectics - with its obstinate pretension of explaining "everything with everything" -has always been the torment, the unhappy conscience for the illiberal flatteries of the nominalist dualism of which contemporary epistemology is so fond [2].

Dialectics, as the theoresis of the willingly unsolved "open contradiction", sees law - considered in an axiological way - as a possibility of regulating and assuming without dissolving the contrasting couples: earth-space; bond-freedom; loyalty (publicity)-privacy; state-individual.

On the other hand, the requirements of "widespread reproduction" determine the oscillations of the margin of freedom the system can tolerate if it is not to enter a crisis.

A. Madec describes the strategic sectors of advanced economies, based on the transformation of information into goods, as follows: "In post-industrial economies, where the management of information still represents between 40-50% of the added value, it is natural that the international exchange of information is called to play a central rôle" [3]

Law, whether national or international, has the difficult task of placing a limit on the system itself, as a corrective to the awareness that there is a fundamental conflict between industrial society (regardless of the implied ideology) and privacy-freedom, both in term of personal autonomy and the right to the privacy of peoples.

Remote sensing is at the cutting edge of information technology and has further sharpened the public-private dichotomy, laying bare how the dualistic attitude - leitmotiv of the post-modern mentality - produces that "blinding connection" which distorts - or even worse - destroys reality.

In going beyond this dualism, one will see how privacy is a concept which is evasive in appearance only. I am, therefore, convinced that a critical use of dialectics (holistic or negative) lays the basis for the wording of a concept of right to privacy of individuals and peoples, which despite its inconsistency can never be eliminated.

I - TECHNOLOGY AND INDIVIDUAL PRIVACY

1.1 The Nature of Technology

The axiological zero setting of technology in the name of pure efficacy is itself enough to conceal the ambiguity of the neutral nature of technology. I say ambiguity, as the concept of neutrality easily lends itself to the disguising of the "will-to-power", which is all the same a constitutive part of its content.

Heidegger - with reference to the estranging moment of technology, expression of an epistemic statute which interposes between man and the object of his/her activity - had already intended to point out its close bond with power. To set out a correct and well-balanced relationship with it, both a right-wing reading (C. Schmitt) and a progressive one (empiricism/positivism) are misleading.

For Schmitt, technology is neutral as "it can be revolutionary and reactionary, it can serve freedom and oppression, centralization and decentralization" [5].

No less insidious - and we shall see it in connection with the retrieval and informatic treatment of data effected in a system of complete freedom, outer space, with remote sensing at the highest resolving power (about 3 m) - is rationalist and positivist optimism, for which science and technology in general and computer science in particular are "intrinsically good", functional in maintaining order in the growing complexity of society. The principle of identity and non contradiction, fruit of contemporary science, at last seems to have materialized in the computer and in all developments linked to it, so as to make it an instrument "good in itself". The problem in these two formulations is not only a matter of recovering the distinction between use/abuse in technology and computer science, but also of the persisting limit (also for future generations of computers, those equipped with the so called artifical senses) peculiar to the sensistism dualic tied up with the linear schematism of cause and effect which incapable of the explanation of "everything with everything" - loses sight of the dialectic connection of individual and society, with the consequent reduction to "second nature" to dominate.

The evidently negative spin off of all this can be seen in the lack of importance which the principle of privacy protection enjoys in domestic and international legal systems. Another consideration on use and abuse: just where does the boundery line between these two concepts lie, especially in a situation of structural crisis where natural (environmental risks and protection) and induced emergencies (widespread creation of wealth and recovery of resources) increase and abuse becomes normative?

1.2 The Risks for (Individual) Privacy

I shall now try to clarify how remote sensing, with its capacity for observing objective or neutral data is a paradigmatic example of the fact that technology cannot be defined as neutral.

As a union of prima facie activities (i.e. aseptic and impersonal assumptions) - the translation of images of objects into digital data [6] as the product of a high-definition optical process - remote sensing conceals the more strictly political implications of the matter, which directly assail the sphere of privacy and its limitations.

All this is in some way arranged by the structure of the computer itself, which is still based around the principle of "discreet logics", in clear oppostion to the requirements of freedom and reservedness appertaining to the right to privacy [7].

Automatic sensing and treatment of data - as it is carried out at present - inhibits and blocks public and private dialectics.

An open society, which is attentive to individual autonomy in its material concrete form, "cannot and must not tend to the construction of a public... totally alienated in a transparency which progressively eliminates room for private and personal autonomy, both in terms of an orginal negation of freedom in research and error of unpredictable deviation, in the shape of a reduction of the person to predicates of the (social) system, to the model and the requirements of the system itself" [8].

Nevertheless, in the increasing expansion of the analytic (finished) some authors see a reversal into the analogical (unfinished) as the real self-contradictory potential of the computer [9].

So, together with the more flexible generations of the so called "dialectic and neuronal computers" [10], we should welcome any signs of optimism, although we must not forget that in the dialectics of subject-object acting as the field of interacting forces, it is only the critical conscience of the subject, which can prevent the adeguatio intellectus rei (the thing, the computer, ever if it is dialectic too).

II - REMOTE SENSING

That the non neutrality of technology is in itself equivalent to an irreversible prior determination, is something which remote sensing and connected techniques demonstrate with even greater frequence. Experience teaches that the risks and limits for the destiny of freedom and the future of civilization, brought about by new technologies, are compensated for by innumerable benefits. Take, for example, an increase in wealth through resource location and the forecasting of the weather and possible natural or human (nuclear) disasters, not to mention the stabilizing effect on domestic and international crises deriving from the preventive control on military balances of power.

I will, therefore, speak about ambivalence of technology, a concept that - if it is not inserted into a historic-materialistic reference framework - risks being deflated on the previously rejected one of "alternative use".

To do this, it cannot break away from pure abstraction as an oriented variant, the absolute neutrality; but, by holding fast to the premise of the intrinsic connotation of technology as the will-to-power, it can make a sort of reversal of the same. This reversal appeals solely and exclusively to the contradiction induced in the knowing subject.

Ambivalence will then be total, because it will revolve around concrete and well restricted data that only the specificity of a determined historical situation can indicate [11].

If the innumerable social advantages resultant upon remote sensing are indisputable, the ambivalence proper to this technique emerges fully with the indivduation and the denunciation of the alarming spin offs in negative terms for the destiny of personal freedom, as a fact immanent of a context positive in itself.

An example: to succeed in an aim unequivocally progressive and useful for humanity, the second generation of satellites launched for earth observaton, SPOT 4 and 5, equipped with (medium infrared band) HRV sensors to measure the humidity of earth vegetation, have reached such levels of perfection as to obtain stereoscopic images for the reconstruction of reliefs with 20-metre definition in a multispectrum way, and with 10-metre definition in panchromatic.

Besides guaranteeing the most efficient means of achieving the aim, this means that these miracles of efficiency and precision can easily be used for the observation of living reliefs: people reduced to an inanimated dimension analysable through a physicalist process.

With regard to the juridical aspect, even if internationl law is quite unanimous in rejecting the qulification of remote sensing as an act of espionage - since it cannot be assimilated to war espionage as mentioned in the Aia Convention of 1907, art. 29 [13] - I do however think that, even though with a different connotation, espionage can be spoken of also under a juridcal profile [14], since electronic surveillance, "activity substanzializing in the retrieval of a heterogeneuous bulk of information, and, if necessary, implying the simple supervision of the ordinary development of civil relationships within a country" [15], implies a persistent situation of hostility (no more against the State but against the individual) thus legitimating a form of unfriendly espionage.

Some more notes on the deceit which is hidden in the presumed neutrality of the so called data surveillance of remote sensing in its two constitutive moments: physical observation and information processing.

At first sight, it looks like a transparent (innocent) data organization. The technique of the inferential relational retrieval, as the putting into sequence of isolated and apparently insignificant data, allows the puzzle of complex individual and social activities to be reconstructed.

The advantage is evident: the classic espion animated by animus spiandi, who can be discovered while following a person's movements, is substituted by the input (formatting) into the data bank of all those loose clues which one takes no further notice of.

As a conclusion of the connective process operated by public or private subjects called to it, the photograph of a person will be complete not only under the external profile but, what is more important, also under the internal profile, i.e. political, religious, sexual and health opinions. In a word, the rate of adaptability to the pre-established loyalty [16].

2.1 Technical Notes

With the English wording of remote sensing the COPUOS has meant the definition of an activity whose characteristics can be synthesized as follows: "the essentially unobtrusive observation and gathering of identificable facts through the use of space objects and their component parts".

Remote sensing can then be effected with the help of more specific activities, which the COPUOS Subcommittee has schematized as follows: data retrieval (by satellites and space station); 2) reception of data (antenna and receiver); 3) pre-working out of data; 4) filing and reproduction of data: 5) analysis of data (interpretation and data processing); 6) use of information (pratical application on the part of users) [17].

First of all, it is necessary to remember some of the differences between telecommunication and remote sensing satellites. The first ones (GEOSAT) move onto a circular orbit (geostationary) located vertically above the equator at about 35,000 km; the function of this satellite is to work as a fixed antenna, by maintaining its angle speed constant, which enables it to be always perpendicular above the same point.

Remote sensing satellites (LANDSAT 1, 2, 3, the French series SPOT and in the future the ERS 1) [18] operate on heliosyncronous orbits at about 700 km in height, and their function is to keep almost the whole earth under control.

This capability of observation is guaranteed by maintaining constant the angle formed by the conjunction of sun-earth centre and planet orbital plane. The satellite passes by at the same latitude at the same hour, thus making it possible to compare the images in temporal sequence.

2.2 Lawfulness

It is my intention to put to one side the analysis of the vexata quaestio concerning the lawfulness or unlawfulness of remote sensing effected through military satellites, even if for peaceful purposes [20], and to draw attention to what is suggested by the so called functional criterion.

In conformity with such criterion, the problem of lawfulness in remote sensing (for non-military purposes) is not defined so much with reference to location, as with regard to the activity carried out and the possible damage to material interests [21].

This has led to much heated discussion: on the one side, the industrialized countries press for the complete liberalization of all activities, including those which are more strictly scientific (meteorological, environmental monitoring, etc.) and those of a more economical bent (land-use situation, discovery of ore deposits, etc.). On the other side, the developing countries appeal to the principle of sovereignty and the inalienable right to dispose of their own resources.

This dispute has found a point of mediation in two limits to the divulgation of the data obtained: 1) interdiction from Third Party access to data obtained via remote sensing without the consent of the sensed State; 2) interdiction from Third Party access to information on economic resources.

2.3 Advantages

"Les possibilités de la télédétection dans le domaine de l'identification des récoltes agricoles (état de végétatioin, espèces cultivées, degrés de maturité, etc.) des ressources biologiques terrestres et aquatiques et des gisements minéraux dans le sous-soil ont été démontrées de façon suggestive par l'expérience aérospatiale combinée effectuée à l'aide de l'avion RB-571 et le vaisseau spatial "Apollo-9" en 1969"; and then "U.S. aid has supported many demonstrations of the usefulness to developing countries of LANDSAT data for resource management, to determine optimum land-use approaches, or measure existing crop acreage, to detect changing stream and flood patterns, to monitor forest cover and changes, and rates of soil erosion or loss of productive grasslands" [22]. We have here a wide range of the different applications of remote sensing in the civil field: from forecasts to the control of the disastrous results of natural calamities; from environmental monitoring to the detection of resources in the subsoil or in the oceans.

This is an incomplete but meaningful picture of the results obtained and of the possibilities in fieri for the increase in benefits which remote sensing can grant.

The following examples provide a hint as to the concrete achievements of this technique: 1) the possibility of water resources in a desert zone (the Sahara) has been revealed thanks to the identification of trails of a waterway which had flowed thousands of years before; 2) the spotting of a minimal route under ice-fields which enabled the SIBIR submarine (Sovietic Navy) to reach the North Pole speedily.

I should also like to recall the results obtained in the geopolitical field thanks to satellite intelligence, either with regard to the strategic balance between the two superpowers, or to the countermeasures taken against countries and régimes that had tried to violate the agreements on non-proliferation of nuclear weapons [23].

An example: in 1977, diplomatic pressure led to the dismantlement of a nuclear power-plant which was being built in South Africa, after retrievals carried out by a Soviet surveillance satellite [24].

2.4 Limits

Even if principles formulated by COPUOS [25] represent a considerable step forward towards the control of a situation of de facto regulation, the provision of an organic normative system on the above mentioned limits and risks of remote sensing is of immediate concern, despite the awareness that there is a lack of power balance working wholly to the advantage of the growing power of technology.

I will attempt a brief analysis, paying particular attention to the truly weak subject of the situation: the individual and his increasingly weakened right to privacy.

Contrary to the communis opinio [26] which minimizes, or even ignores, the dangers of remote sensing for individual freedom, submitting it to the competence of international law under the heading of "Human Rights", I think that the lack of control over the undisputed benefits of remote sensing penalizes as "weak" subjects, not so much the non-space-active countries, the so called developing countries [27], as the individual, on whom moral and material interests converge.

It is now clear that even in the case of remote sensing as for any different technique or science in general, the logic of instrumental reason imposes a reciprocal dialctic reversal of means and ends, but I feel duty-bound to advance a further clarification of the matter.

Well, if it is true, as I said before, that remote sensing is not eo ipso to be considered as espionage tout court, it is no less true that as a technical means for the retrieval of informatic data, it works with the utilitaristic end of the control of "anomaly and difference". But the anomalies/differences that the menas senses with its apparent neutrality, are not only inanimate anomalies (meteorological, environmental, agricultural, etc.) but also human and personal ones [28].

Nobody would be so unwise as to charge a non-human activity (i.e. remote sensing) with an exclusively human responsability: the non-acceptance of anomaly/difference, an acceptance which only a society organized differently in economic and cultural terms (and maybe in legal terms too) can realize.

All this leads again to a consideration of the inextricable nexus of progress/regression which constitutively connotes science and technology.

A last, somewhat bitter consideration: negative-dialectic rigour obliges me to undertake the unpleasant rôle of the critic à outrance. It is paradoxically the case of remote sensing: freedom is fatally overturned into its contrary, or at least into something which has little to do with it.

Just to avoid Utopian and useless forward flights on the plain of "must be", reality increasingly draws our attention to the fact that a normative total prohibition in collecting information - as the only way to protect personal freedom - is something which cannot be proposed, a conclusion consolidated by the increasing tendency to the widespread method of the "permanent census realized through data retrieval directly from data banks" [29].

As was mentioned earlier, because modern remote sensing technologies operate in a régime of complete freedom in outer space, which is not subject to the claims of sovereignty [30], they paradoxically reinforce - being free and unlikely to be identified - the forms of control imposed by States on personal freedom through information technologies.

This does not imply a reinforcement of the principle of State sovereignty, which has too often shown itself in the course of modernity to be an agile instrument in the hands of the ruling class to control and repress freedom [31]. On the contrary, it means that in addition to its many positive aspects, remote sensing threatens the citizen's capacity to identify clearly and clash juridically with the public subject in front of him/her: the State.

I hope that the following examples will clarify the above

considerations: 1) I wonder how dissuasive a citizen of a "sensed state", investigated by a different State via remote sensing, might be, when it is so difficult even in his/her State [32] to refuse permission to collect information about him/her; 2) what effect could "information objection" have as a form of protest - now developing in several countries - in order to deny replies to statistical surveys, irrespectful of privacy, if the interlocutor to whom one should make the protest, cannot be recognized? [33]).

Remote sensing makes self-protection even more difficult also because data banks are obliged to communicate the existence of files to the interested persons, but easily avoid this obligation because the distinction between primary and processed data implies (with regard to primary data, i.e. electromagnetic waves intercepted and then sent back to earth) vulnerability to information piracy by persons or groups, who can intercept and keep for themselves reserved information. Indeed "the national legal systems do not consider decoding an unlawful access to secret information" [34].

There is a futher complication to the protection (and self-protection) relating to data processing. In this phase there is a proliferation of organizations which manage the satellite and have the juridical power on the data checking set (similar to the logical system or programme) necessary to the translation of primary data [35].

Finally, the praiseworthy intention expressed in the VI draft principle (district-based agreements on the grouping of operative centres for data retrieval and infomation processing), as well as in the first paragraph of XII principle (the right of the sensed State to have access to identified information) de facto vanishes as the second paragraph of XII principle expressis verbis excludes the obligation of dissemination of the information processed by private subjects (commercial enterprises), thus annulling de facto the declared purposes which could have safeguarded even privacy.

A further blow to the protection of privacy in both sensing and sensed States especially with regard to sanitary matters, comes from the use of formula such as: the requirements of scientific research; reasons of national security, which lend themselves to a broad interpretation, in conformity with the porosity of the concept "interests" contained in IV Principle.

Contrary to what happens for private-law economic subjects, the infringement of secrecy by public administrations makes the conceptual distinction of public/private weaker and weaker.

III - THE CRISIS OF SOVEREIGNTY

3.1 Economy and Technology: Crisis Factors

The conquest of outer space which took place during the 1960s and the epochal events of the 18th century which saw England become ruler of the waves, have many things in common.

Just as yesterday the seas offered themselves up to England, today outer space is for industrially advanced economies: in both cases there is a challenge which joins open spaces (sea and universe) to the free world market, under an idea of freedom whose protagonist is the network of unrestrained technlogy and economic growth on a world scale [36].

This network has radically changed the relationship between man and the earth, that justissima tellus, ruled by sovereign States which the jus publicum europeum in its two components, State and international law, had tried to govern with its own nomos: "The era of State is at last reaching its end: it is not necessary to waste words on this. With it the whole superstructure of concepts relating to the State - once raised to a Eurocentric science of State and international law - fails" [37].

The category of sovereignty, the juridical and political axis of Staatsrecht and Völkerrecht - a domestic and international order - comes to crisis under the blows of a new sovereign order: that of technology and internationalization of capital [38]: "(...) State and sovereignty are the foundations underlying the limitations of international law, imposed, up to now, on war and hostilities" [39].

The loss of the possibility of building and maintaining order through political/juridical sovereign act, took place when - as C. Schmitt himself understood [40] - Europe lost its world political rôle.

The theorists of sovereignty (Kelsen, Schmitt and now Forsthoff) missed and continue to miss the planetary rôle of the technology-market connection, which cannot be reduced to a mere instrument for the achievement of political/juridical ends.

Neither H. Kelsen nor C. Schmitt could have imagined of contrasting the power of economy and technology: Kelsen with his hypothesis of an order governed by international law which ought to have gained the upper hand over the legal system of individual States, thus emptying the concept of sovereignty of all meaning; Schmitt with his longing for a new nomos and a motorisierter Gesetzgeber, which could shape unrestrained technology just as the Rechtsstaat had done previously for civil war and Nomos der Erde for wars between States [41].

It is clear for everybody how the international economy has had the better of the national legal systems, with the consequence that State sovereignty succumbs to financial markets, as a result of large-scale holdings imposing themselves on State politics.

The foundation and resistance-staying power of the concept of sovereignty is therefore overwhelmed by "the requirements of political/industrial efficiency" [42].

The requirement of production, or rather the process of the widespread reproduction of capital which flourishes in a market independently of political sovereign power [43], is therefore the most suitable ground for an unlimited development of science and technology [44].

It is time to go back to the leading idea: the contradiction immanent of technology, i.e. remote sensing and its more or less mediate implications for the twofold category of the concepts of sovereignty and personal and/or peoples' privacy.

Outer space, as the new dimension of space and the paradigm of post-modern age, clarifies how the contrast between freedom and bond implies the "concrete dialectics" of clashing material interests: those that see the individual opposing the State and those concerning the opposition between advanced and developing countries.

Remote sensing, as I have already said, is at the cutting edge of information and space technology and it is perfectly functional to the interests of the industrial powers. Paradoxically, it plays a double rôle which goes now towards freedom, now towards its negation.

The international law system, in its formulations of jus speciale (space law), communis opinio, the principles worked out by the United Nations (COPUOS Principles) and last but not least, jurisprudence, shows it has accepted , if not the emptying, at least the weakening of State sovereignty decreed by the binomial power technology/economy, and of which remote sensing is a clear example:

1) Article II of the Outer Space Treaty of 1967 underlines the res communis juridical nature of outer space and affirms complete freedom of exploration and use against the claim of sovereignty usque ad sidera of developing countries.

The subsequent corollary is that the launching of satellites for remote sensing is not to be considered as a form of interference in other States' sovereignty [45].

2) The almost total disagreement on the conditions of lawfulness and on the principles governing remote sensing is the reason for the lack of opinio juris on the matter.

On the contrary, with regard to internaional law, the usual contrast between developing countries and hi-tech countries is still present. The latter, thanks to the enormous technological gap, support the lawfulness of remote sensing, notwithstanding the lack of previous consent of developing countries or international regulations.

On the other hand, developing countries, appealing to the sense of embarrassment on the part of industialized countries for the disadvantageous situation, oppose the principle of sovereignty over natural resources, and consider remote sensing activities as intrusive and injurious with regard to the international right to privacy (the so called privacy of peoples).

Finally, the principles of international cooperation, peculiar to the new world order, are intended to deny the lawfulness of remote sensing without the permission of sensed States [46];

3) The draft principles relating to remote sensing, elaborated by the juridical subcommittee of the United Nations Committee on the peaceful uses of outer space (C.O.P.U.O.S.), reflect the trend of the sensing countries which give little importance to the limitations of that activity in observance of the sovereignty right of sensed States. Any reference to the principle of "prior to the consent of sensed States" is missing. There is only a reference to sovereignty in principle IV: "respect for the principle of full and permanent sovereignty of all States and peoples over their own wealth and natural resources with due regard to the rights and interests in accordance with international law";
4) Clear economic reasons (from the interests of holdings to the pressure of transnational groups) lie at the heart of recent trends in international law, whose authoritative representatives (inclusing Vlasic and Christol) naturally belong to industrialized countries. Their theories try to prove the ideological nature of permanent sovereignty over natural resources (Principle IV of COPUOS) supported by developing countries, as well as the defence of their national interests disguised as legitimate rights of States, the right "to be let alone", a true and proper attribute of international personality relating to sovereignty and the right to privacy which cannot be infringed by remote sensing activities [47].

On the basis of these argumentations, Vlasic doubts the principle of permanent sovereignty over natural resources, considering it an escamotage to legitimize the nationalization of foreign interests: "(...) was enunciated to facilitate and legitimize the nationalization of foreign interests in the natural resources of States so as to prevent their falling under alien control in the future and thus to reduce the possibilities of the economic exploration on one nation by another" [48].

It is clear that these authors, claiming the limitless freedom and lawfulness of remote sensing activities, protect group interests in a specular way to the theorists of sovereignty and using the same ideological means.

On the other hand it is necessary to recall with Aristotle that the nature of an action is its end or aim.

The nature of what starts up an economic capitalistic activity is not the common good but the production of profit [49].

Nevertheless - and this is the dialectic aspect of the matter to be analysed in a crtical way - objectively speaking, capitalism and the exploitation it brings with it, has proved itself able to distribute wealth and welfare, and contribute to the development of pluralist States or tolerant and democratic governments, where it has developed.

One last conclusion: Third World Weltanschauungen are praiseworthy and to be shared, and thanks to them juridical concepts like international personality, the right to peoples' privacy, sovereignty over natural resources, legitimate rights and interests for the protection of developing countries, have begun to arouse interest within the United Nations. But one must not forget that developing countries are generally ruled by authoritative or liberticide régimes, and that protecting sovereignty and natural resources by limiting remote sensing acivities does not guarantee an internal process of distributive justice.

In the light of the above considerations I do not feel up to sharing the opinion of those authors [50] who, with regard to remote sensing [51], have attributed an ontologic foundation to an overly abstract and elusive concept as that of "privacy of States and peoples", thus identifying an almost functional bond between the economic information taken from and the damage caused to any sensed State without previous consent.

CONCLUSION

I hope that these occasionally rhapsodic considerations have revealed the intention and concern of my research: to draw attention to the increasing fragility of the individual in a modern and technological mass society [52] and the reasons for the weakness of law and the traditional notion of privacy.

A law which has non concrete links, where everything, even information, becomes "goods", the moloch of a capitalism which has led to the overcoming of the centrale rôle of property [53], by replacing possessions with even greater possessions [54].

In the debates on the transformations which are taking place in advanced societies thanks to modern technologies together with informatic and remote sensing, the fact that the empire of information and communication has upset the dialectic couples of individual/society; public/private; having/being, has often been overlooked.

Since they have contributed to information circulation (as goods generating wealth) from space to earth in actual time [55], it is quite obvious that remote sensing techniques have further helped to accelerate the process of transformation from a local, property-centred capitalism to a multinational and dynamic one, which substitutes the individual entrepreneur: the "new property" corporations, where financial operators and managers make money and goods circulate speedily all over the world.

This affects so deeply the individual in his/her psychic and mental structure that it is necessary to reformulate the concept itself of "right to privacy" in such a way as to protect even the interiority of man through a "right to an inviolate personality".

From the bourgeois, individual owner [56], where individual autonomy protected by law is constituted by the impassability of his/her property, one proceeds to the mass individual, continually exposed to an intrusive public in the information circuit, where the thinking faculty is substituted by the communicating one [57].

In this way social and economic transformations have changed the nature itself of the right to privacy, through the shift of individual autonomy from "having" (property or conract) to "being" (psychic integrity) [58].

On the other hand, in a period where the separation of public and private fails and where the private disappears in the transparency of the public (which is loyalty or conformity to predominant values), the right to privacy has a chance of succeeding only if it joins up with a radical critique of the "blinded connection" where being is having: one is if he has (and one can have only according to the imperative of having more and more).

This is, I think, the deep meaning of the "right to be let alone" in the technological and informatic society [59]. It is a right which, in order to be protected from technology and remote sensing, is, and must remain, strictly individual.

Actually, my doubts concerning the possibility of a de jure condendo construction of any normative hypothesis on "international privacy - right to the privacy of peoples", cover more the philosophical, political and juridical aspects of sources than the hermeneutic ones.

On the whole international law does not help to build up a unifying and organic conceptual category [60].

On one hand, with regard to international law, the formulation of a "privacy of peoples" principle seems a really mechanic transposition of juridical schemes relating to "domestic private law" which "do not suit a community of subjects unless they are individuals" [61]; on the other, the same sources of international law explicitly underline the "attempted threats against the personality ..." [62], thus consenting an extensive interpretation of the concept of "privacy", once the strict connection between "privacy" and "personality" has been accepted (as most doctrines do) [63].

Consequently the problem is not the degree of correlation between privacy and personality [64], but where the limit to what constitutes personality should be set, without losing sight of the purposes of individuation and protection of the weak subject. I, therefore, think that a dialectic consideration, which has this problem at heart, should remain within the sphere of historical materialism.

A materialism which cannot indulge in the forward flights of idalism and/or ontologism, even if concealed by law. In this persepective the sole true, weak, concrete subject, the referent of the right to privacy is the individual in flesh and blood, in his recomposed unity of body and mind, and not abstract or evasive entities, like the concept of peoples [65].

The abstract/concrete distinction cannot be solved through an "intellectual imperious act" of nominalist derivation.

It implies the constant, mutual involvement of the two terms of the problem, which however demands clarification. The result is the inadequacy of the sole distinctive criterion of physical person and juridical one.

The drive towards the protection of individual personality in its unity of body and mind, which brings up the value of human interiority, cannot be considered as something decreeing the way out from a realistic philosophy.

The extension of the value of "privacy", i.e. the principle of habeas mentem [66], is the qualitative leap classical materialism is forced to undergo by law, with the shift from the protection of the individual owner to the owner in his/her comprehensive unity.

Finally, in order to eliminate the economic gap between States in observance of the principles of a new economic international order [67] by balancing the technological potentialities of sensing States with the legitimate rights to information and sovereignty over natural resources of sensed States [68], it would be useful to assimilate the trend of international law, which suggests the introduction of the principle "claim for unjust enrichment": a claim "based on the private law concept of unjust enrichment might be attempted before an international tribunal if it could be proved that the operating State used the information to its economic advantage and that no benefit or compensation was received by the claimant State" [69].

Professor Martino propases (and his proposal is truly worthy of being shared) an action for the claim of unjust enrichment based on equity and to be considered as compensation to developing countries: an action for unjust enrichment which, in order to obtain the greatest effect, should be recognized as "sine causa enrichment" without being bound to tort committment [70].

The greatest potentiality could, moreover, be obtained by introducing the Anglo-American "torter" discipline which imposes very high and easily obtainable compensations, in such a way as to prevent the economic normalization of violations.

REFERENCES

1. On the bond of everything with everything, see Eraclito: "Only one thing is wisdom: to understand the reason why everything is ruled through everything", I Presocratici, fragment 41, vol. I, Laterza, Bari (1986), p. 205.
   The dialectics, which Th. W. Adorno formulated in the negative ("the expression negative dialectics clashes with what is betrayed"), is nothing other than an attempt to reconstruct a general theory of controversy. See P. Sloterdijke, Critica della ragion cinica, Garzanti (1993), p. 307 and fol.
2. I am referring to both the so called "hard sciences" and the more recent social sciences.
3. A. Madec, Les Flux trasfrontieres des donné: Vers une economie internationale de l'information?, La documentation française, Paris (1982), p. 12.
4. M. Heidegger, La questione della tecnica, in Saggi e Discorsi, Mursia, Milano (1976).
5. C. Schmitt, L'epoca delle neutralizzazioni e delle spoliticizzazioni, in Le categorie del politico, Il Mulino, Bologna (1972), p. 178 and fol.
6. The United Nations has adopted a terminology to distinguish primary data from processed data. The former are electromagnetic waves sent forth by objects on the earth, captured by a satellite, transofrmed into digital signals and sent back to earth. The latter, which are the result of man's intervention, are those to be protected. The data of the satellite are the physical "material core" of the computer proposition, a data base, but they cannnot be protected. Forms of indirect protection could be built up through domestic laws, which consider the data given out by satellite as already being covered with secrecy and their retrieval as a form of illegal access to a secret. But it is important to underline that primary data are in themselves incomprehensible and, therefore, secret, up until the time they are decoded. Seee for details: G. Catalano Sgrosso, Aspetti giuridici del telerilevamento, in Diritto dello spazio, Cedam, Padova (1994).
7. See for details J. Lyotard, La condition postmoderne, Les Editions de Minuit, Paris (1979); Italian transl. Feltrinelli (1981) p. 11 and fol. Lyotard charged computerization with the responsabilty of the exclusion from legitimate knowledge of any epistemology not founded on languages translatable into "machine" language.

8. See S. Scoglio, La dicotomia liberticida, in Hermeneutica, nr. 6, Quattro Venti, Urbino (1986), pp. 363-364.
9. See G. Lolli, Matematica e calcolatore, Quaderno de "Le Scienze", n. 14, Feltrinelli (1981). He speaks of the "finito grande" as the possibilty of "exploring an unknown domain".
10. See. D. Meadows, The Electronic Oracle: Computer Models and Social Decisions, Wiley (1985); D. Hofstadter, Godel, Escher, Bach, Basic Books (1979); Italian transl. Adeplhi (1984).
11. S. Scoglio, op.cit., p. 364 and fol.
12. For more details on the technical carachteristics of these and other remote sensing satellites see: G. Catalano Sgrosso, op.cit.
13. See for more details: R. Martini, I principi di diritto internazionale applicati in materia di telerilevamento, in Il regime internazionale dello spazio, Giuffré, Milano (1993), pp. 179-180.
14. See A.R. Miller, The Assault on Privacy, University of Michigan Press (1971); A. Westin, Privacy and Freedom, N.Y., Atheneum (1967).
15. See R. Martini, op.cit., p. 191.
16. See S. Scoglio, op.cit., p. 349 and fol.
17. See S. Martini, op.cit., note 1, p. 169.
18. For the technical characteristics see R. Martini, op.cit., note 65, p. 196; or G. Catalano Sgrosso, op.cit., pp. 145-147.
19. See G. Catalano Sgrosso, op.cit., p. 144.
20. First, the Outer Space Treaty with the partial obligation of demilitarization of space itself and then the SALT 1 agreements of the ABM Treaty between the USA and the ex-USSR containing the acknowledgement of the "technical means of verification", saw (most of) the doctrine leaning to the first solution. See T. Ballarino, S. Busti, Diritto areonautico e spaziale, Milano (1998), pp. 202-203.
21. See R. Martini, op.cit., p. 179.
22. See Marcoff, Traité de droit international public de l'espace, Fribourg (Suisse), 173; Thacher, Space Technology and Resource Mangement, Journal of Intenational Affairs, 39 (1985); R. Martini, op.cit., note 2, p. 170.
23. On "Arms Control Verification" see Krepon, in Headline Series of Foreign Policy Association, Sept/Oct. (1984); Dahlitz, Nuclear Arms Control, Allen and Unwin (1983).
24. T. Ballarino - S. Busti, op.cit., p.203.
25. See supra 2,1.
26. See also for R. Martini, op.cit.
27. Ibid, pp. 181-185.
28. Very sophisticated remote sensing techniques can get information on objects (as small as a number plate and even on persons).
29. S. Scoglio, op.cit., p. 367.
30. Art. 2 of the Outer Space Treaty of 1967 says: "Outer Space, including the moon and other clestial bodies, is not subject to national appropriation by claim of sovereignty by means of use or occupation or by any other means".
31. Via the realization of remote sensing without borderlines technology in general and remote sensing in particular have greatly contributed to putting the dogma of sovereignty in a critical position: international law - favoured by developing countries - has tried to maintain it, giving room to a de jure condendo hypothesis as a real right to privacy of peoples and States.
32. "Equality of consent has always been the mask of class and power inequality" in S. Scoglio, op.cit., p. 368.
33. The German Contstitutional Court was one of the first Courts in Western countries (1983) that recognized and legitimated the radical principle of information self-determination, leaving people with full discretionary power over giving or refusing the concerned information. See ibid, p. 367.
34. G. Catalano Sgrosso, op.cit., p. 165.
35. Ibid.
36. See C. Schmitt, Terra e mare, pp. 75 and fol.; Dialogo sul nuovo spazio, Giuffré, Milano (1986) pp. 100-101. On the close connection between oceanic space and industrial development see also G.W.F. Hegel, Lineamenti di filosofia del diritto, paragraph 247, Laterza (1974), pp. 232-233.
37. C. Schmitt, Le categorie del politico, Il Mulino, Bologna (19), pp. 90 and fol.
38. See E. Severino, La Bilancia, pp. 190-198, Rizzoli (1992) and Il declino del capitalismo, Rizzoli (1994).
39. C. Schmitt, Le categorie del politico, cit.
40. See C. Schmitt in the Introduction to the Italia an Anthology of his Works, 1972.
41. See C. Schmitt's conclusion to Dialogo sul nuovo spazio, cit., pp. 108-109.
42. Expression used by M. Losano in Il diritto pubblico dell'informatica, Torino (1985).

43. This is the position of F.A. von Hayek who affirms that a market working in full autonomy is itself a law producer in a condition of freedom, in contrast to the thesis of the political production of law supported by the theorists of sovereignty, which might contain the germs of totalilitarism in nuce.
44. See E. Severino, Il declino del capitalismo, cit.
45. The proposal of the equatorial States (Declaration of Bogotà) has not received the attention of the international community. They have claimed sovereignty over those orbital parts above their country where geostationary satellites are in operation.
46. See R. Martini, op.cit., p. 172.
47. This is what Christol affirms in The Modern International Law of Outer Space, New York (1982) p. 720: "(...) Other have assumed that a proper interpretation of national sovereignty allows for the right to maintain a closed society. Thus it has been suggested that unhibited remote sensing with its accumulation of primary data as well as with the subsequent dissemination of such primary data or its analysed information has cast a cloud over national privacy the sovereign right of State to be let alone".
48. Vlasic, Principles relating to Remote Sensing of the Earth from Space, Manual on Space Law, New York (1979), p. 311, cited in R. Martini, op.cit., note 64, p. 196.
49. See E. Severino, Il declino del capitalismo, cit., pp. 153 and fol.
50. Bleckmann, Das Souveranitätsprinzip, in Völkerrecht, Archiv des Völkerrechts, 1985/4.
51. See R. Martini, op.cit., pp. 194-196.
52. See Th. W. Adorno, Società di massa, in Dialettica dell'Illuminismo, Torino (1974).
53. See P. Barcellona, Il capitalismo come puro spirito, Editori Riuniti (1990).
54. G. Scoglio, op.cit., p. 371.
55. Dalfen: The Internationl Legislative Process, Direct Broadcasting and Remote Earth Sensing by Satellite, in Canadian Yearbook of International Law, (1972), p. 92; see also Cohen, International Communication, An Intercultural Approach, Cooperation and Conflict, 1987, 2, p. 63; and Sfiss, Grenzüberschreitende Massenkommunikation und Internationale Sicherheit, I.P.V. Berichte, 1987/4, p. 10.
56. P. Barcellona L'individualismo proprietario, Boringhieri (1987).
57. Th. W. Adorno, Libertà di pensiero, in Minima Moralia, Torino (1979), p. 70.
58. S. Scoglio op.cit., p. 173.
59. The expression "right to be let alone" was first used by S. Warren and L. Brandeis, The Right to Privacy, in Harward Law Review, 1890/4, pp. 193-220, cit. in S. Scoglio, op.cit., note 5, p. 335.
60. R. Martini, op.cit., pp. 182-185.
61. ibid
62. See the Declaration of Rights of International Law relating to the friendly relationship among States, United Nations Treaty Series, (1993): "(...) all other forms of interference or attempted threats against the personality of the State or against its political, economic and cultural elements are in violation of international law".
63. The interdependence between "privacy" and "personality" is now admitted in international law and has been included in most Italian doctrine in the conflictual dimension of the "negative implications of public life": "Privacy can be defined as the way of being of a person: it excludes the possibility of a third party knowing anything relating to the person him/herself", in De Cupis, Nuovissimo Digesto, vol. XVI, p. 115; Cicù Messineo, I diritti della personalità, in Trattato di Diritto Civile e Comerciale, vol. I, Milano (1959); see also A. Baldassarre, Privacy e Costituzione; Sul diritto alla riservatezza nella costituzione italiana, in Prassi e Teoria, Milano (1976); Mazzotti, Diritto alla riservatezza e Costituzione, in Giurisprudenza Costituzionale (1970), p. 1530.
64. Although reservedness has a more wide ranging meaning in the U.S.A., it is still more limited in meaning than privacy.
65. R. Martini has a different doctrinarian opinion, which I do not share for the reasons cited above. He partly believes in the validity of a de jure condendo hypothesis of peoples' privacy belonging to their own governments.
    In this he is moved by the realistic realization that in a world of social unrest, whole communities are disadvantaged because of a tehcnological gap and it is, therefore, necessary to find juridical instruments to overcome these advantages. See R. Martini, op.cit., p. 183.
66. A. Baldassarre, Privacy e Costituzione, cit., p. 468.
67. See Nuovo Ordine economico internazionale, Principi 3 e 4, Lettera C, E, U.N.G.A. RES 3201 (S. VI) in International Legal Materials, (1974), p. 715.
68. "De nombreux Etats pensent, cependant, que le principe de la souveraineté nationale de chaque Etat sur ses richesses naturelles doit non seulement comprendre ses droits souverains sur l'utilisation et la diffusion des donnée concernant ses richesses mais, également, les droits

souveraines sur l'acquisition de ces données" in Matte, Droit, p. 161, cit. in R. Maritni, op.cit.,note 48, p. 189.

69. See Dalfen, The International Legal Process, cit., p. 195, see also F. Francioni, The Compensation for Nationalisation of Foreign Property: The Bordeland between Law and Equity, in "International Comparative Law", Quaterly 1975, vol. IV, p. 272, cit. in R. Martini, op.cit., note 53, p. 190.
70. See R. Martino, op.cit., p. 190.

ISSUES PERTAINING TO THE LEGAL DEFINITION "SPACE OBJECT" *

PROF. STEPHEN GOROVE

Chairman, Ed. Bd. JOURNAL OF SPACE LAW and Professor of Law Emeritus, University of Mississippi Law Center; Chairman: Space Law Committee, International Law Association (ILA), American Branch and International Space Law Interest Group, Am. Soc'y of Int'l Law. International Astronautical Federation (IAF) and ILA representative to the U.N. Committee on the Peaceful Uses of Outer Space; Honorary Director and Former Vice President, International Institute of Space Law (IISL); Associate Fellow, American Institute of Aeronautics and Astronautics; Honorary member, Japanese Society for the Study of Law and Policy on Space Utilization; Author/Editor: Space Law: Its Challenges and Prospects (1977), The Space Shuttle and the Law (1981); The Teaching of Space Law Around the World (1986), Developments in Space Law: Issues and Policies (1991); United States Space Law - National and International Regulation (1981-94).

Abstract : *The purpose of this presentation is to focus solely on some significant issues and related policy considerations pertaining to the notion of "space object" and associated with the five U.N. drafted space treaties. The term "space object" is central to the international law of outer space and the policies and laws relevant to its application will become more crucial with the anticipated expansion of space activities in connection with the building of the US/International Space Station and the contemplated moon and Mars missions in the next century. Following a brief introduction, the presentation deals with the use of the term "space object" as well as related "terms other than space objects" in the space treaties and issues of their interpretation. The conclusion provides a single text indicating many possible alternatives, the variations being reflected by additions or omissions of words in square brackets and parentheses.*

INTRODUCTION

Notwithstanding the remarkable achievements of the U.N. Committee on the Peaceful Uses of Outer Space (COPUOS), as reflected in the drafting of five main space treaties, commonly known as the Outer Space Treaty,[1] the Rescue Agreement,[2] the Liability [3] and Registration

Conventions [4] and the Moon Agreement, [5] a number of crucial concepts and terms like "space object," "outer space," and "launch" remain only partially clarified while many others, including "space debris," "astronauts," "personnel" and "procurement" remain undefined.

There can be little doubt that a clarification of concepts and phrases used in major space agreements and other international instruments constitutes an important aspect of legal development which will have to be addressed by lawyers and policy makers in order to allay legal uncertainty, a serious potential impediment to the prudent involvement of private enterprise in space activities.

The purpose of this presentation is to focus solely on some significant issues and related policy considerations pertaining to the notion of "space object" and associated with the aforementioned five space treaties. The term "space object" is central to the international law law of outer space and the policies and laws relevant to its application will become more crucial with the anticipated expansion of space activities associated with the building of the US/International Space Station and the contemplated moon and Mars missions in the next century. Individuals and organizations, both public and private, engaged in space activities will have to know what their rights and responsibilities are when dealing with objects in outer space. They will have to know whether to regard a particular object in a given set of circumstances as a space object because significant legal consequences, particularly with respect to liability for damage, follow from such determination.[6]

I. USE OF THE TERM "SPACE OBJECT" IN THE SPACE TREATIES

While the major space law treaties frequently use the phrase "space object," unfortunately, only a partial definition may be found in the Liability and Registration conventions, both of which state that the term "space object" includes "component" parts of a space object as well its "launch vehicle" and "parts" thereof.[7]

The fact that the partial definition of "space object" refers back to itself when speaking of "component parts" of a "space object" and "its" launch vehicle leaves the fundamental issue of what is or is not a space object or under what circumstances an object becomes or ceases to be a "space object" and the question of the applicability of the relevant space treaty provisions unanswered, thereby necessitating a systematic analysis of the various scenarios in which the issues may arise.

Before embarking on such a detailed review of the notion of "space object," it may be appropriate to address ourselves, first of all, to two initial issues raised by the partial definition, namely: (a) the question of its applicability to all of the main space treaties and (b) the question of the meaning of "component parts" and "parts" with special attention to the issue of what constitutes "space debris."

1.1. Applicability of the Partial Definition of "Space Object" to all of the Space Treaties.

The first question that comes to mind in connection with the Liability and Registration conventions' partial definition — that a space object includes its component parts as well as the launch vehicle and parts thereof — is whether such a definition is also applicable to the other space treaties. The answer to this query is likely to be in the affirmative. For one thing, there is no indication in the Outer Space Treaty and the Rescue and Moon agreements or in their travaux préparatoires that the launch vehicle and parts thereof or the component parts of a space object would not be regarded as space objects. For another, it may be pointed out that the Outer Space Treaty speaks of liability for damage caused by a space object or its "component parts," [8] thereby implying the inclusion of such parts in the notion of space object in the particular context. Also, the Rescue Agreement refers to component parts [9] and it would appear untenable, for instance, for a state party to argue that the launch vehicle is not a space object and deny its return in a given situation on that ground.

1.2. Meaning of "Component Part" and "Part" with Special Attention to Space Debris.

It is the authoritative choice by policy makers that determines in what situations damage caused by space objects will entail international liability and the scope and extent of such liability. Thus a determination of whether to regard an object in a given set of circumstances as a space object or a part of it and impose liability for damage caused by it reflects an antecedent policy choice which should be kept in mind throughout the ensuing discussion.

Reducing the policy considerations to a textual legal analysis, since a space object under the partial definition includes its "component parts" as well as the "launch vehicle and parts thereof," it appears necessary to determine what can be regarded as a component part and a part. Pieces, fragments and other substances of a space object, would normally be regarded as parts of that object. Thus the basic issue arising from the partial definition of a space object is whether the phrase "component part" is to be equated with the term "part." Can any part of a space object be regarded as a component part, or, to put it differently, are all parts of a space object necessarily component parts?

Admittedly, the term "component part" has a distinct meaning and it may be legitimately argued that the drafters of the Liability Convention by their definition regarded only "component parts" and not all "parts" of a space object as being subject to the constraints of the Convention. However forceful this argument may be at first sight, the fact remains that the quoted definition itself speaks not only of component parts but also of parts when it makes reference to the launch vehicle and "parts" thereof. It would appear unsound and unworkable within the context of the Liability Convention to regard any "part" of the launch vehicle as a space object and, at the same time, to assert that only a "component" part and not just any "part" of a spacecraft is to be taken as a "space object." [10] There is no indication that the drafters ever intended to make such distinction when they formulated the partial definition of a space object. This conclusion is also reinforced by state practice to date. To the knowledge of this writer, whenever there was a question of liability arising from the fall of space debris on earth, the issue of whether the debris was a component part or just a part of a space object with the idea of possibly denying liability in the latter case has never been given consideration.

Of course, as a practical matter, it is highly unlikely that the state of registry or launching authority would request the return of worthless fragments of a space object, particularly since such a party would have to bear the expenses associated with the recovery and return of such fragments. At the same time, it is quite conceivable that a request would be made for the return of a valuable component part.

The Liability and Registration conventions' provisions that the space object includes its component parts has also brought to the fore an important question in connection with the US/International Space Station. The issue that policy makers faced was whether such a station should be conceived of as a single space object, with the various elements being regarded as the object's component parts, or whether it should be taken to constitute a cluster of different space objects requiring separate registration. The latter had notable relevance in connection with the exercise of jurisdiction and control. This matter was settled in the US/International Space Station Agreement which provides for separate registration of each of the flight elements supplied by the partners.[11]

An issue closely associated with component parts and parts is whether debris, which may result from the break-up, deterioration, loss or abandonment of a space object, is a space object. If space debris is regarded to be a space object or a component part of such an object or happens to be its launch vehicle or a part of it, under the Liability Convention the launching state's international liability would be absolute in case damage was caused by it on earth and would be based upon fault if damage occurred in space [12]

To date, there has been no general agreement in the scholarly literature on the issue whether space debris is to be regarded as a space object. Some notable space law writers maintain that space debris is not to be considered a space object or a part of it [13] This would mean that a space object which malfunctions or cannot be controlled any more, like a broken part, would no longer be regarded as a space object or a part of it and, as a result, any damage caused by such debris would not fall under the provisions of the Liability Convention. Such a position appears to run counter to the intention of the drafters of the Liability Convention and can hardly be supported by rational arguments. As Bin Cheng quite correctly observes, fragments of a space object are treated as space objects both in the Liability Convention and the Rescue Agreement.[14]

While it is difficult and somewhat risky to attempt to provide a workable definition of space debris, the latter may be looked upon as a no longer functioning, no longer controlled, non-useful or abandoned space object or a part of such an object, when no change can reasonably be expected in these conditions for the foreseeable future. [15] Under such a definition, every bit of space debris is a space object or a part of a space object but every space object is not necessarily space debris. However, as with any definition, here also care should be exercised in its application. For instance, should a space object be branded immediately as "space debris" — with whatever legal consequences may follow from such determination — when a loss of radio contact and control occurs? Most likely not. Also, it should be noted that the space treaty provisions which are currently applicable to space objects do not appear to place any limitation arising out of the kind and size of such an object, whether controlled or uncontrolled.

The issue of the nature of space debris is also significant because, under the Outer Space Treaty, ownership of objects launched into outer space is not affected by their presence in outer space. [16] Currently, there is no right to remove no longer functioning (uncontrolled) and even useless space objects without permission, unless legally justified under the rules of international law governing self defense. It is doubtful that a potential (not actual) threat to one's own functioning space object or one's space activities, would be considered as sufficient justification for such a removal. However, if space debris is not a part of an object launched into outer space, including objects landed or constructed on a celestial body, the ownership provisions of the Outer Space Treaty would not be applicable to it, though conceivably ownership rights could still be asserted, albeit with less legalistic justification.

The conclusion that emerges from the foregoing discussion is that the Liability Convention is clearly applicable to damage caused by space debris and that de lege ferenda, the international community should address the all-important issue to determine in what situations and under what conditions could space debris be lawfully removed from outer space, bearing especially in mind the Outer Space Treaty's stipulation that, in the absence of contrary agreement, the state of registry retains jurisdiction and control over an object launched into outer space and that ownership of such objects is not affected by their presence in outer space.

II. USE OF RELATED TERMS OTHER THAN "SPACE OBJECTS" IN THE SPACE TREATIES

Apart from "space object," the treaties also use such related phrases as "objects launched into outer space" [17] or "into earth orbit or beyond," [18] or placed "in orbit around the Earth," [19] or "around or other trajectory to or around the moon" [20] and other celestial bodies within the solar system. [21] They also speak of "objects landed or constructed on a celestial body, "[22] and an occasional reference may also be found to "man-made space objects" [23] and a variety of other "objects."

The panorama of additional phrases dealing with objects in the five space treaties and the possible future scenarios that they may imply, call for a consideration of a number of significant legal and policy issues, a clarification of which may shed light on the notion of "space object" and its applicability in the context of the space treaties. These issues and possible interpretations may be conveniently discussed under the following headings: (a) the relevance and purpose of "launching," (b) the pre-launch and landing phases, (c) outer space, (d) objects landed or constructed on the moon or other celestial bodies, (e) extraterrestrial objects, (f) stations and habitats in free space, and (g) the notion of an "object."

2.1. Relevance and Purpose of "Launching"

The space treaties occasional allusion to "objects launched" or the "launching" of an object makes one wonder whether the act of launch or launching is an essential prerequisite for an object to be regarded as a space object. The space treaties do not define "launching" or "launch" apart from a stipulation in the Liability Convention that the term "launching" includes "attempted launching."

This issue will assume particular relevance in connection with the advent of the aerospace plane which is expected to take off as a conventional airplane without being launched and may reach outer space. Would such a vehicle have to be "launched" to be regarded as a space object? Should the fact of launching make a difference? Is the meaning of "launch" crucial? Should the aerospace plane be regarded as a space object throughout its flight or, more precisely, should the Liability Convention's provision be applicable to the flight of the aerospace plane in the airspace or in outer space?

Obviously, in the absence of an authoritative determination, several conjectures may be envisaged. The policy choice may well be not to apply space laws to an aerospace plane and adopt the functional approach if the vehicle is used in the course of a point-to-point transportation on earth even though, during its flight, it may reach the fringes of outer space. Another possibility would be to apply provisions of space law while the plane is in outer space. Admittedly, such a solution would require a clarification of the boundary line where outer space begins and the line where airspace ends. [24]

If the term "launch," that is, the manner in which the object ascends, is not crucial in determining whether to regard the aerospace plane as a space object, one may use the term take-off or "lift-off" which could conceivably be applied to both the aerospace plane and the shuttle. What appears important, however, is that the act of launching in the sense of lift-off or take-off or its "attempt" must in fact take place before an object may be regarded as a space object, assuming of course that the purpose of the intended activity was to put the object in orbit around the earth or beyond and there was a realistic expectancy of achieving it. If, under such circumstances, the launch or lift-off is attempted but fails and the object does not reach outer space, the respective space treaty provisions regarding liability for damage and the return of space objects would still be applicable. For the same reason, sounding rockets which are not launched with this required purpose would not be regarded as space objects.

Launching may take place from land, water, or even from the airspace as recently demonstrated when a B-52 released a Pegasus rocket in the air carrying a satellite into outer space. Karl-Heinz Böckstiegel suggests among several possible alternatives that one might consider the start of the airplane already as the beginning of the launch so that the state from whose territory this start was effected would be regarded as the launching state. [25] However, it is more likely that, in the absence of contrary understanding, the state from the airspace of which the object was launched by the airplane would be the launching state.

Another issue which may be raised in connection with launching is whether the launch from a celestial body or from free space would entail the application of the discussed space treaty provisions. Inasmuch as the provisions of several space treaties refer to objects launched "into" outer space, strictly speaking there would be no such occurrence since the object would be launched "in" and "from" outer space and not "into" outer space. Would damage done by such objects on the moon and other celestial bodies in the course of human intervention call for the application of the space treaties? The answer to this question may not be as significant as it appears to be. Liability in such a case under the provisions of the Liability Convention would be predicated on fault and such liability would likely exist even without the provisions of the space treaties. Of course, whatever advantages a recourse to the Liability Convention may carry would be lost if the objects are not considered to be space objects.

2.2. Pre-launch and Landing Phases

Another aspect of the definitional issue of space object relates to the pre-launch and post-landing or disembarkation phases, that is, the relevance of time and place. Does the location of an object or the time element make any difference with respect to the occurrence of damage caused by the object in determining liability? To put it differently, at what point in time and place does the Liability Convention's provision become operational or at what point should we regard an object, such as a launch vehicle, to have become a space object or have ceased to be one for purposes of the Convention? Should it always be regarded as a space object and damage caused by it always entail liability irrespective of where and when the damage occurs? Should it make any difference whether the damage causing object is in a manufacturing plant, or in a test facility, or in the process of being transported to the launch site, or being assembled but not installed there as yet? An additional question is whether such flight includes the space object's ascent and descent through the airspace,

Under the definition of a proposed Draft for a Convention on Manned Space Flight, an object with a human being on board intended to be launched into space would be regarded from the point of embarkation through the launch, in orbit, deorbit, reentry, landing, and disembarkation phases as a manned space object. [26] While the use of the phrase "embarkation" and "disembarkation" may be questioned, [27] the only query that arose during the drafting process was in relation to the possible use of the term "post-landing" rather than "disembarkation." However, "disembarkation" appeared to be a better term considering that reference to a "post-landing" phase might have implied an extension of the time period after landing without any specific limitation.

For certain purposes, such as the exercise of jurisdiction and control, the fact of embarkation and the closing of the doors may be significant as provided, for instance, in the application of U.S. territorial jurisdiction and control. Nonetheless, there is no indication that the drafters intended to have the space treaty provisions apply to objects prior to a launch from earth or an attempted launch. Thus an abortive fire on the launch pad, even after embarkation, prior to an attempted launch, would appear to preclude the application of the Liability Convention. Acts preparatory to the launch, including the embarkation and count-down, by themselves, would not be regarded as an attempted launch. Only when the engines fire and the lift-off is endeavored would it appear to be an attempted launch. In this connection, it may be important to stress that for an act to qualify as an attempt, it must be intended; it cannot be absolutely impossible of commission; it must involve "perpetration" or "execution" rather than mere "preparation;" it must come "close to success" and the "means" used must be adequate.

If space law were to be applied to the flight of an aerospace plane, it would have to be determined at what point in time (take-off, closing of the doors, etc.) such laws would apply to it. Short of an authoritative determination to the contrary, most likely a lift-off would have to be attempted in order to have relevant space treaty provisions applicable to it.

In sum, from the viewpoint of the lex lata, it may be better to regard the launching and attempted launching rather than the embarkation and closing of the doors as the crucial element in the determination of international liability. Such a position appears to be more in line with the space treaty provisions which only refer to launching and attempted launching and make no reference to embarkation, disembarkation, closing of the doors and similar expressions. [28]

In light of the preceding considerations, it would also appear that, prior to a launch or attempted launch, the launch vehicle and its parts should not be regarded as space objects, just as an object or component parts of an object would not become space objects prior to a launch or an attempted launch. They would not qualify for such characterization in the manufacturing plant or test facility or on their way to the launch site or at any time, prior to an attempted launch.

Does it make any difference if the damage occurs at the moment or shortly after the object's return to earth or in the course of its subsequent refurbishment? At what point in time and place does the Liability Convention's provision cease to be operational or at what point should we regard an object, such as a launch vehicle, to have ceased to be one for purposes of the Convention? The Outer Space Treaty and the Rescue Agreement speak of "landing" of astronauts and "return" of objects to earth and it may be assumed that such landing and return was intended to serve as the cut-off point. [29] De lege ferenda another more specific event, like "opening of the doors," or perhaps the Draft's provision designating disembarkation [30] as a point of termination for manned space flight may be considered for possible adoption by international policy makers.

2.3. Relevance of Outer Space

When the space treaties speak of objects "launched," they occasionally add the phrase "into outer space" or "in orbit around the earth," or "into earth orbit or beyond," [31] and, thereby, they raise the question whether it is necessary for the object to reach "outer space." In other words, should one draw the conclusion that, for an object to be regarded as a space object, it must have reached outer space or be or remain in orbit around the earth? This would mean that, if the object is no longer in orbit, it would cease to be a space object and, as a result, the relevant provisions of the Liability Convention and the Rescue Agreement would not be applicable to it. Such position would appear to run contrary to the intention of the drafters of the space treaties which speak of liability for damage caused by a space object on the surface of the earth. How could there be international liability if the object would have ceased to be a space object upon its return to earth?

Some space law commentators, invoking Article II of the Registration Convention, appear to suggest that an object is not a space object unless it is already in earth orbit or beyond. [32] However, the particular provision was not intended to define the space object but only to determine what objects were subject to the requirement of registration. An object not only can be but is in fact a space object during its flight from earth to outer space and back throughout the indicated phases and it remains a space object during its flight in outer space. Such an object would not be subject to the requirements of registration until it is in earth orbit or beyond. The Registration Convention stipulates that when "a space object is launched into earth orbit or beyond" the launching state shall register the object. [33] This statement makes it clear that the object to which the registration applies is already a space object, otherwise reference would have been made to an "object" and not to "a space object." Also, the reference in Article IV to space objects concerning which the state of registry has previously transmitted information, and which have been and are no longer in earth orbit, [34] implies that such objects remain space objects irrespective of whether they are in orbit or not, and so also during their phase of descent.

Another issue which has relevance to outer space, in connection with the characterization of objects and the attendant policy alternatives, is whether there can be any objects which are launched from earth into outer space or reach outer space as a result of human intervention and are not to be regarded as space objects. Are personal belongings which accompany an astronaut during the flight into and in outer space considered space objects? Possibly so. [34a] Does damage caused by such objects make it subject to the application of the relevant provision of the space treaties and do such objects have to be returned to the launching authority under the Rescue Agreement? The space treaty provisions do not appear to shed light on these issues and in the absence of contrary provisions, it would appear that such objects would be regarded as space objects.

2.4. Space Objects Landed on the Moon or Other Celestial Bodies

In connection with the moon and other celestial bodies, some of the space treaties make occasional references to a variety of other objects such as, for instance, "objects landed or constructed on the Moon," "vehicle" [35] and "space vehicle," [36] "supplies," [37] "equipment," "installations," "facilities," [38] and "manned and unmanned stations." [39] The envisaged scenarios that may be associated with these references raise a number of additional issues which have a bearing on the notion of "space object."

First and foremost, the question arises whether an object launched from earth would lose its legal characterization as a space object upon its landing on the moon, or Mars or another celestial body? Would a moon rover or other movable objects, equipment or supplies originating from the earth cease to be space objects and would the relevant space treaty provisions not be applicable to them following such landing? Or would such objects continue to remain space objects and, if so, for how long?

While there is a temptation to argue that such objects should no longer be regarded as space objects after their landing or during their stay on a celestial body, it is somewhat doubtful that, in the course of a mission to Mars, a temporary landing is likely to be regarded as a sufficient justification for taking the objects out of the operation and application of the relevant provisions of the space treaties. A contrary position would have to come to grips with the issue whether such objects following their landing would again become space objects after their relaunch. The Draft left open the question whether a space object remains a space object following its landing on the moon or another celestial body. [40] Logically, and by definition, a flight would come to an end after landing if flight is understood in the conventional sense of the word. Also, a space flight from a celestial body would presumably involve similar phases of embarkation, launch, in orbit, deorbit, and disembarkation, as a flight does from earth to outer space. Notwithstanding the logic of this reasoning, looking at this issue from the vantage point of an earthly perspective, states parties to a convention similar to the Draft who participate in a manned expedition to Mars, may not regard a temporary stopover on the moon or another celestial body as necessarily suspending the operation of the convention. For purposes of both the uniform application of the law as well as reason and logic, it would appear preferable to regard such objects as space objects during such stay.

In general, two things may be emphasized which may be important for the policy choice : the length of time during which space objects are utilized after their landing on the moon or another celestial body and the associated preferability of making a break with what may likely entail an endless extension of the Liability Convention's application to situations for which it was originally not intended. Admittedly, authoritative policy makers would have to make a determination.

Another issue that arises apart from the landing and length of stay of a space object on a celestial body relates to the possibility of a moveable space object being made into or becoming part of an immovable structure in the form of a station or facility on a celestial body. In such a case, it is doubtful that the policy choice would be to continue to regard such objects as space objects.

2.5. Extraterrestrial Objects.

The Moon Agreement envisages the eventual exploitation of "natural resources" of the moon and other celestial bodies under the auspices of an international regime [41] and also makes reference to "samples of mineral and other substances" [42] which may be used for scientific purposes. The possible use of extraterrestrial materials may also be envisaged by the Outer Space Treaty's reference to objects "constructed" on celestial bodies. [43] If such natural resources and other extraterrestrial materials cause damage at the time of their collection or at a later stage in the course of their use in support of a space mission or upon their return, should liability attach on the basis of the application of the provisions of the Liability Convention or the Outer Space Treaty? Should such objects be required to be returned to the launching authority in appropriate instances under the provisions of the Rescue Agreement? Briefly put, should they be regarded as space objects in given contexts?

As to objects not originating from the Earth, there is some doubt that the provisions of the major space treaties applicable to space objects could be properly invoked. The Outer Space Treaty, the provisions of which were further developed by the Liability Convention, speaks of the "return" of space objects to earth, thereby implying that the objects had to be on the earth beforehand to be regarded as space objects. Up to now, the issue has not been pressing but it could assume significance in future scenarios involving the exploitation of natural resources in space and the use of extraterrestrial materials in support of a space mission.

In the course of the construction of objects on the moon or another celestial body, the question may also arise whether such movable materials made in part of terrestrial and in part of extraterrestrial materials will remain space objects or whether they will lose their legal identity? If the notion of space object were extended to cover extraterrestrial objects handled by humans as space objects for purposes of the relevant space treaty provisions, then all such composite materials would automatically be regarded as space objects. [44] In the absence of such policy determination, the likelihood is that the space treaty provisions relating to space objects would not be applicable to them.

2.6. Stations and habitats in free space.

The reference to "stations" in the Outer Space Treaty and to "manned and unmanned stations" in the Moon Agreement [45] relate to such structures located on the moon or another celestial body. However, legal and associated policy issues may further present themselves in conjunction with the building of space stations and habitats in free space in which both terrestrial and extraterrestrial materials may be used.

There is little doubt that the US/International Space Station built of terrestrial materials would be regarded as a space object or cluster of space objects or parts of a space object to which the space treaty provisions pertaining to such objects would be applicable. This conclusion is fully substantiated by the relevant provisions of the US/International Space Station Agreement. [46] Even if some extraterrestrial materials were used, the likelihood is that this fact alone would not change the outcome.

Permanent habitats in free space, as in L-5, which could conceivably be built in whole or in part from materials not originating from the earth may require further scrutiny and consideration by policy makers, This is not a far-fetched possibility in the 21st century. For instance, Professor O'Neil from Princeton University worked on the design of a mass driver to be located on the moon for the purpose of hurling lunar materials into space which could be used when building a habitat in free space, possibly for manufacturing solar power satellites. [47] Even if the materials originate exclusively from the Earth, the long term extension of the applicability of the Liability Convention may have to be reexamined.

2.8. Notion of an "Object".

Finally a word may be added relating to the connotation to be attached to the word "object." This term in every day usage refers to a person or material thing that can be seen or touched and is stable in form. [48] Seen in such a context, solar energy, electromagnetic impulses, cosmic and other forms of radiation, as well as nontangible biological or chemical agents, are not regarded as objects. [49] At the same time, installations, equipment, materials, payloads, fragments and debris would be included in the category of objects or parts of objects.

Notwithstanding the seeming simplicity of the foregoing differentiation, questions may arise, for instance, with respect to the applicability of the Liability Convention's provisions when damage is caused by atomic radiation, or solar energy. Under the relevant stipulations, damage must be caused by a space object [50] and if nuclear or solar energy does not qualify as an object how can liability arise? In such cases, the object regarded to be causing the damage is the nuclear power source from which the radiation emanates, or the solar power satellite which transmits solar energy to earth via microwave or laser.

CONCLUSION

The preceding overview and analysis is intended to shed light on the multifaceted issues presented by a single but central notion in the space law literature, that of "space object." The discussed scenarios underscore the need for further delineation of the term, especially in situations when different people can legitimately and, in some cases with equally strong logic, maintain divergent views. Definitional clarifications involve policy choices which decision makers will have to make in light of their value judgments on behalf of the countries they represent.

From among many possible alternatives, space limitations only permit a few examples of definitions, the variations being indicated by additions or omissions of words in square brackets and parentheses. Within the context of the main space treaties, a space object may be defined as "an object launched or attempted to be launched in orbit around the earth or beyond [and includes stations, installations and other objects (whether terrestrial or extraterrestrial) constructed or used by humans in outer space, including the moon or another celestial body]. Such object [or a part of it] is a space object [or a part of it] from the time of its launch or attempted launch, through its ascent from earth to outer space or while in outer space, as well as during its orbit, deorbit, reentry and landing on earth. [In case of a manned space flight, a space object (or a part of it) is a space object (or a part of it) from the time of embarkation (closing of the door) to that of disembarkation (opening of the door) on earth.] [A space object (or a part of it) landed on the moon or another celestial body which becomes part of an immovable structure ceases to be a space object (or a part of it).]

Associated with the definitional clarification is the issue whether to split the single legal notion of "space object" into several well defined categories, such as "space station," "space object" in the narrower sense, and "space debris," as suggested by Vladimir Kopal. In his view, such a split will become sooner or later inevitable. [51]

Even with the adoption of one of the intimated definitional alternatives, there will be many remaining issues which may need further consideration. Of course, it is well to keep in mind that there is no fool-proof definition to take care of all possible scenarios which may arise in the future. Nonetheless, even a limited removal of some of the uncertainties associated with the notion of "space object" would go a long way in allaying concerns of private entrepreneurs when undertaking space activities. In addition, a clarification of some of the authoritative policy choices would also help in eliminating possible sources of disputes.

The clarification of definitional issues will be ever more pressing as we expand the horizons of space exploration and use in the 21st century. In light of the reduction of international tensions and disappearance of the cold war psychology, the unique opportunities of world-wide international cooperation make the objective of achieving consensus on the scope and meaning of undefined or only partly defined legal terms less difficult to achieve. It is this writer's belief that the time has arrived to advance suitable proposals to clarify key notions and phrases of space law for consideration in a multidisciplinary effort by national and international institutions and policy makers. It is hoped that strides in this direction will continue unabated and will eventually lead to positive results.

REFERENCES

* The topic of "definitional issues of space law" was first discussed at the Montreal Colloquium in 1991. This was followed up by a decision of the IISL Board at its 1992 meeting in Washington, D.C. to establish a Working Group, consisting of Professors S. Gorove (chair), K.-H. Böckstiegel, and V.S. Vereshchetin, for consideration and development of the subject matter. While this paper deals with the notion of "space object," it also touches upon issues of "space debris" as well as those of "launching." At this same time, Böckstiegel's papers (IISL-94-IISL.2. 828(a) and (b) address the concepts of "appropriate State" and "launching State," whereas the contributions prepared by Dr. E. Kamenetskaya (IISL-94-IISL.2 826) and Dr. E. Zhukova (IISL-94-IISL.2 827) in cooperation with Prof. Vereshchetin, focus on the terms "astronaut," "personnel," "crew," "passenger," and "space debris."

The present paper, which was presented at the 1994 IISL Colloquium in Jerusalem and is expected to be included in its Proceedings, is a slight variation of the first version which appeared in the Journal of Space Law [vol. 21 (1993), pp. 11-26], insofar as it makes an occasional reference to the other papers of the collaborative effort and considers comments received in the exchange of views among the members of the Working Group..

1. Treaty on Principles Governing the Activities of States in the Exploration and Use of Outer Space, Including the Moon and Other Celestial Bodies, Jan. 27, 1967, 18 U.S.T. 2410, T.I.A.S. No. 6347, 610 U.N.T.S. 205 (entered into force for the United States Oct. 10, 1967) [hereinafter "Outer Space Treaty"].
2. Agreement on the Rescue of Astronauts, the Return of Astronauts, and the Return of Objects Launched into Outer Space, April 22, 1968, 19 U.S.T. 7570, T.I.A.S. No. 6599, 672 U.N.T.S. 119 (entered into force for the United States Dec. 3, 1968) [hereinafter "Rescue Agreement"].
3. Convention on International Liability for Damage Caused by Space Objects, March 29, 1972, 24 U.S.T. 2389, T.I.A.S. No. 7762, 961 U.N.T.S. 187 (entered into force for the United States Oct. 9, 1973) [hereinafter "Liability Convention"].
4. Convention on the Registration of Objects Launched into Outer Space, opened for signature Jan. 14, 1975, 28 U.S.T. 695, T.I.A.S. No. 8480, 1023 U.N.T.S. 15 (entered into force for the United States Sept. 15, 1976) [hereinafter "Registration Convention"].
5. Agreement Governing the Activities of States on the Moon and Other Celestial Bodies - adopted by the U.N. Gen. Assembly on December 5, 1979, opened for signature on Dec. 18, 1979, entered into force July 11, 1984 (not in force for the United States), U.N. Doc. A/RES/34/68 (1979) [hereinafter "Moon Agreement"].
6. While occasional definitions of the notion of "space object" may be encountered in the scholarly literature since the dawn of the Space Age, it was not until the Montreal IISL Colloquium in 1991 that, under the chairmanship of this writer, a special session was devoted to "Definitional Issues in Space Law" which inter alia also touched upon issues relating to space objects. See Bin Cheng, "Space Objects," "Astronauts" and Related Expressions, 34 Proc. Colloq. L. Outer Space 17 (1992); HE Qizhi, Review of Definitial Issues in Space Law in the Light of Development of Space Activities,

id. at 32; Vladimir Kopal, Issues Involved in Defining Outer Space, Space Object and Space Debris, id. at 38; William B. Wirin, Space Debris and Space Objects, id. at 45.

7. Liability Convention, art. I(d), Registration Convention, art. I(c).
8. Outer Space Treaty, art. VII.
9. Rescue Agreement, art. 5.
10. For a discussion of this issue, see Stephen Gorove, Developments in Space Law - Issues and Policies 151-52 (1991)
11. See art. 5 of the Agreement Among the Government of the United States of America, Governments of Member States of the European Space Agency, the Government of Japan, and the Government of Canada on Cooperation in the Detailed Design, Development, Operation, and Utilization of the Permanently Manned Civil Space Station, Signed September 29, 1988, in Washington, D.C. (hereinafter US/International Space Station Agreement). For a text of the Agreement, see United States Space Law - National and International Regulation, Sec. II. A. 22 (Stephen Gorove ed., 1982-1994).
12. Liability Convention, arts. II and III.
13. See, for instance, HE Qizhi, Review of Definitial Issues in Space Law in the Light of Development of Space Activities, supra note 6, at 35. In Wirin's view, "space objects and components" should be distinguished from small pieces and fragments of debris which are not capable of reentering the atmosphere and should not be regarded as space objects or components. See William B. Wirin, Space Debris and Space Objects, supra note 6, at 50.
14. Cheng states that "Fragments of a space object that fall on the earth are certainly treated as parts of that space object, and are given exactly the same status as the whole object, were the object to come back in one piece.... Nothing suggests otherwise, or that shattered fuel tanks or flakes of paint from space objects in outer space should be treated any differently." Bin Cheng, "Space Objects," "Astronauts" and Related Expressions, supra note 6, at 24. Prof. I.H. Ph. Diederiks-Verschoor in her recent book "An Introduction to Space law" (1993), at pp. 117-18, appears to support this position which is also shared by E. Zhukova. See her paper (IISL-94-IISL.2 827) on "The term 'Space Debris': An Attempt of Definition" prepared for the Jerusalem Colloquium.
15. Stephen Gorove, Space Debris in International Legal Perspective, 32 Proc. Colloq. L. Outer Space 97 (199); Cf. International Academy of Astronautics, Committee on Safety, Rescue and Quality, Position Paper on Orbital Debris, August 27, 1992, at 1.
16. Outer Space Treaty, art. VIII.
17. Id, at arts. VII and VIII; Registration Convention, Preamble.
18. Registration Convention, art. II.
19. Outer Space Treaty, art. IV, para. 1.
20. Moon Agreement, art. 3, para. 2.
21. Id. at art. 1, para. 1.
22. Outer Space Treaty, art. VIII.
23. Moon Agreement, art. 3.
24. For a detailed discussion of the legal and policy choices associated with the aerospace plane, see Stephen Gorove, Legal and Policy Issues of the Aerospace Plane, 16 J. Space L. 147 (1988).
25. Karl-Heinz Böckstiegel, The terms "Appropriate State" and "Launching State" in the Space Treaties - Indications of State Responsibility and Liability for State and Private Space Activities, 34 Proc. Colloq. L. Outer Space 15 (1992).
26. See art. I, para. 2 of the Draft for a Convention on Manned Space Flight (hereinafter "Draft") which was prepared by three leading institutions in Germany, the former Soviet Union, and the United States and submitted by this writer to the U.N. Committee on the Peaceful Uses of Outer Space in 1991 on behalf of the International Institute of Space Law. For a text of the Draft, see 18 J. Space L. 209 (1990).
27. See comments of Judge Guillaume, in Manned Space Flight - Legal Aspects in the Light of Scientific and Technical Development, Proceedings of an International Colloquium, Cologne, May 20-22, 1992, at 201 (Karl-Heinz Böckstiegel ed., 1993)
28. Cf. Stephen Gorove, Environmental Risks Arising from Space Activities, in Environmental Aspects of Activities in Outer Space - State of the Law and Measures of Protection 130 (Karl-Heinz Böckstiegel ed., 1989).
29. E.g. Outer Space Treaty, arts. V and VIII; Rescue Agreement, arts. I, 2, 4, and 5.
30. Draft, art. I, para, 2.
31. E.g. Outer Space Treaty, arts. IV, VII, VIII; Registration Convention, art. II.
32. HE Qizhi, Review of Definitial Issues in Space Law in the Light of Development of Space Activities, supra note 6, at 32.

33. Registration Convention, art II.
34. Id. art. IV, para, 3.
34a. In his written exchange of views, Vereshchetin noted that this idea is "rather disputable."
35. Moon Agreement, art. 12.
36. Outer Space Treaty, arts. V and XII; Moon Agreement, arts. 8 and 12.
37. Moon Agreement, art. 12.
38. Outer Space Treaty, arts. IV and XII; Moon Agreement, arts. 3, 8, 10 and 12.
39. Moon Agreement, art. 9.
40. Cf. Draft, art. I, para, 2.
41. Moon Agreement, art. 11, para. 5.
42. Id. at art. 6.
43. Outer Space Treaty, art. VIII.
44. Bin Cheng favors an expansion of the definition of space objects to include "Stations and Installations constructed by humans in outer space or on the moon or other celestial bodies." See Bin Cheng, "Space Objects," "Astronauts" and Related Expressions, supra note 6, at 24.
45. Outer Space Treaty, art. XII; Moon Agreement, art. 9.
46. See, e.g., US/International Space Station Agreement, supra note 11, at arts. 2, 5 and 17.
47. See G.K. O'Neil, The High Frontier, Human Colonies in Space (1977); idem, The Colonization of Space, Physics Today 32 (Sept. 1974); For a comprehensive assessment of the international legal implications of solar power satellites, see Stephen Gorove, Satellite Power Systems - International Agreements (U.S. Dept. of Energy, 1978). For a recent discussion, see Eilene Galloway, The Legal and Regulatory Framework for Solar Power Satellites, in Solar Power Satellites - The Emerging Energy Option 183 (Peter E. Glaser et al. eds., 1993).
48. Cf. The Random House Dictionary of the English Language 993 (unabridged ed. 1966).
49. Referring to Prof. L. Perek, E. Zhukova notes in her paper (IISL-94-IISL.2. 827) that only "dust particles, molecules and gaseous components" are not regarded as debris.
50. Liability Convention, arts. II and III.
51. Vladimir Kopal, Issues Involved in Defining Outer Space, Space Object and Space Debris, supra note 6, at 41.

LOOPHOLES IN LIABILITY? ASPECTS OF LIABILITY FOR DAMAGE SUSTAINED IN THE COURSE OF SATELLITE TELECOMMUNICATIONS ACTIVITIES

FRANS G. VON DER DUNK

Co-Director, International Institute of Air and Space Law, Leiden University; Assistant Professor, Department of Public International Law, Leiden University; Coordinator Dutch NPOC/ECSL; Member IISL, ECSL, AIAA, ILA, NVIR.

I - SATELLITE COMMUNICATIONS, SPACE LAW AND LIABILITY.

Satellite communications today is one of the most dynamic areas of space activities. Although its relative relevance within the larger field of telecommunications is constantly shifting, for reasons predominantly of rapid technological developments both on the terrestrial and on the extra-terrestrial front, in absolute terms it is still rapidly gaining in importance. Inevitably therefore, disputes relating to these activities will gain in importance as well, which in turn makes that the status of the law becomes of increasing practical importance.

In this respect it should be noted that any analysis of the legal aspects of satellite communications operations should start with space law. Satellite communications first and foremost is still a space activity, despite being part of the field of telecommunications activities as well, and hence also falling under pertinent telecommunications law, focused as that might be on earthbound legal realms.

An important dichotomy arises here between the earthbound operations involved and the operations of the space segment orbiting high above the earth. Legally speaking as well as in practical terms, all terrestrial activities related to and connected with the satellite operations as such have a clearly distinct character from the truly space-bound parts thereof.

This article will only deal with the satellite part of satellite communications activities, since terrestrial activities basically fall within the respective jurisdictions of the states on whose territories they take place. While of course these states may have taken international steps to provide for some rules amongst themselves as to how their national jurisdictions are to be exercised in this respect, this still has to be viewed from the perspective of the national sovereign rights of every state to structure its telecommunications activities as it wishes. [1]

States are not bound to adhere to any international regime such as that agreed upon for instance in the framework of EUTELSAT except by their own free will. Up to a point, this even holds true within the European Union where something like one territory for the purpose of telecommunications is going to be created, since membership of the Union is still a voluntary affair, and individual member states at this point still some the possibilities through the Council of Ministers to obstruct full application of the relevant principles to their own territory.
So what concerns us here is the space part of satellite communications, since it takes place in the area of outer space which is free of national sovereignty and can not be appropriated by any state [2]. Space law further provides that outer space shall only be used for peaceful purposes [3] ; but other than that, it remains free for exploration and use [4]. Therefore, there can be no question as to whether satellite communications as such is allowed as long as remaining within the bounds provided by international space law reigning the realm of outer space.

On the other hand, it must be acknowledged that the rules of space law by applying to communication satellites not only in a practical sense have an impact on the terrestrial operations. They also in a legal sense sometimes may extend into the national realms of sovereign states by imposing international obligations excluding discretion of those sovereign states as to the way in which they deal in law with the terrestrial part of space communications.

Liability for damage as a consequence of space activities is the most important example thereof, already in the sense that such damage incurred on earth falls under the space law liability regime, and is not a matter of unrestricted national discretion to deal with by means of law.
When it comes to the issue of disputes of a legal nature in respect of satellite communications, it must therefore be realized that the most substantive issues to become the subject of such disputes, as a

consequence of the ultra hazardous nature still clinging to space activities, will relate precisely to such questions of liability for damage caused by space activities.

Also, it must be noted that because of the unity of space as a legal realm, such liability is dealt with by space law in a unitary fashion, applying not only to satellite communications operations, but also to all other activities involving satellites or their launch vehicles. This is logical and appropriate in view of the fundamental aim of the liability regime - and indeed space law as a whole - to protect innocent third parties, for whom there is no fundamental difference between damage caused by a communications satellite and damage caused by a remote sensing satellite or a launch booster.

It is the liability regime of international space law which this article will make a few observations on. For while it may be undisputable that so far at least international satellite communications activities have not led to legal claims with respect to liability, inside or outside of the courts,[5] it remains to be seen whether the present system of liability as established by international space law will hold its own once it is really tested in disputes between entities involved in international satellite communications. Such disputes may arise sooner rather than later, in the light of the rapidly growing importance of satellite communications in the wake of telecommunications in general, and this moreover in two distinct fashions.

II - RECENT DEVELOPMENTS IN INTERNATIONAL SATELLITE COMMUNICATIONS AND THE RESULTING QUESTIONS.

There are two recent, distinct but interrelated developments which may challenge the liability system as presently provided. They make any perceived flaws in the body of international space law of vital importance when it comes to dealing with satellite communications. In terms of liability, these developments essentially relate to two respective aspects which may be detected: the structural aspect and the substantive one.

2.1. The Growing Involvement of Private Enterprise in Satellite Communications.

The structural development presently changing the environment for satellite communications to an almost revolutionary extent relates to the shifting focus from states and state-centred public international organizations to private entities and international consortia, in consequence of a perception shifting from telecommunications as a public utility to telecommunications as a commercial activity and prime motor of economies.

The heart of the problem from a legal perspective concerns the exact place, role and status of private enterprise within the public body of laws regulating telecommunications at the international level. Thus, for instance within the International Telecommunications Union (ITU), a public international organization with states as its exclusive members, discussions are raging as to how it should adapt to the growing involvement of private companies, national or transnational, in the increasingly profitable business of space telecommunications. [6]

Another important element concerns those international bodies undertaking satellite communications themselves. Since the rise of private entities doing the same they have found themselves to be in some sort of competition with the form, while remaining in essence intergovernmental, bureaucratic operating organizations intended to undertake activities thought to be of a public utility-nature. To what extent should for example EUTELSAT, the European Satellite Telecommunications Organization, be privatized, and be torn loose from a rigid structure of having one, usually public, bureaucratic and monopolistic partner for every national territory? [7] Obviously, the European Union with its battle-cry of free competition plays a dominant role here. [8]

Thus, more fundamental questions raise their heads. Whilst international organizations have their legal rights and obligations relatively clearly outlined in the space treaties, even if only by continuous reference to states, [9] the question becomes perhaps even more pertinent: what is the place of private entities within this larger scheme? What particular legal structure is in place to solve actual disputes on questions of damage and compensation involving private enterprise therefore? And finally of course, is what exists a satisfactory structure, or are there flaws and loopholes to be detected, and rather important ameliorations to be achieved?

2.2. The Increasing Scramble for Satellite Communication Orbits and Frequencies.

Ultimately, all the aforementioned issues relate back to one underlying current: the growing profitability of telecommunications in general and satellite communications in particular. This is the prime motivating factor for private enterprise to become increasingly involved in any activity, but it also is a strong inducement for intergovernmental organizations to undertake such activities, in addition to or in extension of their public tasks.

This brings us to the complement to these structural developments: the substantive developments in international satellite communications today. In view of the scarcity of the available resources, increasing profitability and commercial viability of satellite communications are leading to an increasing scramble, especially for orbital positions and frequencies in the geostationary orbit, amongst those interested in providing satellite communication services - with the accompanying and exponentially increasing risks of disputes on interference and the consequent damages.

Thus, disputes in respect of several orbital slots over East Asia, one of the fastest growing regions of the world when it comes to telecommunications, have already reached the pre-juridical public phase. [10] Moreover, the fact that so far such disputes have remained pre-juridical seems to owe a lot to the continuing uncertainty regarding the exact legal nature of the results reached within the ITU's regulatory framework, and even more to the fact that at any rate no juridical or juridically effective dispute settlement mechanism can be found, either within or outside of the ITU framework. [11]

Once more, the result of this increasing competition for satellite orbits and frequencies must be seen in the light of the fact that profitability and viability take on a rather special character when it concerns space activities. While the earthbound elements of satellite communications activities may have become fairly routine operations, operating uplink and downlink stations as well as the accompanying networks being nothing special when compared to running completely terrestrial networks, operation of the space segment still is quite something else.

Satellite activities, whether it concerns the launch and tracking and control activities or the specific activities related to telecommunications (pointing of antennae, switching of channels etcetera), are still very hazardous activities. They entail both a relatively large chance that some mishap might occur, and a relatively large chance that once a mishap occurs, it will be of catastrophic magnitude.

For that reason, probably more than with any other private activity undertaken for profit, the way in which the pertinent law deals with these questions of damage, both from the victim side and from the causing side, is of essential importance for any private entity.

On the substantial side of liability therefore a few pertinent questions should be asked as well. What regime of liability applies, if any? What cases of damage are covered; what are the exceptions, and why? How much in the end will one receive, respectively have to pay, in compensation? And once again, finally, to what extent does international space law present a satisfactory legal regime for liability?

III - THE STRUCTURAL ASPECTS OF LIABILITY AND THE LOOPHOLES.

The first aspect of liability is the structural one, since it deals with the question of the allocation of any material claim for damage and the resulting liability between the entities involved. Here, the place of the private entity is defined with respect to questions of liability, where in the state-oriented body of international space law both on the victim-side and on the inflictor-side the state legally speaking is the relevant entity.

3.1. The Victim-Side of Damage.

Thus, on the victim-side it is the state which has an exclusive right to claim even if the damage was actually suffered by private entities. [12] The question of allocation here is answered by providing exactly the specific categories of private victims on behalf of whom a particular state may claim.

Thus, a state has the right to claim primarily in respect of damage suffered by its nationals, including private juridical legal persons with its nationality. Secondarily, it may claim in respect of damage suffered on its territory by nationals of another state if the latter state does not claim itself. And tertiarily, it may claim in respect of damage suffered by those permanently resident in its territory, as long as the state of nationality of the victim respectively the state where the damage actually was sustained do not use their right to claim themselves. [13]

It should be added that, in line with general public international law, those states are not internationally bound to compensate the actual private victims if these claims are satisfied. That depends upon national law respectively comity and decency of the governments in question, balanced against other interests of the state in not recompensing.

Private entities are not principally deprived of any possibility to forward legal claims themselves for compensation of damage suffered, but these possibilities also depend upon national legislation [14] and not upon international space law. To this extent, the liability arrangements are still fragmented: national sovereign legal systems have to be used in the absence of an overriding international legal system. [15]

The Liability Convention in this respect merely stipulates, that in case such a private claim has been brought before a national court, the procedure provided by the Liability Convention can not be used at the same time by the pertinent state to present a claim in respect of the same case of damage. [16] The other way round, a private entity still would have the option to sue in a national court despite concurrent state-to-state claims under the Liability Convention.

In practice, it would mean that private entities could indeed try to claim compensation on their own accord. Most likely options here are the courts of their own state of nationality - which still has disadvantages however in terms of evidentiary matters and actual execution of a judgement - or the courts of the state of nationality of the causing entity (if that is not a state itself) - with all the difficulties going with prosecution in foreign courts and under foreign procedures. In addition, to the extent the entity accountable for the damage is a foreign state or government body, such private claims might run up against the state immunity potentially invoked in court.

Compare this to the status of international organizations of space law. While, once more, states - read: the member states - are the entities to claim if an international organization is the entity actually suffering the damage, [17] the inclination of states to assert a claim on behalf of an organization which they have joined out of their own will usually will be much larger than to assert a claim on behalf of a particular private entity, even if such an assertion is not urged or obliged under the organization's constitutional document.

Moreover, whereas in theory one state acting in that way on its behalf would suffice for an international organization, usually in contrast to private entities it could count on a number of states principally interested: INTELSAT currently has 135 member states, INMARSAT 76, and EUTELSAT 46. [18] With these entities partially operating in the same fields as private enterprise, this will lead to discrepancies on the practical consequences of liability and the consequent need to insure, which can not all be traced back to the public utility function of the international organization.

3.2. The Inflictor-Side of Damage.

On the other end of liability, it is - at least under international space law - always a state or a number of states which are liable, even if the activities causing the damage as such are privately undertaken. To see how the question of allocation is answered on the inflictor-side, one has to focus on the issue of the 'launching state' of the space object accountable for the damage, since it is the launching state (or states) which is are liable for that damage. [19]

The launching state is defined in four different alternative ways. [20] If the space object causing the damage is launched from the territory of a state, from a facility of a state, launched by a state or if its launch is procured - that is basically: paid for - by a state, such a state becomes liable. This construction results in the possibility of four (or more) different states being liable in respect of the same case of damage. [21] Once more, the position of international organizations, though secondary to that of states, is relatively clear. Damage caused by a space object which actually is owned and operated by an international organization under normal circumstances simply results in the member states of that organization, or at least those member states involved in the launch of the particular space object by means of contributing funds, being liable as states co-procuring the launch in question.

Internal settlement of compensations paid will be dealt with by internal procedures of the organization, if provided for, but at any rate will not result in a larger financial burden than those member states want it to shoulder. This is interesting to note, since one way or another those member states in the end collectively will have to make it possible for that organization to shoulder the share of the burden thus decided upon, complementing their own shares to be shouldered directly. Here, the public character of these intergovernmental organizations as instruments of states for more efficient use of their public funds becomes apparent.

Even in the exceptional cases of the European Space Agency (ESA) and EUTELSAT, which have produced declarations accepting the rights and obligations under the Liability Convention, [22] this merely means that claims for liability for damage caused by space objects in respect of which ESA or EUTELSAT qualify as 'quasi-launching-state' could be directed in first instance to the organization in question. If such claims are not settled in that mode, in the end however the member states involved as launching states themselves will have to pay.[23]

For partly or completely privately conducted launches resulting in some damage invoking the Liability Convention, however, this regime results in a more complicated system of allocation. If a private entity undertakes the launching activity and the satellite thus launched causes damage, there is no state which can be held liable under this criterion. At the same time however, the private entity itself is not liable either, at least not directly under the Liability Convention.

Mutatis mutandis, the same holds true in respect of two other criteria for becoming a liable entity under international space law. If the launch at issue is procured solely by one or more private entities, respectively undertaken from a completely private facility, there is no state liable - and hence no entity at all - under these criteria.

This makes the remaining criterion of essential importance for the question of allocation. Only states possess territory in the international legal sense of the word, and this means that under this criterion launches which in other respects are completely private, through the territory where they take place will always result in some state being liable under international space law. [24]

In this way, international space law in essence provides for a system of allocation of private entities for the purposes of liability focusing on territory. Consequently, it is basically the state of the territory of the launch which will have to take special care that those launches from its territory which are private in character are under its legal control.

It can do so in two ways: by establishment of a national space law providing for obligatory licensing in respect of launches undertaken from its territory, including an obligation to reimburse the state concerned for any international liability claims, [25] or by simply obliging the existing launch providers operating from its territory to accept such reimbursement obligations, [26] leaving it to them to further derogate compensations to be paid to the other entities involved, might the case arise.

3.3. The Loopholes.

So far, this system has provided a satisfactory legal means of dealing with international liability claims related to space activities - at least in theory. [27] The victim-oriented nature of the system, as evidenced by the broad fourfold definition of the states to be held liable and the three bases on which states may claim on behalf of individual victims, was considered its greatest asset in this respect.

However, in view of the increasing number of private entities which become involved in satellite communications, a few flaws perceived merit greater attention. Firstly, the large gap arising between the legal status of international intergovernmental organizations, even when acting partly for commercial purposes, such as INTELSAT, INMARSAT and EUTELSAT, and private entities trying to enter the same or similar markets, comes to mind.

While it may be that such a gap is inevitable due to the differing starting points - international organizations still proceeding from the public utility-perspective, private entities basically in it for the money - and consequently differing legal status - international organizations public bodies created by states, private entities private bodies created within states - to the extent that they operate in the same commercial or semi-commercial markets preferably ways should be found to deal with these increasingly irritating discrepancies wherever they actually arise. Admittedly, however, this problem oversteps the boundaries of the narrower issue of liability to be dealt with here, while at the same time the solution as to liability depends upon the larger solution with respect to the fundamental problem. For that reason, for example bodies such as the European Union or the newly founded World Trade Organization should, as the convenient and appropriate fora for these issues, deal with the larger problems first, which will then almost automatically lead to solutions on the inequality and resulting loopholes to abuse the prerogatives of international organizations for plainly making money in terms of liability as well.

That brings us to the second flaw becoming apparent in the present liability system, which is more of a real loophole still. Whereas the liability of international organizations is internationally dealt with, liability of private enterprise is not. The resulting loopholes at least in theory can be filled easily by national legislations derogating the consequences of international state liability by the states so concerned to the private entities actually to be held accountable.

Practical reality is a different matter however, presenting not just a patchwork but one with many holes as well. Those national space laws in existence which deal with the issue of derogation operate in different manner and with different scopes, [28] and this leads to different conditions for private entities from different states. Other states have no visible or understandable regulation of private space activities at all, but nevertheless are already or might soon become involved.

Such a patchwork with many loopholes could even result in license shopping, respectively 'non-license shopping', and a phenomenon of 'launching states of convenience' arising, similar to the phenomenon of 'flags of convenience' well known in the law of the sea [29] - with all due consequence for the safety of space activities and space communications. With ongoing technological developments, the day may be not far off any longer when small and difficult-to-detect launching installations may become available for interested private entities, making it no longer viable for states to trust in their general ability to detect these activities before harm might result.

A closely related fundamental problem is hidden in the focus of liability on the territory of the launch, with the aforementioned result that only states with launching sites on their territory have established national legal regimes dealing with liability by providing for licensing schemes. With satellites orbiting for many years, and as a consequence opportunities arising of sales-in-orbit or leases-in-orbit [30], the logical legal link between the entities involved in the launch and any damage caused by a satellite becomes increasingly less logical.

While of course contracts between the states permitting their territory to be used for the launch and the actual satellite owners and operators can deal with this problem by means of obligatory derogation clauses to be included in any future contracts of these satellite owners and operators with buyers or lessees of the satellite, the result might be a string of legal texts in which uniformity and logic can easily get lost.

Also, the victim-orientedness of such legal constructions threatens to disappear into the distance. Every link in the chain could make for extra time before any compensation would actually be paid, in spite of the remaining firm obligation of the (original) launching state to pay. [31] Such a state might have considerable interests in stalling payment to the victim, in order to take care politically of its interests in derogation further down the chain successfully first.

Finally, the possibility of private entities launching from outside any territorial jurisdiction from private launch sites becomes larger the more financial interests are at stake. Plans have already been formulated to create private spaceports in Cape York, Australia, Florida and other places, alternatively commercialize the Swedish launch site Esrange at Kiruna. [32] Also, off the coast of Kenya in international waters the San Marco-platform belonging to the state of Italy has been used for launch activities. [33] How far will the day be off, before private enterprise will start adding one and one, with as a result that no one could be held liable under international space law if damage occurs?

In short: it seems that from the viewpoint of logic and expedience, as well as from the viewpoint of victims and the general safety of space activities, in particular with reference to space communications, the present system of liability is shown to present some loopholes which might be used sooner than desired.

IV - THE SUBSTANTIVE ASPECTS OF LIABILITY.

Moving now to the second category of aspects of liability which is important in view of the recent developments in international satellite communications, here concern is with the substantive consequences of damage once it has occurred. A crucial role is played in this respect by damage, and as a consequence also by the definition of damage.

4.1. Liability, Damage and Compensation for Damage.

As already mentioned, damage under the Liability Convention entails the liability of the launching state or states of the space object which caused the damage. Of crucial importance here is the distinction made between damage caused on earth or to aircraft in flight, where absolute liability applies, and damage caused to another space object in outer space, where the extent of liability for the damage concerned depends upon the extent of the fault on the part of the launching states of the respective space objects. [34]

Exoneration of absolute liability is possible, to the extent that the victim state or its nationals have contributed to the damage in a grossly negligent fashion. [35] Such an exoneration however does not apply if the damage in question is due to an activity which itself is not in conformity with international law; if a satellite causes damage on earth or to an aircraft in flight while in violation of an international legal obligation, absolute liability applies even in face of contributory gross negligence on the victim side. [36]

Whether absolute liability or fault liability applies, in either case damage is of course necessary in order to trigger liability. Damage is defined as including at any rate all direct material damage. [37]

As to immaterial damage, the ruling interpretation of "other impairment of health" includes therein immaterial personal damage, in other words mental and psychological damage, in deference to the authoritative definition of health as "a state of complete physical, mental and social well being" offered by the World Health Organization. [38] Other immaterial damage, for instance the violation of a state's sovereignty, is not included in the term "damage" for the purposes of the Liability Convention.

More importantly is that the question whether indirect damage is included has never been answered satisfactorily. [39] Thus, whether for instance the loss of revenue resulting from the inoperability as a consequence of a collision will be compensated under the Liability Convention, will ultimately depend upon the particular Claims Commission before which such a claim is brought. [40]

Of general importance finally is the fact that under the definition provided the damage must have occurred to someone's property, whether that of a state or of a natural or juridical person.[41] Damage to a 'res communis' or a 'res nullius' can not result in liability under international space law.

Once liability is established, of course the next question will be to what extent the damage for which it is established is compensated. The Liability Convention is quite clear on this issue: the compensation required should be of such magnitude as to restore the situation which would have existed if the damage had not occurred. [42] In other words: there are no limits to the compensation to be paid under international law, other than those potentially established by "the principles of justice and equity" [43].

4.2. The Loopholes.

These provisions of the Liability Convention also have been defended as being rather victim-oriented, in view of for instance the application of absolute liability to damage on earth and the absence of limits to compensation, and therefore as satisfactory.

To what extent the recent developments in satellite communications mentioned however necessitate a reevaluation in this regard, remains to be seen. The overriding question for private entities after all is: which liabilities are indeed covered by international space law, and to what consequences do they lead in material terms?

Once again, in my opinion a few flaws should be noted here, which maybe as of yet never have led to serious disturbances in the international satellite communications field, but nevertheless better already at this point should be evaluated and, preferably, dealt with.

To begin with, the issue of immaterial damage should if possible, be settled at the international level, and thereby in a uniform sense. The prevailing uncertainty under international law allows states to deal with this problem, once it arises in a concrete fashion within their jurisdictions, at their own discretion - and hence results both in uncertainty among entities active in outer space as to the damage they would be able to be compensated for or, alternatively, have to pay, and in forum shopping.

While these phenomena are of course not peculiar to the satellite communications business, in view of the inherently international character of these activities and the enormous sums involved, they would be of much greater consequence here than in most other areas of economic activity. Nevertheless, it is probably beyond the possibilities of international space law to solve these problems.

Another loophole to be noted by contrast is of a typically space-related character, pertaining to the inflictor-side of the damage. The Liability

Convention can be invoked only in case the damage is caused by a space object. [44] While detailed discussion on what constitutes a space object can be foregone at this point, [45] it is clear that only tangible hardware comes under the definition of space object, that is basically satellites and launching devices, and component parts of any of these.

Such a definition does not exclude damage caused for instance to a communications satellite other than by physical impact, since radio or other signals emitted from another satellite to the extent they cause damage to the first satellite, still result in damage "caused by a space object" [46].

However, such a definition does exclude damage sustained as the consequence of radio or other signals emitted from earth stations. In view of the rapidly growing amount of satellites not just in the geostationary orbit but also in other orbits, the risks of interference with satellite operations by means of earth stations, whether by accident or on purpose, are exponentially increasing. [47]

This loophole may become the more relevant to the extent that the growing commercial interests involved vis-à-vis the continuing scramble may lead to an irresistible temptation to take short cuts and ignore the ITU regulatory system of coordination, with its uncertain legal but certain bureaucratic lack of expediency, [48] which might in the end result in simple power politics. This would hardly seem to bode well for the future development of satellite communications both as a public utility and as a commercially attractive driving force of the economy.

V - TOWARDS SOME PROPOSED SOLUTIONS.

It was argued that some of the flaws of the liability system are too general and too comprehensive to be solved by space laws as such, let alone in the framework of the liability system. This also holds true as it applies under present-day international space law to international satellite communications, and increasing private participation therein as well as increasing chances that disputes on damages of any sort incurred by those activities arise.

Amongst such problems more or less beyond the scope of international space law proper, one should also count the questions regarding direct legal effect and effectiveness of the ITU and its regulatory framework. Dealing with much more than just the regulaton of space communications, that framework must be seen and dealt with as a whole when such problems are to be solved.

That leaves us however with some loopholes which are of a typically space-related character, where filling the loopholes seems to be a matter of international space law indeed. Basically, they are four in number: the focus on launching for purposes of liability which is loosing its logic and efficiency when it comes to satellite communications operations as such; the focus on territory which may become circumvented by daring entrepreneurs starting space operations from the high seas; the focus on damage caused by a space object which would leave a whole category of damage of increasing potential reality outside the grasp of international liability arrangements; and the focus on direct material damage, which leaves the uncertainty as to whether indirect damage such as loss of revenues would be recoverable under these arrangements.

As to the last loophole, the juridical solution is relatively simple: an amendment to the Liability Convention's definitional Article [49], or if that is deemed undesirable or inappropriate, general agreement to interpret damage in such a way as to include indirect damage would suffice. Since especially in the case of satellite communications for example loss of revenue might be reasonably easily determined, on the basis of costs of satellites, transponders and their operations, average lifetimes of satellites, amount of transponders, average 'load factors' and average lease or rental fees, a secondary option may lie in using the ITU for creating global acceptance of such an interpretation of damage under the Liability Convention.

With the remaining three loopholes, the issue would seem to lie more complicated juridically speaking. Nevertheless, it is submitted that it is indeed up to international space law to solve those problems, and even better, that international space law itself already provides for a nucleus of guidelines showing the road to follow in order to arrive at such solutions. This nucleus is formed by the principle of state responsibility as it operates specifically with regard to space activities.

VI - A FEW IDEAS ON RESPONSIBILITY AND LIABILITY.

States are internationally responsible for national activities in outer space, including national activities of non-governmental entities, if such activities are in violation of the rights of other states .[50] Such private national activities moreover are to take place under the authorization and continuing supervision of the "appropriate state". International space law itself does not further elaborate how responsibility for such national space activities should operate in practice, for instance with respect to the consequences in law once responsibility has been established.

This makes it necessary to turn to general public international law in order to define the further ramifications of state responsibility for private space activities. [51] While there are some important differences to be noted between the two fields - under general public international law states can only incur due care state responsibility in respect of purely private activities, whereas under the Outer Space Treaty states are fully accountable for such private activities just as if it concerned their own activities - the essence of the state responsibility doctrine with regard to the consequences in law thereof should be deemed to apply to outer space.

These consequences in law, captured by the notion of reparation, basically may be considered to be threefold. [52] In logical order, the first form reparation can take is that of *restitutio in integrum*, reestablishment of the situation as it would have existed had the violation of a state's right not occurred. Violating a duty not to interfere with a communications satellite of another state under applicable ITU regulations for instance, if considered to be legally binding, would thus need first of all putting an end to such interference.

Where *restitutio in integrum* is impossible, or does not repair all harmful consequences of the violation, the second form of reparation is at issue: that of compensation, usually in monetary form. In view of the material nature of compensation, it is the most logically and most often invoked form of reparation in cases of material damage, whether direct or indirect. In our example, the indirect damage incurred by the inoperability of the communications satellite during the period it was interfered with, consisting of the loss of revenue and related losses, could conveniently be repaired by invoking such monetary compensation.

The third form of reparation is basically the counterpart of material compensation for material damage, namely a kind of immaterial compensation named satisfaction for immaterial damages such as the violation of sovereignty or moral damages. This form of reparation is of little consequence for the field of satellite communications, however.

It may have become clear already, that responsibility, if invoked in this way, presents an interesting option to deal with the third loophole detected, which concerned the focus on damage caused by a space object. The phrase "national activities in outer space" of Article VI of the Outer Space Treaty indeed seems to include interference with space objects from the ground, and any violation of another state's right to use outer space freely for peaceful purposes such as telecommunications activities by one's own earthbound activities would logically entail the duty to compensate for the damage, such as the loss of revenue, which went with the violation in question.

Responsibility would at least also cover the loopholes existing in geographical terms, where liability at present only deals with territory under a state's jurisdiction, respectively caused by the focus of liability on the launching, in a convenient, consistent and logical fashion. It must be said that the discussion on the meaning of "national activities" has not yet led to a final and conclusive definition thereof. Some authors prefer reading into it 'activities for which a state would qualify as a launching state', to align responsibility to liability, others figure "national activities" to be 'activities of nationals', still others combine the two. [53] Reference to general public international law however provides us with an interesting further option: the national activities for which a state is to be held internationally responsible should encompass those activities undertaken by its nationals as well as those undertaken from its territory. Thus, the jurisdiction of a particular state, which is in general exercised with respect to its nationals as well as to occurrences on its territory, would be perfectly matched by those activities for which a particular state can become responsible.

Under that doctrine, a private entity buying or leasing a satellite in orbit and operating it henceforth, would incur the responsibility of its state of nationality as well as territory (if these are not one and the same), which states thereby also would be held to compensate damages as a consequence of these activities. This would do away with situations where the original launching states, which need have nothing more to do with the satellite after the sale, are still held to compensate, as is presently the case under liability.

Similarly, any private entity undertaking space activities from outside any state's territorial jurisdiction, would by means of its nationality still fall under some state's jurisdiction - and thereby under some state's international responsibility. Consequently, under reparation, any material damage which is the consequence of such an activity, would obligatorily be compensated by at least one state. This state in turn can take care to derogate such claims by means of national space laws.

The only problem with having responsibility 'take over' in cases of damage from liability in the way the latter is presently dealt with admittedly is a huge problem. While responsibility as such is an established notion to a certain extent even in space law, the existence of liability in a much more elaborated fashion under the Outer Space Treaty and the Liability Convention seems to have barred the way for the effective use and practical application of the responsibility-mechanism.

At present, as *inter alia* analysis of the national space laws, which take care of the international obligations of states in those respects at the domestic level, shows, every state, organization or entity is focused primarily on liability as it is presently arranged. Little if any attention is paid to the potential legal consequences of responsibility in those regards.

The juridical evaluation of this diagnosis is rather difficult. Constant usage, by states and other entities undertaking outer space activities as well as by authors and documents, of the liability-mechanism as the sole mechanism dealing with cases of damage could mean that those elements of responsibility relating to damage as the consequence of space activities should be seen indeed to be regulated by liability - despite the fact that no such link has ever been explicitly established, and despite the fact that liability and responsibility have different scopes when it comes to private participation. [54]

On the other hand, it should really be questioned whether a principle such as that of responsibility, as an age-old doctrine of public international law, can be denied application to outer space activities, with all their remaining links with the body of general public international law in spite of their distinct character, by merely keeping silent on such application in theoretical discussions and general documents. It should be reiterated here, that so far no actual international third party-liability dispute has ever reached a court, arbitral tribunal or Claims Commission, national or international.

It is the opinion of the present author, that on the contrary good use should be made of the principle of responsibility for the purpose of legally regulating private participation to the greatest benefit both to those private entities and to society in general, since because of the arguments submitted above, it seems a more logical and comprehensive tool to do so than liability presently presents - not in the least with regard specifically to satellite communications. Liability, with its focus on launching states and damaging space objects, would therefore need to be abolished effectively as a separate notion, or simply be redefined as the aspect of responsibility where actual material damage is involved, and be refocused on the 'nationality' of space activities while not confining itself to damage caused by space objects proper.

CONCLUSION.

The problem of practical dimensions involved in changing space law to the extent desired being what it is, perhaps the aforementioned arguments may lead at least to a reevaluation and reconsideration of the issue of liability. To the extent that these arguments are particularly relevant for satellite communications moreover, in view of the increasing involvement of private enterprise and the increasing scramble for the limited natural resources available in outer space for these activities, for example in the framework of the ITU special attention should be given to these principles of responsibility and liability and their operation vis-á-vis private enterprise.

In the end, both the general safety of space activities as protected by the liability system of international space law and the justified interests of private enterprise participating in satellite communications activities in the operation of such a liability system will benefit from such attention. Even while many questions have not been addressed yet, it may well be, that the law's greatest contribution to the prosperous development of international satellite communications for the next few years could be realized on this issue.

REFERENCES

1. And indeed, as the first volume of the Telecommunications & Space Journal (1994), has shown, many national laws on telecommunications have included at least some aspects of space communications in their respective scope, e.g. most explicitly Germany (K.J. Kraatz, *The Regulatory Framework of Satellite Communications in Germany*, 175-81), Italy (M. Gioscia, *Recent Developments in Telecommunications in Italy: Privatization and Liberalization*, 184, 194) and Brazil (W.D. Stuber & M.C. Semionato, *Investments in Telecommunications in Brazil*, 257 ff.).
2. See Art. II, Treaty on Principles Governing the Activities of States in the Exploration and Use of Outer Space, including the Moon and Other Celestial Bodies (hereafter Outer Space Treaty), of 27 January 1967.
3. Cf. Art. IV, Outer Space Treaty.
4. Cf. Artt. I, II, Outer Space Treaty.
5. Actually, the only dispute which ever came close to necessitating a really juridical solution, the dispute concerning the fragmentation of the Soviet nuclear powered satellite Cosmos 954 over Canadian territory in 1978, in the end was solved by diplomatic negotiations resulting in a protocol of settlement which on purpose did not refer to the Liability Convention. See e.g. P.L. Meredith & G.S. Robinson, Space Law: A Case Study for the Practitioner (1992), 68; also S. Gorove, Developments in Space Law - Issues of Law and Policy (1991), 239-48.
6. Cf. e.g. the report of the Secretary-General of ITU of June 1993, C93/49, on "Participation of entities and organizations other than Administrations in the activities of the union"; second draft discussion paper of the Dutch government on active participation of 'small m-members', of February 1994, CEPT ECTRA/ERC AHWG PP94(94) 01 rev. 1, on the subject of Resolution 4, "Participation of Entities and Organizations Other than Administrations in the Activities of the Union", as proposed to the ITU Plenipotentiary Conference of September/October 1994, in Kyoto.
7. Cf. amongst the mass of literature on this subject e.g. The Implications of the CEC Green Paper on Satellite Communications in Europe, Proceedings of the First Dutch NPOC/ECSL Workshop (1992); E. Ducasse, L'Europe des Telecommunications par Satellite: Entre Liberalisation et Cooperation [The Europe of Satellite Telecommunications: Between Liberalization and Cooperation] (1993), 54-7, 111-9; P. Malanczuk, *Ten Years of European Telecommunications Law and Policy - A Review of the Past and of Recent Developments*, 1 Telecommunications & Space Journal (1994), 41-2; the authors's *Satellite Communications in the European Community: The Tide is Changing Wave by Wave*, in Issues in International Air and Space Law, and in Commercial Law (1994), 314 ff. Also, Issue No. 14 of ECSL News, of November 1994, has been devoted to discussion of these issues.
8. Cf. also: Towards Europe-wide systems and services: Green Paper on a common approach in the field of satellite communications in the European Community, COM(90)490, of 20 November 1991; Communication [by the Commission] to the European Parliament and the Council on satellite communications: the provision of, and access to, space segment capacity, COM(94)210, of 10 June 1994.

9. Cf. Artt. VI, XIII, Outer Space Treaty; Art. 6, Agreement on the Rescue of Astronauts, the Return of Astronauts and the Return of Objects Launched into Outer Space, of 22 April 1968; Art. XXII, Convention on International Liability for Damage Caused by Space Objects (hereafter Liability Convention), of 29 March 1972; Art. VII, Convention on Registration of Objects Launched into Outer Space (hereafter Registration Convention), of 14 January 1975; Artt. 4, 11, 16, Agreement Governing the Activities of States on the Moon and Other Celestial Bodies, of 18 December 1979.
10. See e.g. P. Seitz, *Firms Request Tongasat Sanctions*, Space News, 6-12 Sept. 1993, 3, 21; and B. Manasco, *The Rise of Tonga's Kingdom in the Sky*, 9-6 Via Satellite (June 1994), 18, 20, on the first dispute to arise in that respect; or on more recent issues, P. Seitz, *Intelsat Gets Key Orbital Position*, Space News, 13-19 February 1995, 18.
11. Cf. e.g. F. Lyall, *The International Telecommunications Union Reconstructed*, Proceedings of the Thirty-Sixth Colloquium on the Law of Outer Space (1994), 79; H.M. White, *Space Communications Law and the Geostationary Orbit*, 3 American Enterprise, the Law, and the Commercial Use of Space (1987), 71-2, 83-8; R.S. Jakhu, *International Regulation of Satellite Telecommunications*, in Legal Aspects of Space Commercialization (1992), 80-6; M.A. Rothblatt, *New Regulatory Ideas and Concepts in Space Telecommunications*, 20 Journal of Space Law (1992), 27-8.
12. See e.g. wording of Art. II, III & VIII, Liability Convention.
13. Article VIII(1),(2) & (3), Liability Convention, respectively.
14. In particular, this regards the applicable national laws of the states accountable for the activities causing the damage, as far as the exercise of active personal jurisdiction is concerned, the states of nationality of the victims, as far as the exercise of passive personal jurisdiction is concerned, and the states where the damage was actually sustained, for purposes of exercising territorial jurisdiction.
15. That an international public system for dealing with private legal claims for damage need not be an illusion or oxymoron, is shown by the neighbouring field of air law: both the Warsaw Convention for the Unification of Certain Rules Relating to International Transportation by Air, of 12 October 1929, dealing with liability for damage sustained by passengers and cargo, in its Art. 28(1), and the Rome Convention on Damage Caused by Foreign Aircraft to Third Parties on the Surface, of 7 October 1952, in its Art. 20, provide for harmonized extra possibilities for private parties to claim compensation form the damage concerned.
16. Article XI(2), Liability Convention.
17. Art. XXII(4), Liability Convention.
18. With the proliferation of new states in recent years, figures are constantly changing; the figures cited come from the issue of Space News of 6-14 February 1995; p. 11 as to INTELSAT, p. 3 as to INMARSAT, and p. 6 as to EUTELSAT.
19. Cf. e.g. Artt. II, III, IV & V, Liability Convention.
20. See Art. I(c), Liability Convention; cf. also Art. VII, Outer Space Treaty; Art. I(a), Registration Convention.
21. Of course, they will not each have to pay the full amount of damage awarded; Artt.IV & V, Liability Convention, point out how such multiple liability would have to result in the victim being paid his rightful compensation and nothing more.
22. These declarations were effective as of 23 September 1976 and 25 January 1988 respectively; cf. K.H. Böckstiegel & M. Benkö, Eds., Space Law - Basic Legal Documents, A.III.2.1.
23. Cf. Art. XXII(1) resp. (2), Liability Convention. Note that even in case ESA or EUTELSAT are completely accountable for the launched space object, after a period of six months individual member states' liabilities can be invoked; whereas in case there are states qualifying as launching states next to ESA and EUTELSAT, this possibility already exists from the start; cf. Art. V(1) in conjunction with Art. XXII(1).
24. The exceptions here relate to launches undertaken from launch facilities on the high seas, Antarctica (or at least its unclaimed parts) or from within outer space itself. Such options at the moment may be labelled as theoretical, especially if they are to be exercised by private entities. See however my remarks further infra, text at notes 32 and 33.
25. This road has been taken especially by the United States, by means of its Commercial Space Launch Act of 1984, as amended 1988, providing in Sec. 16(a)(1)(A) for obligatory reimbursement of the US government for international third party liability claims up to an amount to be determined by the US Secretary of Transport, but by no means higher than US $ 500 million. See further e.g. P.D. Nesgos, *Commercial Space Transportation: A New Industry Emerges*, 16 Annals of Air and Space Law (1991), 405 ff.; V. Kayser, *An Achievement of Domestic Space Law: U.S. Regulation of Private Commercial Launch Activities*, 16 Annals of Air and Space Law (1991), 372 ff.In a more summary fashion, the national space laws of Sweden (1982), the United Kingdom (1986), South Africa (1993) and Russia

(1993) also provide for licensing obligations including explicit or implicit possibilities for the relevant government authorities to impose reimbursement conditions. See respectively: J. Reifarth, *Nationale Weltraumgesetze in Europe [National Legislation Relating to Outer Space Activities in Europe]*, 36 Zeitschrift für Luft- und Weltraumrecht (1987), 3-16, or the author's *The Swedish and British Space Acts and Private Commercial Enterprise under Public International Space Law*, Memoria, Conferencia Espacial de las Americas (1991), 336-42; on South Africa: I. de Villiers Lessing, South Africa: Recent Developments in Space Law, 1 Telecommunications & Space Journal (1994), 139-49; on Russia: E. Kamenetskaya, V.S. Vereshchetin & E. Zhukova, *Legal regulation of space activities in Russia*, 9 Space Policy (1993), 121-3, and M. Tarasenko, *Russia's place in space: a home view*, 10 Space Policy (1994), 118.

26. This road has been followed by France, which in respect of its launch complex in Kourou, in French Guyana, run jointly by the French Centre Nationale d'Etudes Spatiales (CNES), the European Space Agency and the private launch services provider Arianespace, has provided for a rather complex regime. The net result as far as important at this point is that Arianespace is obliged to reimburse the French government up to an amount of FF 400 million (approximately US $ 70 to 100 million at present rates) for international third party liability claims; see e.g. M.G. Bourély, *La Production du Lanceur Ariane [The Production of the Ariane Launcher]*, 6 Annals of Air and Space Law (1981), 293 ff.
27. Once more, it should be noted that the Liability Convention has never been invoked in a juridical dispute settlement procedure; see *supra*, note 5.
28. See the examples provided *supra*, in notes 25 and 26.
29. Cf. e.g. V.F. Wodié, The High Seas, in M. Bedjaoui, Ed., International Law: Achievements and Prospects (1991), 888-90; I. Brownlie, Principles of Public International Law, 3rd ed. (1979), 424-6, 494-5.
30. See e.g. S. Chenard, *Lean Times for France?*, 8-5 Space (Oct.-Nov. 1992), 26; Space News, 22-28 March 1993, 8; Space News, 19-25 July 1993, 30; *Space industry: the scene in 1992*, 9 Space Policy (1993), 239.
31. Cf. once more the unequivocal obligations of launching states in this regard under Artt. II, III, IV(b) & V, Liability Convention.
32. *Commercial spaceports - a window of opportunity?*, 9 Space Policy (1993), 66-8.
33. F.J. Mathy, *Weltraumpolitik Italiens [Space Policy of Italy]*, in K. Kaiser & S. von Welck, Weltraum und internationale Politik (1987), 353; The Cambridge Encyclopedia of Space (1990), 121, 129.
34. Art. II resp. III, Liability Convention. For damage caused to a communication satellite, in essence only absolute liability will be at issue, whereas for damage caused by communications satellites, both types of liability may become involved.
35. Art. VI(1), Liability Convention.
36. Art. VI(2), Liability Convention.
37. Art. I(a), Liability Convention, enumerates "loss of life, personal injury or other impairment of health, or loss of or damage to property of States or persons, natural or juridical, or property of international intergovernmental organizations".
38. Preamble, Constitution of the World Health Organization. See further e.g. C.Q. Christol, The Modern International Law of Outer Space (1984), 97-100; Gorove, Developments in Space Law, 242.
39. See e.g. Meredith & Robinson, 63.
40. In case diplomatic negotiations on disputes on international third party liability have not led to an agreement thereupon, the Liability Convention provides for the establishment of a Claims Commission; see Artt. XIV-XX, Liability Convention. It must be noted, however, that such a Claims Commission can only issue a binding decision if both parties to the dispute agree that it should be binding; cf. Art. XIX(2).
41. Cf. Art. I(a), Liability Convention.
42. See Art. XII, Liability Convention.
43. Art. XII, Liability Convention. Compare however notes 25 and 26, supra, on how individual states can domestically apply caps on liability and effectively take the burden of remaining amounts to be compensated.
44. Cf. e.g. Artt. II, III, IV(1) & V(1), Liability Convention.
45. In this respect reference can be had e.g. to S. Gorove, *Toward a Clarification of the Term "Space Object" - An International Legal and Policy Imperative?*, 21 Journal of Space Law (1993), 11-26.
46. E.g. Art. III, Liability Convention.
47. Cf. e.g. W. Pritchard, *Satellites in non-geostationary orbits*, 9 Space Policy (1993), 199-208.
48. See e.g. *supra*, literature at note 11
49. I.e. Art. I(a), Liability Convention.
50. Cf. Art. VI, Outer Space Treaty.

51. Art. III, Outer Space Treaty, provides for general application of international law to space activities, in essence as a lex generalis to the extent that the lex specialis of space law does not itself provide for clearly defined rights and obligations.
52. Cf. e.g. Brownlie, 457-54; M.N. Shaw, International Law, 3rd ed. (1991), 496-9.
53. Cf. e.g. H.A. Wassenbergh, Principles of Outer Space Law in Hindsight (1991), 22-31; B. Cheng, *Space Activities, Responsibility and Liability for*, 11 Encyclopedia of Public International Law (1989), 299-303; H.L. van Traa-Engelman, Commercial Utilization of Outer Space - Legal Aspects (1989), 44-48; M.G. Bourély, *Quelques Réflexions sur la Commercialisation des Activités Spatiales [Some Reflections on the Commercialization of Space Activities]*, 11 Annals of Air and Space Law (1986), 176-8; K.H. Böckstiegel, *Reconsideration of the Legal Framework for Commercial Space Activities*, Proceedings of the Thirty-Third Colloquium on the Law of Outer Space (1991), 5; B. Cheng, *The Commercial Development of Space: the Need for New Treaties*, 19 Journal of Space Law (1991), 36-40.
54. As discussed *supra*, liability of a state applies to private activities, in case these activities cause damage in respect of which that state can be considered a "launching state", i.e. basically if its territory has been used for the launch of the space object (ultimately) causing the damage, whereas responsibility of a state applies for private activities which can be qualified as its "national activities". See for a more detailed analysis further the author's *Liability Versus Responsibility in Space Law: Misconception or Misconstruction?*, Proceedings of the Thirty-Fourth Colloquium on the Law of Outer Space (1992), 363-71.

LEGAL REGIME OF HUMAN ACTIVITIES IN OUTER SPACE LAW

CARLO GOLDA

University of Genoa and Studio Legale Pini, Via Brignole de Ferrari 6-10, 16125 Genova, Italia.

Abstract : *Current developments in space activities increasingly involve the presence of humans on board of space crafts and, in the near future, on the Moon, on Mars, on board Space Stations, etc. With respect to these challenges, the political and legal issues connected to the status of astronauts are largely unclear and require a new doctrinal attention. In the same way, many legal and political questions remain open in the structure of future space crews: the need for international standards in the definition and training of astronauts, etc.; but, first of all, an international uniform legal definition of astronauts. Moreover, the legal structure for human life and operations in outer space can be a new and relevant paradigm for the definition of similar rules in all the situations and environments in which humans are involved in extreme frontiers. The present article starts from an overview on the existing legal and political definitions of "astronauts", moving to the search of a more useful definition. This is followed by an analysis of the concrete problems created by human space activities, and the legal and political responses to them (the need for a code of conduct; the structure of the crew and the existing rules in the U.S. and ex-U.S.S.R.; the new legal theories on the argument; the definition and structure of a code of conduct; the next legal problems in fields such as privacy law, communications law, business law, criminal law, etc.).*

INTRODUCTION

Since 1961, which marked the onset of human space flights with Yuri Gagarin, and until today, no more than 200 individuals the world over can boast the title of astronaut. Almost all of these are either US or ex-Soviet citizens and most are career military officers [1].

On the other hand, a period of significant changes in human space flight has already begun. First of all the "Freedom" space station program (continuing, although behind schedule) will lead to the establishment of communities in space, on board special vehicles of a totally new type and, more importantly, of greater size [2].

Additionally, ten years of experience with the American S.T.S. (the Space Shuttle) have demonstrated that, although not without difficulty, it is possible to exploit extra-atmospheric space commercially for activities that require man's physical presence (scientific experiments, manufacture of special chemical and biomedical substances, satellite repair, etc.). Then one should consider that both cost and domestic/foreign policy issues have led to an increasing internationalization of activities in space, with the consequence of an increasingly varied and multifaceted composition of flight crews, both in the West and in the ex-communist countries.

Lastly, it should be underscored how an easier and, within limits, more "economical" access to extra-atmospheric space has motivated private organizations (led by the large multinational corporations) to gain access to space for a variety of motives, preferably with their own personnel included as crew members [3].

I - WHO IS AN ASTRONAUT

Consequently, the time has come for jurists to focus on man, the astronaut, in order to confront the mass of legal issues apparently generated by human space flight. This need, which is felt by Americans in particular, was recently addressed by international diplomacy by the COPUOS [4] legal sub-committee.

On this issue, the situation today is still quite confused, so much so that not even a univocal legal definition of the term "astronaut" has been agreed upon [5].

The international law sources, which in some way and for various ends recall the concept of "man operating in space", provide vastly different indications.

First of all, art. 5 of the Outer Space Treaty [6] as well as the tile and preamble of the Rescue Agreement [7] utilize the term "astronaut", art. 8 of the Outer Space Treaty utilizes the term "personnel", articles 1 and 4 of the Rescue Agreement refer to "crew", art. 12 of the Outer Space Treaty mentions "representative", articles3 and 5 of the Liability

Convention [8] mention "persons aboard a space object", and, finally, art. 5 of the Outer Space Treaty proposes the classical definition of "envoys of mankind".

1.1. Astronauts as "envoys of mankind"

First of all it is appropriate to discuss the validity of this latest definition that, repeated and used on several occasions in the political and diplomatic arenas, is apparently an expression of the most widely accepted classification of personnel operating in outer space [9].

In reality, and beyond its rhetorical value, the definition "envoys of mankind" is totally devoid of any juridical utility and today is obsolete, just as the Rescue Agreement itself [10].

This treaty, ratified in 1968, was the product of a world comprising only two space powers engaged in a cold war and involved in infrequent and very costly space activities. Therefore the Western and Communist blocks felt a need, including a military one, for reciprocal reassurance that any men and equipment that might fall into "enemy" hands would be returned [11].

In essence the intent was to guarantee a sort of diplomatic immunity to the astronauts (the boundaries of which were all but well defined), sanctioned by the foremost supranational authority in light of the benefits, indeed international, that derive from human activities in space and in light of the risks that crews voluntarily agree to take.

The dissolution of the communist empire, on one hand, and the extreme ambiguity of the cited definition, on the other hand, compel us to reject it in the context of this paper, even though the Rescue Agreement has been ratified by more than 80 countries.

If anything, the latter datum can assist in directing the hermeneutic research of the emerging principles of international public law in the matter of research and rescue during space activities.

1.2. Astronauts as "persons on board a space object"

If we examine all of the previously cited definitions, the most juridically useful one is apparently that of astronauts as "group of persons on board a space object [12]".

This definition must be accepted as the one least likely to be criticized since it is directly traceable to the principles of navigation law of the Scialoja School.

If indeed all human activities in extra-atmospheric space must take place within a truly autarkic vehicle and if every role or function of man in space can in all cases be traced to the supporting vehicle and its navigation, then the above mentioned definition seems the most fitting because it makes it possible to use (and quote) all the doctrinal and normative efforts devoted so far to similar cases in maritime law, including ancient and aeronautical law.

Furthermore, it is also necessary to point out that the proposed definition is independent of the substantial and diverse activities that can be conducted in space (piloting, experiments, etc.) as well as of the duration of space flights, the composition of the crew, etc. Obviously, E.V.A. [13] (i.e. space walks) must be included within the expression "on board...".

It cannot be ruled out that, in the near future, the definition of choice will become inadequate to cover the spectrum of human extraterrestrial activities.

In particular it may become necessary to distinguish between astronauts and the normal passengers of extra-atmospheric flights.

However we can affirm that such a distinction would effect certain aspects of international private law (in the areas of trade, compensation, etc.) more than international public law and that, as of today, a development of such magnitude cannot be anticipated. More important is the issue concerning human settlements on the surface of other planets [14].

Obviously, the definition proposed must be integrated to include persons who, already members of a crew of astronauts, would then remain on a celestial body, whether or not becoming separated from the rest of the crew [15].

In this respect the definition contained in the Moon Treaty (the last of the space-related treaties) could be used. Art. 10 of this treaty defines astronauts as "any person on the Moon.....".

This concept integrates perfectly with that of "person on board a space object" as the first expresses the static phase while the second expresses the dynamic phase of human activities in space.

Furthermore, this definition does not exclude the preceding one and allows both concepts to apply at the same time, referring each time either to one or the other, depending on the juridical nature and the characteristics of the problems to be confronted.

1.3. Current situation with international norms

Moving on to review the current regulatory evolution of the issue of the juridical qualification of astronauts, it is necessary to note that two documents have already been presented to the U.N. in 1987 (more precisely the C.O.P.U.O.S. Subcommittee), both of which concern the matter under scrutiny. The first document regards accidents and critical situations in space flight, the other the legal status of the crew and rescue operations.

The first is the working document (No. A/AC-105/C-2/L 159 dated 27/3/87) presented by the United Kingdom to promote the broadening of international cooperation in the event of an accident or critical situation on board an inhabited space station. The second working document (No. A/AC-105/C-2L 161 dated 1/4/87) was presented by Czechoslovakia to bring the issue of the juridical status of the crew of a space object and the matter of rescue operations of such a crew to the attention of the COPUOS juridical subcommittee [16].

II - THE SPECIFIC PROBLEMS POSED BY HUMAN ACTIVITY IN SPACE

Having exhausted the discussion on the general picture of international public law concerning the actual definition of "astronaut", it becomes necessary to investigate the specific problems posed by the expansion of human activities in space. The debate over the definition can be actually circumvented by a homogeneous juridical discipline under various profiles of law (private, penal, etc.) capable of leading to an "a posteriori" definition of the astronaut concept.

Given the vastness of the issues before us, it is helpful to identify some of the specific questions that will and, at least in part, already face us2

2.1. The Code of Conduct

It seems that the issue of a "Code of Conduct" for astronauts is the most urgent one, also given the premises. Actually, strong internationalization of space activities will increasingly involve the selection of heterogeneous crews impossible to govern by applying specific national laws or the usual conflict norms of international private law [17].

a). The crew's roles

Prior to a discussion on the "code of conduct" there is the question of the definition of the roles of the individual crew members of a spacecraft. In fact, the roles of the individual astronauts constitute a technical and juridical framework for the ensuing behavioral norms, as the latter must necessarily blend in with the technical role entrusted to each astronaut.

The order of the stated ideas is then confirmed in practice by today's activities in space.

Both the US and the ex-USSR have indeed devised international norms (at times involving simple administrative acts) in order to differentiate between the various roles, functions and authorities among the crews.

The definition contained in the NASA documents is particularly useful and clear [18].

Within a crew, the American Space Agency identifies the roles of captain, pilot, mission specialist and payload specialist [19].

The Captain on board is a NASA career astronaut. During the flight he is entrusted with the absolute power to order all actions that he deems necessary to ensure discipline and the safety of the spacecraft and the astronauts.

To achieve this end the captain may avail himself of all means at his disposal, including physical coercion.

The captain's authority extends over all persons on board, regardless of their nationality, and also includes transferees from other vehicles, even only temporary ones, including personnel involved in Extra Vehicular Activities. In addition, in the event of a descent on a celestial body, the captain assumes all duties of the leader of the landing party, regardless of whether or not he is concretely involved in the activities on the surface of the celestial body.

The Pilot is also a career astronaut and his duties are of an exquisitely technical nature (even though he remains under the authority of the captain even as far as piloting the spacecraft is concerned).

Most important is the role of Second in Command that the pilot assumes every time the captain is otherwise occupied and/or prevented from carrying out his duties.

The Mission Specialist is equally a "professional" astronaut deeply involved with achieving the results of the individual mission. Consequently, the mission specialist is involved in planning the mission and is responsible for coordinating the use of the spacecraft and the payload.

The Payload Specialist is the only crew member who is not necessarily a career astronaut. He generally is not part of the flight crew (and may be a national of a country other than the US) and can even come from the private sector (industry, universities, etc.), rather than space agencies [20]. The payload specialist does not take part in any extra-vehicular activity and his duties are limited to activating all of the instruments that pertain to the payload as well as obtaining the scientific and experimental results for which the latter was planned.

b). New doctrinal tendencies in defining crew roles

In light of the fact that flight crews in the ex-USSR are structured in a similar way, it is immediately apparent that crews have been designed on the basis of a limited number of members with a tendentially homogeneous level of training.

The future of space colonization, on the other hand, will carry with it the need for increasingly large crews (communities) with various backgrounds (physicians, scientists, etc.) whose level of preparedness for permanence in space may not be homogeneous and in any case whose level of inner conflict (due to ethnic, religious, personality or other reasons) could be higher.

However, for the most part it will be impossible for the captain to exercise his authority in person and the chain of command will necessarily involve some degree of subdivision to administer particularly vast and complex vehicles.

Over the last few years, the best doctrine, American in particular, has raised the issue of delegating the powers of the captain and, consequently, of the different organization of the crew, focussing in particular on the differences between the S.T.S. and the Space Station [21].

The most significant and original fruit of this speculation has been the statement of two possible approaches to the problem: the "Spacecraft Commander Approach" and the "Delegation Approach" [22].

In essence, this latter approach stems from the premise that it is impossible, on the part of the captain, to personally manage the life of the vehicle and crew, and, as a solution, proposes the creation of Personnel Specialists. These would be individuals assigned to the Space Station for a certain period of time where they would carry out the tasks delegated to them by the captain to ensure order and discipline, provide for the safety and the well-being of the personnel on board, solve controversies within the crew and protect crew members and equipment on board the Space Station. Their curriculum must include training in psychology, law, public relations, medicine, etc., however without excessive specialization and they may also be other than NASA career astronauts and have only part-time duties on board.

The proponents of this theory cite an historical precedent for this role identified as the "consul" figure in the "Codice Amalfitano" (Amalfi

Code) a person who, on board, acted as a judge to quell disputes among the crew members [23].

The captain would maintain the "ultimate responsibility" in governing the vehicle as well as for the order, safety, well-being and integrity of the spacecraft and its passengers. Finally, the captain would maintain the right to order the use of force, however with the possibility of delegating such use to one or more Personnel Specialists.

Although the approach illustrated briefly above is certainly original and presents some valid items, we feel that it is unacceptable despite being supported authoritatively and diffusely in American doctrine.

Indeed, the creation of a figure such as the Personnel Specialist presents a number of uncertainties.

First of all there is the fact that the captain is deprived of certain control and intervention faculties (for the most part typical of military command) on the life of his crew, with the substantial risk that he may be increasingly perceived as detached from the live of his crew members, more a controller than a chief.

Furthermore, at the juridical level the captain would retain the ultimate responsibility for decisions made by others, the curriculum and abilities of whom he could not directly ascertain in numbers that may be substantial enough to create a real and true counterpower, especially on board large and very large spacecraft. This would actually pose a great threat to order, discipline and safety on board. We fail to understand, in the event of errors or abuse on the part of Personnel Specialists, how the captain could be called to answer in person, especially in the more serious cases, without artificially creating a sort of objective responsibility against him, totally incompatible with the subtlety and difficulty of his duties [24].

Having ruled out that they would only be American citizens (if nothing else for obvious political reasons), it is clear that personnel specialists could not belong to a single, well-defined, ethnic or religious group, least they lose their credibility in front of the other crew members with a different origin and culture. It is unrealistic today to imagine the existence of such individuals.

Finally, and still at the juridical level, the supporters of this theory have failed to explain where Personnel Specialists should be placed in the chain of command on board of the spacecraft, whether under normal circumstances or during emergencies.

In particular, there is a risk that these individuals, somehow authorized to supervise and refer on the psychic health of the crew, could constitute a sort of internal lobby and dispose of the power to influence and to move accusations against any individual that is totally disproportionate and incompatible with any guarantee of the astronaut's freedom.

Furthermore, in the event of an emergency, it is unclear whether they would be the immediate subordinates of the captain or of the second in command/pilot, or if they would indeed maintain an independent capacity to intervene.

In this case, the proposed model would seem to lead to a "community" management of life on board that would be particularly harmful to the efficiency of the vehicle and unfit for space activities, at least those that can be hypothesized today [25].

The "Spacecraft Commander Approach" is substantially different. Based on a more traditional paramilitary model of the crew (as illustrated above) and supported by current international (art. 8 of the Outer Space Treaty) and national (US) regulations (42 U.S.C. par. 2473, first section of the US Code) this model contemplates the total subordination of each crew member to the captain, without any regard to individual roles or nationality [26].

In the case of non-American astronauts, these would in any case be subject to both American law and the laws of the countries they are nationals of, on the basis of diplomatic agreements made on a case by case basis [27]. Obviously, hierarchic subordination becomes even more rigid in the case of a military crew and it is precisely to the military origins of NASA personnel (at least during the Sixties) that supporters of the delegation of powers doctrine refer to prove the outdatedness of the approach just illustrated. In particular, the most valid criticism is based on the captain's physical impossibility of supervising every function "without becoming a space bureaucrat" on one hand and without creating command vacuums on the other.

In effect, this objection is well taken where it states that "the Spacecraft Commander Approach" could be beneficial for small traveling communities, for a limited amount of time, formed by homogeneous members (the Shuttle is a typical example), whereas it would be inadequate for larger communities.

Actually, we believe that the latter approach is absolutely preferable for a variety of reasons.

First of all, it has strong foundations in historical maritime and aeronautical law. Whether military or civilian, the hierarchical distribution of roles on board "autarkic sailing vehicles" has always postulated the existence of a sole person responsible for the life of the vehicle as well as the crew, with the consequent attribution to the same of at times limitless powers. Moreover, selection of this model makes it possible to fully exploit all of the doctrinal and jurisprudential developments that have occurred to date in the field of maritime law, and in uniform international norms in particular, which, in this field, are considerably vast, especially at the level of general principles.

Then, the problem of the captain's actual ability to carry out all of the functions assigned to him and the problem of the risk of abuse, facilitated by the magnitude of the powers assigned to the him, can be solved without recourse to delegation of power within the crew (as would occur by establishing Personnel Specialists) by delegating said powers outside the crew.

In essence, it would be necessary to establish a regime by virtue of which the captain, at least during non-emergency situations on board, would be subjected to higher Ground Control Authorities, on the basis of the suggestions or, if necessary, orders of which he would act to regulate life on board. Such a system was effectively tested in the past through the "aeronautical director", and the commercial flight management system in particular, where most decisions concerning non-emergency flight are made by ground controllers and, for the most part, even the captain's decisions on board are made on the basis of precise and extremely detailed operations manuals that contemplate practically all possible flight situations. A similar system can be hypothesized for the relationship between Ground Control Authorities and the captain of a spacecraft, with the additional advantage that such a vehicle could be managed entirely through automatic and radio controls from earth, making it possible to exclude the captain should he no longer prove sufficiently reliable.

At the political level of international law, such a command chain would be the most favorably accepted by all Countries involved in an international mission, as it would allow for true international control over the life of the vehicle and its crew and, more importantly, it would situate the discussion and decisions involving the primary theoretical and juridical issues well before the beginning of the mission.

c). Concrete definition and structure of a Code of Conduct

Having exhausted the review of hierarchy within the crew it is possible to analyze the issue of a "code of conduct".

Even today this regulation of the astronauts' life is already a problem that will become a most impelling one in the near future [28].

In fact, crew internationalization, their increasingly varied composition and the size of the vehicles and duration of permanence in space will increasingly configure human activities there as those of a traveling community, in part isolated, and on an almost totally autarkic vehicle.

In this framework, the code of conduct must become a sort of international "general law" for all human activities in space, containing reference norms that, perhaps modified on a case by case basis, can guarantee a real and true legal regime of astronaut life.

With reference to the most recent doctrinal elaborations (I.S.U. 91), we can very briefly state that the code of conduct should be inspired by current international law on the matter of space activities and by the new requirements that will surface as space stations are commissioned and lengthy interplanetary voyages are undertaken.

Therefore it will probably have to avoid norms of substantial and procedural penal law, since this branch of the law is the strict domain of individual countries (this would also hasten the acceptance of the Code of Conduct by the various Governments involved in space activities) [29].

The main issues that the code of conduct should regulate are the jurisdiction and definition of the roles and positions on board the spacecraft.

Then, at the formal level, the code of conduct should be assembled as a series of uniform international law norms, whenever possible avoiding conflicting norms that would be difficult to apply, either diplomatically or juridically [30].

As far as jurisdiction is concerned, suffice it to state that, beyond applicable conventions [31], jurisdiction for a violation of the norms of the code of conduct should be maintained by the country of origin of the perpetrator of the violation (in case of crimes requiring complete penal proceedings for an evaluation, after return of the astronaut to his country of origin), while so-called "simple insubordination" may be disciplined directly by the captain of the spacecraft.

As far as the different roles on board the spacecraft and the related chain of command are concerned, with reference to earlier statements, the division of roles among Captain, Pilot (and First Officer), Mission Specialist and Payload Specialist are acceptable. In particular, the duties and powers of the captain should be defined as follows:

The captain is the highest ranking authority on board and/or on the surface of a planet. His powers include that of using or ordering the use of physical force against crew members, but only in case of an immediate and absolute necessity, to guarantee the safety of the vehicle [32]. The captain may consult with his first officer or all of the crew as far life on board is concerned and must in any case consult with the first officer before administering punishments and taking steps against any member of the crew.

Emergency situations are those in which there is an immediate and obvious risk of damage or destruction of the spacecraft or its equipment, as well as those in which the physical health of the crew is even simply at risk.

d). New juridical problems

Beyond the structure of the code, as examined above, there are issues of substantial law concerning the life and activities of the astronauts.

For the sake of this discussion it is necessary to propose a particular definition of astronaut, capable of distinguishing the latter from a possible passenger of space flights, to underscore the technical and juridical peculiarities of space work, separate from the contractual relationship of mere transportation which, in the future, could be the setting of the juridical positions of passengers.

In this sense, accepting the recommendations of the best international doctrine [33], a "crew" is defined as "persons conducting professional activities during a space flight".

Having clarified this issue, let us now identify the areas of the most urgent intervention.

First of all, at the civil law level there are personal rights [34]. With reference to the situation on earth, the astronaut lives and works in an environment the limited space and continuing forced cohabitation of which emphasize the need of ensuring some form of privacy [35].

The right to privacy, one of the issues that has developed the most in the field of personal rights over the last few years, is yet to be established in terms of extra-atmospheric activities.

It implies the possibility for astronauts to practice their religion, to take their personal objects with them, to organize and use their own "personal" space within the spacecraft, to have the freedom to organize, albeit minimally, their work and rest schedules.

Within the privacy sphere there is the issue of communications and information, with features that are markedly different from earth.

Astronauts must be guaranteed a possibility to communicate individually with Earth, whether with the ground control structures (GCA, etc.) or their families and employers as well as the reciprocal right to receive all the information that they may wish to receive, at least in the family and personal safety areas.

Then, in particular, it will be necessary to reach a compromise between the possibility for the individual to use private radio frequencies (even if at given time intervals) and the need for the entire life of the space vehicles to be under control and also in order to ensure military and scientific secrecy. Another privacy issue is the limits to investigations of the life of men in space (both physically and psychically) and to the transmission of news concerning any individual on the part of other crew members [36].

Other pressing issues surface in the sphere of work and welfare law.

There is definite proof that, in the absence of gravity, the human body experiences physical damage of significant entity (lack of calcium in the bones, cardiocirculatory problems, alteration of reflexes, etc.).

This damage cannot be prevented beyond a certain limit and implies the right of astronauts to a special safeguard, both medical and compensatory.

Moreover, internationalization of crews in the current situation where each Country provides individually for recruiting and training personnel (at times with differences between military and civilian personnel) creates an astronaut coverage problem.

Therefore it will be necessary to establish homogeneous standards for recruiting, qualifying, training and determining the attitude for flight and lengthy sojourns in space, granting licenses and professional certification in this field.

Still in the astronautical field, but with implications that involve all the other fields of law, there is the problem of the right to safety. This right will have to constitute the basis of every juridical elaboration concerning the legal regime of astronauts and extend to all processes involving transportation, sojourn in space, scientific experiments, handling of dangerous substances, definition of parameters for clothing (intra and extra-vehicular), etc.

Once again, special attention must be paid to radio communications in order to ensure the use of protected emergency frequencies (along the lines of existing aviation law) [37].

Then it will be necessary to provide a safeguard of the intellectual property rights that derive from human scientific activities in space in order to protect and encourage investments on the part of private companies and individual countries in this area, on one hand, and to insure the exploitation and dignity of the work of the men on the other. These men, individually, accept the risks and discomfort without which similar scientific experiments could not take place [38].

Finally, in the perspective of astronauts from many different backgrounds (national and international space agencies, military, commercial enterprises, research institutes, etc.) it will be necessary to guarantee uniformness of economical treatment in the wider sense of the term, establishing a single or very homogeneous labor contract (at least for extended flights), that could possibly be included in the code of conduct [39].

From the perspective of civil liability of the astronaut it will be necessary to provide a transfer mechanism of the individual astronauts' liability and, secondly, the liability of the countries of which they are nationals. This would ensure a sort of "immunity" to the astronaut that could compensate the hardship, including physical hardship, in which they work, and at the same time avoid depriving the subject who has suffered a loss of an equitable reimbursement for the damages sustained. At the penal level it will be necessary to deeply rethink the limits of guilt with regard to the astronauts' new and special living conditions.

In particular it will be necessary to establish precise limits to the extension of criminal behavior to include a greater number of "involuntary" actions that could be perpetrated in space [40].

A different approach must be adopted for concepts such as total, partial and mental infirmity. Finally, for behavior defined as "simple insubordination", it will be necessary to establish a sanctioning system capable of ensuring respect of the norms on the part of the individual astronauts, also out of fear of punishment, but without involving sanctions that would deprive the crew of the contribution of the single members whose role is often technically irreplaceable. These sanctions must therefore be essentially economic in nature and/or effect the professional career perspectives of the astronaut [41].

REFERENCES

1. The record for the longest time in space belongs to the Russian L. Kizine with 375 days, 16 hours and 19 minutes of space flight.
2. See A. Farand "La Station Spatiale et son regime juridique", Min. des Aff. Ext. - Dir. Gen. des Aff. Jurid. Ottawa, 1989; Anajiom, Goldman, Meeks Space and Society: Challanges and Choices, Am. Astronautical Soc, Publ., 1984.
3. See I.S.U. "International Mars Mission Design Project", Toulouse, 1991, pp. 32 et seq. where the initial theoretical basis for the creation of an international organization dedicated to future space activities was laid. Young A.J., "Law and Policy in the Sp. St. Era", Utrecht, 1989, M. Nijhoff Pub., pp. 98 et seq., where the profile of space activity internationalization and privatization, although partial, is examined in greater depth.
4. C.O.P.U.O.S. Committee for the Peaceful Use of Outer Space, a permanent UN Committee established in 1959 by Resolution 1472 (14) adopted on 12/12/59 "International Cooperation in the Peaceful Use of Outer Space", by the UN General Assembly
5. Among the various authors who have addressed this problem, see: G. LAFFERANDERIE, "Pour une charte de l'Astronaute", Annual of air and Space Law, vol. XII, 1987, pp. 263 et seq., McGill Univ., Montreal; K.H. BOCKSTIEGEL, "Draft for a convention on Manned Space Flight", in ZLW 40. Jg. 1/1991, pp. 5 et seq.; Goldman, "International Space Law 1957-1979", Part II, pp. 74 et seq.
6. Treaty on Principles Governing the Activities of States in the Exploration and Use of Outer Space Including the Moon and Other Celestial Bodies (the Outer Space Treaty, 1967).
7. Agreement on the Rescue of Astronauts, the Return of Astronauts and the Return of Objects Launched into Outer Space (the Rescue Agreement, 1968).
8. Convention on International Liability for Damage Caused by Space Objects (The Liability Convention, 1973).

9. See the Outer Space Treaty, art. 5: "States Parties to the Treaty shall regard astronauts as envoys of mankind in outer space...."
10. See Goldman, cit. p. 76
11. This is the framework in which some of the COPUOS member nations signed the Agreement, however reserving the right of granting political asilum to foreign astronauts that were to make such a request.
12. See in this respect: N.M MATTE, "Space Activities and Emerging International Law", Centre for Research of Air & Space Law, McGill Univ., Montreal 1986; SZALSO, "Air Crew and Space Crew - A modest Analysis of Analogies from Air and Space-law rules", document presented at the seminar of the International Institute of Space Law, Oct. 9-15, 1983, Budapest; DE SAUSSURE, Astronauts and Seamen - A legal Comparison, 10, J. Of Space L. 165 (1982); and others.
13. E.V.A. = Extra Vehicular Activities: all activities performed by the astronauts outside the spacecraft (i.e. walking in space or on celestial bodies).
14. International consensus is increasing on the "Agreement Governing teh Activities of States on the Moon and Other Celestial Bodies" (The Moon Treaty, 1979) containing the general principles that pertain to human extraterrestrial settlements, in accord with the Outer Space Treaty.
15. In this sense also I.S.U., cit., p. 48.
16. As of today, none of these documents has been developed. The en d of the cold war and of the resulting political bipolarism should allow for both of these diplomatic proposals to be resurrected.
17. See, in this respect, I.S.U. cit., p. 49; LAFFERANDERIE, cit., p 271 et seq.; FARAND, cit., p 13 et seq.; BOCKSTIEGEL, cit., p 5 and 6.
18. Presidential Executive Order 1977 establishing the Space Shuttle Program; U.S. Code of Federal Regulations, Title 14 - Chapter V - N.A.S.A. - part. 12/4; N.A.S.A. - NMI 7100-8-1987.
19. See S.F. MARCHI., "Authority of the Space Station Commander: the need for delegation", in Glendale Law Review, Vol. 6, n. 1, 1984, Glendale Univ., page 73 et seq.; LAFFERANDERIE, cit., pp. 270-271; I.S.U., cit., p. 50 et seq.
20. For example, the first Italian astronaut, Franco Malerba, is an engineer, not a pilot, employed by a private industry.
21. S.T.S. = Space Transportation System: official name of the "Shuttle". (Presidential Executive Order 1977 establishing the Space Shuttle program; US Code of Federal Regulations, Title 14 - Chapter V - NASA - part. 12/4; NASA - NMI 7100-8-1987); Space Station = "Freedom" Space Station: an orbital structure under development by the US, the European Countries (through ESA) and Japan, with the aim of creating a large permanently inhabited settlement in earth orbit.
22. See MARCH, cit., p. 74; I.S.U., cit., p. 48; FARAND, cit., p 18-21, etc.
23. The Amalfi Tables contain an ancient Mediterranean Code, dated 1131; see GILMORE & BLACK, The Law of Admiralty (2nd Ed., 1975) p. 5 note 14.
24. Among others, ROBINSON, "NASA's Space Station and the Need for Quantifiable Components of a Responsive Legal Regime", 6 Int'l Law 292, 297-300 (1972) discussing the risk factors of human relations within a space crew; NOTE, "Dispute resolution in Space", 7 Hastings Int'l & Comp. L. Rev. (1984).
25. This is the sense behind the criticism of the "Delegation Approach", es. in I.S.U., cit., p. 50.
26. For a complete definition and analysis of the Sp. Comm. App. see: GORONE, "The Sp. Shuttle: Some of its features and legal implication", 6 Annuals of Air & Space Law 381, 389-90 (1981); MOSSINGHOFF and SLOUP, "Legal Issues Inherent in Sp. Shuttle Operation", 6 J. Space L. 47, 67 (1978).
27. Although regarding other juridical issues, this hypothesis was already contemplated for the Shuttle flights by MOSSINGHOFF, "The Space Shuttle era: International and Domestic Legal Aspects", 72 Proceedings of the Am. Soc. Int. Law 249, 257-58 (1981).
28. The issue of a "Code of Conduct" for astronauts for years has been at the center of the best doctrine. For example, see I.S.U., cit., p. 48 et seq.; YOUNG, cit., p 135 et seq.; BOCKSTIEGEL, cit., p. 6; Maj. Gen. T.B. BRUTON, USAF Judge Advocate General, "The status of Criminal Jurisdiction in Outer space", within the Proceedings of the 24th Conference of the Inter-American Bar Ass. Panama, February 1984 (no page no.), etc.
29. This is the direction in which international doctrine appear to be moving, especially until the end of political bipolarism, given the considerable differences that exist in the definition of crime and sentencing between communist and non-communist regimes.
30. See I.S.U., cit., p. 35 et seq.; YOUNG, cit., p. 67 et seq.
31. Art. 8 of the Outer Space Treaty
32. See, in this respect, YOUNG, cit., p 152-53, MATTE, cit., pp 342 et seq.
33; See, in this respect, YOUNG, cit., p 152-53, MATTE, cit., pp 342 et seq

34. In particular, LAFFERRANDERIE, cit., p 269 et seq.; BOCKSTEGEL, cit., p 6
35. See I.S.U., cit., p 115 et seq.
36. See P. SMITH, "Space Stations Intended Law and Policy", 312-61 (1979).
37. Even during a past Apollo mission there occurred problems with interference, loss of radio contact, etc., that ceased after the use of the "Hot Line" between Washington and Moscow (See LAFFERANDERIE, cit., p. 272).
38. See YOUNG, cit., pp. 164 et seq.; HOOVER, "Law and Security in Outer Space from the Viewpoint of Private Industry", 11 J. Space L. 115, 122 (1983); B. LUZENBERG & G.J. MOSSINGHOFF, "Intellectual Property and Space Activities", 13 J. Space L. 8, 13 (1985); see also on the issue of discovery and patent rights, among others T.M. FOLEY, "U.S. Proposal would restrict European, Japanese Station use", in Aviation Week & Space Technology, 16/2/87, p. 23; and R. O_STERLINK, "Les inventions de l-Agence et leur protection", ESA Bulletin n. 27, Aug. 81, pp. 22-26.
39. FARAND, cit., pp. 23 et seq.; YOUNG, cit., pp. 146 et seq.; MATTE N.M., "Aerospace Law From Scientific Exploration to Commercial Utilization", (1977, The Carswell Co. Ltd.) pp 162-163, (note 24).
40. See BLUTH, "Consciuosness Alteration in Space", Paper presented at the Fourth Princeton/Am inst. of Aeronautics and Astronautics (AIAA) Conference on Space Maifacturing Facilities, Princeton, N.J. 14-17/5/79; AKINS, "Psychological and Psychophisiological Effects on Long Duration Space Flight", in Update on Space 64-81 (Bluth & McNeal ed., 1981).
41. This is the direction apparently taken by the US in applying art. 8 of the Outer Space Treaty, reference to which is made by 2 sections of the US Code (42 U.S.C. 2473); 18 U.S.C. 799); in any case see McDOUGAL, LARSWELL, VLASIS "Law and Public Order in Space", 1973, pp. 668-74.

- IV -

REGULATORY REFORMS

EMBRACING THE FUTURE : RECENT CANADIAN TELECOMMUNICATIONS DECISIONS

RICHARD J. SCHULTZ

Department of Political Science and Centre for the Study of Regulated Industries McGill University.Richard Schultz is Professor of Political Science and former Director of the Centre for the Study of Regulated Industries at McGill University. He has written extensively on Canadian telecommunicatuions policy issues and served as a member of the Sector Advisory Group on General Services for the Canadian Government during the Canada-United States Free Trade negotiations and the GATT Uruquay Round. His most recent publication is "Paradigm Lost: Explaining the Canadian Politics of Deregulation" in Franks et al. (eds) Governance in a Mature Society: Essays in Honour of John Meisel (McGill-Queen's Press, 1995)

Abstract : *In 1992 and especially 1994 the Canadian telecommunications regulatory agency made decisions that, if fully implemented, will make Canada one of the most open and competitive telecommunications markets in the world. This chapter summarizes these decisions and sets them against earlier more conservative decisions. The chapter concludes with a discussion of some of the obstacles that will have to be overcome to give the decisions full effect.*

Key Words : *Telecommunication ; regulation ; competition, Canada.*

INTRODUCTION

Canada was the first country to follow the precedent of the United States in liberalizing sectors of its telecommunications market. Beginning in 1979 with private line voice and public data services and then between 1980 and 1982 with terminal equipment, Canada appeared to be well launched on the road to unlimited telecommunications competition.[1] The momentum towards competition was short-lived, however, as reaction soon set in and Canadian regulators opted for a more modest, conservative approach. Indeed it was an approach that resembled that found in most of the industrialized world outside of the United States. Consequently, the early decisions soon appeared to be anomalies as Canadian regulators resisted further demands and attempts for liberalization.

The last two years have witnessed a dramatic turnaround, however, as once again Canadian regulators have embraced a decidedly pro-competitive philosophy. First in 1992, and especially in late 1994, Canadian regulatory authorities have taken decisions that, if given full effect, promise to make Canada one of the most open and competitive telecommunications markets in the world. The purpose of this chapter is to describe and analyze the dramatic turnaround in Canadian telecommunications regulation and the recent embrace of competition with particular reference to decisions taken in 1994.

I - BACKGROUND

Before turning to more recent decisions it is instructive to take note of earlier regulatory outcomes in order to appreciate the magnitude of the changes introduced by the CRTC in 1992 and especially 1994. The Canadian Radio-television and Telecommunications Commission (hereafter CRTC) permitted the introduction of competition, albeit in a very limited form, only a few years after the American Federal Communications Commission had authorized private line competition in the United States. In 1979, the CRTC gave permission to CNCP Telecommunications (now Unitel) to provide public data and private line long distance services in competition with those offered by the federally regulated carriers, Bell Canada and B.C. Tel. [2] The next year in an interim decision which was followed two years later with a final decision the CRTC matched the FCC's terminal equipment policies by permitting unlimited attachment of customer-owned terminals and other equipment. [3] It is worth noting, in terms of subsequent regulatory decisions and behaviour, that the CRTC in its 1979 decision ominously warned that, if it came to believe that private line competition turned out to be not in the public interest, it would reverse itself. [4] There is no question that behind this warning was the CRTC's fear that Canada might be subjected to the recent American experience where similar competition soon extended, first illegally and then with regulatory sanction, to public voice competition. [5]

The CRTC's underlying commitment to telecommunications competition was tested within five years when CNCP applied for permission to provide public voice long distance competition. Just as it has been said that "judges follow election returns" so it would appear do regulators. In view of the extensive public and especially intergovernmental debate that followed its 1979/1982 pro-competition decisions, concern for the political reactions to its decisions would appear to be the primary explanation for the CRTC's 1985 decision rejecting the application despite its extensive and approving discussion of the benefits that would flow from the extension of competition to public long distance voice services. [6] In this regard it should be recalled that the Federal Government had agreed, prior to the CRTC's decision, to a major intergovernmental study of telephone pricing and the possible negative impact on universal service of competition. The report of the "Federal-Provincial Examination of Telecommunications Pricing and the Universal Availability of Affordable Telephone Service" was issued in 1986. Furthermore the then Federal Minister of Communications had stated only months before the CRTC issued its decision that his first priority was universal, affordable service.

As if recognizing the criticism it would receive from some quarters for reversing its direction on the issue of competition, the CRTC in its 1985 decision sought to appease its critics by two measures. The first was to support, in principle, the liberalization of provisions governing the resale and sharing of telecommunications services as an alternative to the rejected full-blown competition in long distance service. The operative phrase is "in principle" because the Commission established very stringent conditions subsequently that in effect reduced substantially any opening for competition in the message toll (MTS) or wide area telephone service (WATS) markets.

The Commission had earlier proposed to liberalize the so-called enhanced service market and its 1985 decision promised to extend this liberalization to the resale and sharing market segments. [7] However, the Commission was adamant, when it subsequently issued the regulations to govern resale and sharing, that any liberalization in the provision of resale and sharing would occur only in the enhanced services market. [8] To enforce this approach, the CRTC engaged in an ultimately successful battle, at least in the short term, with an obstinate and inventive reseller. [9] As Hancock notes, reflecting its 1985 conservative approach to competition, the Commission "was predominantly focused on the restrictions required to prevent resellers from offering MTS/WATS-like telephone services and containing the growth of competitive service offerings to private lines." [10] She also emphasizes that there "were few (if any) voice services which the Commission would consider non-MTS/WATS." [11]

The primary concern in both the rejection of long distance voice competition in 1985 and the restrictive provisions for resale in 1987, notwithstanding the accompanying rhetoric about liberalization and the putative benefits of competition, was maintaining the existing level of cross-subsidization of local telephone service from the revenues for long distance services. The cross-subsidy system was central to the second measure that the Commission adopted in its 1985 "Interexchange Competition" decision to placate some of its critics.

The CRTC refused in 1985 to undertake any attempt to rebalance rates but it conceded that the extensive cross-subsidy system was the source of potential problems, in particular the diminution of the benefits that even it recognized would flow from the introduction of competition, if it were to be permitted. Consequently, the CRTC announced that, at least as a first step, it would freeze the level of "contribution" or subsidy from long distance to local service at its 1985 amount. This was intended to demonstrate that, notwithstanding the political pressures it faced, the Commission was concerned about this problem and would act on it more directly some time in the future. Despite the promise, the CRTC did not freeze or cap the level of contribution after 1985. Figures filed by Bell Canada establish that the contribution level grew over the next five years. In addition, by emphasizing the issue, the CRTC gave legitimacy to claims of critics of competitive telecommunications that universal service would be threatened by competition. [12]

In light of the foregoing and especially the fact that the application made to it in 1990 was from only two companies, the CRTC's decision in 1992 was truly dramatic. Whereas in 1985, the Commission had ruled against allowing competition primarily because it deemed the financial plan of the specific applicant was suspect and therefore the firm might not prove to be viable, in 1992, the CRTC ruled that the applicant, Unitel, would appear to be viable. More importantly the Commission reversed its previous rather paternalistic approach to declare that even if the applicant proved unable to meet the demands of a competitive marketplace as structured in part by the Commission, "the risk is ... that of the company's shareholders." [13] What was even more pathbreaking was that the Commission went on to declare that it would not impose an economic viability test on any other potential entrant. Rather it concluded that any applicant that was prepared to accept the Commission's conditions governing entry would be automatically approved. [14] In addition the Commission ruled that entry would not only be extended to as many facilities-based carriers as applied, national or regional, but that resale rules would be liberalized such that restrictions on competition within this market and between resellers and facilities-based carriers would be removed. In short both the conceptual and regulatory barriers to competition found in previous decisions were torn down in the 1992 decision.

This is not say that the overall regulatory philosophy was changed. On two very important matters, the CRTC appeared to remain steadfast. First, the CRTC appeared to believe that unrestricted competitive entry raised at best minimal questions about its continued regulatory role. Indeed, shortly after the decision, Keith Spicer, the CRTC Chairman, indicated that the move towards increased competition would merely require "adjustments to regulation." [15] The Commission did not seem to appreciate that a competitive marketplace required a fundamental recasting of regulatory instruments, roles and especially philosophies.

More importantly, the Commission's 1992 Decision provided no new direction on the issue of cross-subsidization of local services from long distance. Indeed, the Commission appeared to set in concrete the traditional social policy objectives linked to those cross-subsidies. The Commission ruled that "competition need not add any significant increases in local rates if contribution payments are adequate" and declared that nothing in its decision would jeopardize "the principle of maintaining local rates at affordable prices." [16] To give effect to these decisions the Commission imposed a subsidy recovery mechanism, namely "contribution payments" on the new entrants that amounted in 1992 to $3 billion.

II - EMBRACING THE COMPETITIVE FUTURE : DECISION 94-19

Despite suggesting that only "adjustments to regulation" would be required in the world of competitive telecommunications, within six months the CRTC launched a general, all-encompassing review of its regulatory framework. As this proceeding developed, the Commission was aided by the passage, after three previous attempts over an eighteen year period, of new telecommunications legislation. [17] Three specific provisions of this legislation were particularly timely. The first was the policy instruction "to foster increased reliance on market forces for the provision of telecommunications services and to ensure that regulation, where required, is efficient and effective." [18] Secondly, where the Courts had denied the Commission the authority to do so under earlier laws, the Commission was explicitly allowed to forbear from regulation. Finally, the Commission was granted permission to employ those regulatory tools that it deemed necessary for effective regulation rather than be limited to traditional rate of return and tariff regulation.

The first striking characteristic of Decision 94-19 is its rhetoric. Although in earlier decisions as suggested above the CRTC made obeisances towards the value of competition, in 94-19 the Commission appears to have found religion. It argues that in a competitive world, what "ultimately emerges will be determined by the demands of users and the willingness of suppliers to take risks . The role of the Commission should be to ensure that the right economic and technical conditions for open access are in place...."[19] In such a world the environment is changing "in ways that outpace the ability of regulators to recognize and define, let alone control." Elsewhere it states that "market forces allow for greater choice and supplier responsiveness and ensure that user applications, not regulators, drive supply considerations." [20] The Commission appears to have rejected its Chairman's earlier suggestion about "adjustments" when it declares that "regulation must also be flexible and responsive to change, unencumbered by objectives based on static definitions of markets or services." [21]

Nowhere is the change in the language and presumably the philosophy of the Commission greater than in the discussion of cross-subsidies. The Commission now concedes that "meaningful regulatory reform cannot be undertaken without a significant reduction in the subsidy that users of long distance services are currently providing in order to keep rates for local/access low." [22] Not only is the principle of cross-subsidization open to scrutiny, it is now to be redefined as both a barrier to "meaningful regulatory reform" and an inequity "for those users who are bearing the burden of a subsidy that is larger than required to achieve universal service obligations". [23] In fact the Commission adopts the language of its critics when it declares that the subsidy is "an inequitable and unnecessary burden ... [that] amounts to a tax on information-intensive enterprises". [24]

Rhetoric, of course, is the easy part of a decision. Crafting the appropriate package of instruments to give effect to the language and philosophy of the decision is the much more difficult task. In Decision 94-19 the CRTC has surprised even its most vocal critics with a highly imaginative, comprehensive, coherent and integrated set of proposals. The starting point for the CRTC was to establish a set of underlying principles to guide its decision. These were :

1. Regulation should focus primarily on monopoly services and the prevention of anti-competitive behaviour... Regulation should be reduced or removed for competitive services.
2. All telecommunications markets should be opened up including the

local telephone market, to competition and barriers should be removed that discourage telephone companies from entering information services markets.

3. Consumers must not be expected to pay artificially high prices for any telecommunications services, particularly Canada's quickly expanding long-distance service. [25]

From these principles were derived the specific aspects of the decision. In order to reduce barriers to entry, the Commission extended its competitive philosophy for long distance markets to local telephone service, both basic telecommunications services and information-based services. It is instructive to note that when the Commission initiated its regulatory framework proceeding, there were no applications before it for entry into local service. Henceforth, anyone can enter this market including presumably cable companies, other telephone companies, resellers and providers of wireless services when they are licensed in the near future.

The second side of this decision was the removal of barriers on telephone company entry into information-based telecommunications markets. Consequently the telephone companies are now free "to provide any telecommunications information service [including] transactional or interactive services such as home banking, teleshopping, distance learning, telemedicine, [and] on-line access to libraries." [26] The Commission went even further than this in permitting telephone companies to deliver broadcast programming on behalf of licensed broadcasters. Furthermore the CRTC announced simultaneously that it would permit the telephone companies to begin trials of the technology to deliver video-on-demand services. What the Commission could not do, however, was to grant the telephone companies licenses to provide broadcasting services directly as this is not permitted under the Broadcasting Act. Extending such permission will require legislative approval and few observers think this will be forthcoming, at least in the near future.

In order to give effect to its principle of reducing or narrowing the focus of regulation, the CRTC developed a package of specific regulatory tools. The first was to split the rate base of the traditionally dominant telephone companies into two components: Utility and Competitive. The "Utility" segment is defined by the CRTC as those "services in the current Monopoly Local and Access service categories, plus a share of certain common costs and local network transport services." [27] This split rate base is to be used for a transitional period while the Commission develops an alternative to rate of return regulation in the form of price caps. This split is reflective of the Commission's desire to focus regulation on monopoly services primarily and its stated goal of preventing anti-competitive behaviour by the dominant firms. Central to the new approach is the decision not to regulate the Competitive segment of the rate base as of January 1 of this year.

Rejecting almost ninety-five years of Canadian telecommunications regulation as no longer appropriate for an increasingly competitive environment, the Commission also announced that effective 1998 rate of return regulation would be replaced by a system of price caps for the monopoly or utility service segment. Echoing language similar to that used in other countries and jurisdictions that have substituted price regulation for rate of return regulation the Commission contends that the former will

- reduce incentives for telephone companies to overinvest;
- reduce incentives and opportunities for telcos to cross-subsidize their competitive services from money earned on their monopoly services;
- increase efficiency by providing incentives for telephone company shareholders to benefit from increased productivity;
- reduce the need for regulatory assessment of investment, earnings and expenses, thereby reducing the complexity and cost of CRTC regulation. [28]

Another measure that the Commission addressed in its attempt to reduce regulation wherever appropriate was regulatory forbearance. The Commission had attempted to forbear from regulation earlier but this effort was ruled unacceptable by the courts. In the new Telecommunications Act, however, Parliament conferred such a power on the CRTC. In an earlier Decision (94-14 August 4, 1994), the Commission exercised its new power to refrain from regulating for some purposes the sale of competitive terminal equipment by Canadian carriers. In 94-19 the CRTC was prepared to extend forbearance for that market for all purposes as well as for the "Competitive Terminal-Other" market, that is the single-line telephone and accessories market segments.

Perhaps as, if not more, important as the actual exercise of its new forbearance powers was the Commission decision to set out the criteria in advance it would employ for any future applications before it for forbearance. In its decision the Commission summarized these as follows:

1. The Commission should forbear when a market becomes workably competitive;
2. A market cannot be workably competitive if the dominant firm possesses substantial market power;
3. Market power is a function of three factors: (a) market share held by the dominant firm; (b) demand conditions affecting responses of customers to a change in price of the product or service in question; and (c) supply conditions affecting the ability of other firms in the market to respond to a change in the price of the product or service;
4. High market share is a necessary but not sufficient condition for market power; other factors must be present to enable a dominant firm to act anti-competitively. [29]

In large part, the Commission appears to have adopted the criteria for determining market power of the Canadian competition authority, the Bureau of Competition Policy and its Director. This is not an insignificant step for it suggests that the Commission in future proceedings will adopt a very focussed definition of market power thereby allowing it greater opportunity to refrain from regulating. Such self-denial, if acted upon, will be admirable and certainly in keeping with the principles the Commission has espoused.

One of the central components of the CRTC's decision in 1992 to permit open competition in long distance services was the imposition of a mechanism to collect "contribution" in order to maintain the subsidy for local services at the existing level. In 1992 that mechanism was a per-access-trunk charge. This was based on a per-minute rate "which is converted to a per-trunk charge based on estimates of the number of minutes that are carried on access lines within trunk groups of various size." [31] At the time the Commission deemed this method to be the least intrusive from a regulatory perspective.

Existing carriers argued before the Commission, however, that the per-trunk mechanism was not regulatory neutral but in fact gave an advantage, through a potential discount, to new entrants. The CRTC was persuaded by this argument and consequently changed, in one of the most controversial aspects of its 1994 decision, the contribution mechanism to a "Carrier Access Tariff (CAT)" designed to result "in greater equity between the telephone companies and competitors". [31] This new charge will be applied to both the originating and terminating points of a call and will "apply to all telephone company voice/data traffic carried over the interexchange portion of the PSTN and to all competitor voice/data traffic accessing and egressing the PSTN via trunk-side connections." The CAT, the CRTC stipulated, will consist of two components: a network access component which includes charges for switching and aggregation, the recovery of start-up costs and other unbundled elements; and a contribution component to subsidize the rates for local telephone service.

The Commission also introduced a new mechanism to guard against anti-competitive or predatory pricing by incumbent, dominant firms. The objective is to ensure that telephone companies recover all relevant underlying costs in the prices they charge. The new mechanism is an "imputation test" based on three elements:

(a) the costs (or average cost per minute) for bottleneck services used by the telephone company in the provision of the service in question, using CAT prices, including that for recovery of start-up coats) for bottleneck services as the cost of the bottleneck service;
(b) Phase II [the CRTC"s Costing System] causal costs, excluding costs for services covered in (a); and
(c) contribution, if applicable. [32]

Any price charged which is less than the sum of these elements will be deemed to be anti-competitive.

The final major aspect of the CRTC's 1994 decision was its initiative on rate rebalancing. As indicated above, the Commission radically shifted both its rhetoric and its perception of the nature and impact of the long-standing extensive cross-subsidy system in telecommunications pricing. The Commission is now of the opinion that

> *Reducing the subsidy from long distance to local service is key to reforming Canada's regulation of telecommunications so that it can contribute to the ongoing development of a modern public infrastructure to meet the information and communication needs of Canadians into the next century.* [33]

As a consequence, the CRTC ordered its first rebalancing by imposing a rate increase of $2/month each year for three years for the prices charged by the telephone companies for basic local telephone service.

The Commission also ordered, however, that the telcos could not use the increases as a revenue generating measure but must make corresponding decreases in basic long distance service prices. Although this increase makes a small dent, at best, in the cross-subsidy system its importance is not to be denied as a most significant symbolic first step by the regulator to address this issue. As noted below, on appeal, the Canadian Cabinet ordered the Commission to review this aspect of its decision. Even if rolled back in whole or in part, I would nevertheless argue that even the attempt by the regulator should be applauded. Reversal should at worst be seen as a delaying tactic rather than a fundamental barrier to change.

The Commission's Decision 94-19 is also significant for two actions it did not contain. The first pertains to "customer-specific tariffs" or special contracts for individual customers. The Commission decided that such tariffs "under current market conditions ... present a greater risk that unjust price discrimination will arise" and therefore ruled that they were not acceptable. [34] As Hudson Janisch has noted "the Commission can't quite bring itself to accept the freedom of contract implicit in a true marketplace." [35]

The second and perhaps more controversial "non-action", at least for some observers, is the Commission's approach to the issue of vertical integration. A number of parties to the proceeding urged the Commission to order structural separation or divestiture. While the Commission agreed that "the telephone companies have both the incentive and the opportunity to use their vertically integrated structures to lessen competition by exploiting control over bottleneck facilities", it rejected the divestiture option. In its view, "there is no compelling evidence to support the dismantling ... of the telecommunications system in Canada. There are better methods...." [36] It concluded that "divestiture could be damaging to the competitiveness of Canada in global markets and could damage the emergence of integrated services evolving from the convergence of the communications, information, computing and entertainment industries." While the Commission expresses confidence that its various regulatory alternatives to divestiture will be effective, other observers believe that this is a question that will remain on the table for some time, especially as some of the competitors face difficult times which they will seek to trace to continued vertical integration.

CONCLUSION

By any criterion, CRTC Decision 94-19 is one of the boldest in Canadian regulatory history. It is well-grounded in an integrated, articulated set of principles. It has a comprehensive, coherent and imaginative set of policy instruments and mechanisms to give it effect.

The overall result, particularly through its forward-looking nature on the issues of diminished and flexible regulation and convergence is to make Canada one of the most open, progressive telecommunications market in the world today and in the foreseeable future. This is truly an impressive accomplishment that is doubly so given both earlier decisions and the apparently well-entrenched regulatory philosophy.

Of course much remains to be done to give it full effect. For its part the Commission must develop the split rate base system, the imputation test and the price cap regime and then implement each of them. Moreover, and perhaps more importantly the Commission will have to navigate through treacherous political waters to give full effect to its decision. Already there are signs not only that this will be difficult but that the Commission may backtrack on some of its more controversial measures.

On one front a coalition of new competitors has already persuaded the Government of Canada to set aside the Carrier Access Tariff because of claims that equity of treatment amongst competitors may be too burdensome for new entrants. Others are asking that they be given additional contribution discounts at least until they become profitable. For their part both new competitors and some consumer associations have successfully petitioned the Government to request the CRTC to reconsider its modest rate rebalancing initiatives.

Perhaps the most serious obstacle to full implementation of the decision as it pertains to open competition in local service and convergence will be the Government's positions, yet to be developed, on these issues. In October 1994, the Government directed the Commission to undertake a full public hearing on convergence and particularly the nature of competition, if any between cable companies and other telecommunications providers. The Government claims to have a "vision of competition" but it is most decidedly a peculiar one inasmuch as the Government has told the Commission that while it is alright for individual firms to fail no sector may disappear. [37] This coming year may witness a fundamental and far-reaching clash of regulatory cultures: the protectionist culture of the **Broadcasting Act** versus the pro-competitive, open market approach as enunciated in CRTC Decision 94-19. Perhaps a subject for next year's **Telecommunications and Space Journal** ?

REFERENCES

1. Canadian Radio-television and Telecommunications Commission (CRTC), "CNCP Telecommunications, Interconnection with Bell Canada", Telecom Decision CRTC 79-11; CRTC, "Bell Canada - Interim Requirements Regarding the Attachment of Subscriber- Provided Terminal Equipment" Telecom Decision CRTC 80-13 and CRTC "Attachment of Subscriber-Provided

Terminal Equipment" Telecom Decision CRTC 82-14.
2. See Footnote 1. The 1979 Decision was made applicable to B.C.Tel in CRTC, "CNCP Telecommunications: Interconnection with B.C. Tel", Telecom Decision CRTC 81-24.
3. See Footnote 1.
4. CRTC, ibid. CRTC 79-11 at p. 243.
5. The American experience is described in Robert B. Horwitz, The Irony of Regulatory Reform (New York: Oxford University Press, 1989) pp. 233-34.
6. CRTC "Interexchange Competition and Related Issues", Telecom Decision 85-19. The controversies are surveyed in Richard Schultz, "Forward to the Past: The Canadian Approach to Telecommunications Reform" in William and Anne C. Averyt (eds.) Managing Global Telecommunications: North American Perspectives (Burlington, Vermont: 1989) and W.T. Stanbury "Decision-Making in Telecommunications: The Interplay of Distributional and Efficiency Considerations" in Stanbury (ed) Telecommunications Policy and Regulation: The Impact of Competition and Technological Change (Montreal: Institute for Research on Public Policy, 1986).
7. CRTC, "Enhanced Services" Telecom Decision CRTC 84-18.
8. CRTC, "Tariff Revisions Related to Resale and Sharing", Telecom Decision CRTC 87-2.
9. For an excellent and insightful review and analysis of this battle and the larger regulatory issues associated with it, see Terry Dawn Hancock "Regulated Competition: Resale and Sharing in Telecommunications" Media and Communications Law Review 2 (1991) pp. 251-296.
10. Ibid. p. 261.
11. Ibid., p. 266.
12. For the details and discussion see Richard Schultz "Old Whine in New Bottle: The Politics of Cross-Subsidies in Canadian Telecommunications" in Globerman, Steven et al. (eds) The Future of Telecommunications Policy in Canada (forthcoming).
13. CRTC,"Competition in the Provision of Public Long Distance Voice Telephone Services and Related Resale and Sharing issues", Telecom Decision CRTC 92-12, p. 111.
14. Ibid. p.117.
15. Keith Spicer, "Responses to Globalization of Telecommunications" Telecommunications Policy, 16 (Dec. 1992) p. 720.
16. CRTC, Decision 92-12, p. 13.
17. For an overview see H.N.Janisch "At Last! A New Canadian Telecommunications Act", Telecommunications Policy Dec. 1993.
18. Telecommunications Act, Statutes of Canada 1993, c.38,s. 7 (5).
19. CRTC Telecom Decision 94-19, p. 33.
20. Ibid., p. 13.
21. Ibid., p. 11.
22. Ibid., p. 14.
23 Ibid., p. 14.
24. Ibid., p. 19. For criticism of the traditional language see Schultz, "Old Whine in New Bottle..." and the references cited therein.
25. CRTC, "News Release" September 16, 1994, pp. 1-2.
26. CRTC, "Fact Sheet: Expanding Competitive Opportunities".
27. CRTC, "Fact Sheet: Glossary".
28. CRTC, "Fact Sheet: A New Price Regulation Framework"
29. CRTC, Telecom Decision 94-19, p. 64.
30. Ibid., p. 125.
31. Ibid., p. 126.
32. Ibid., p. 88.
33. CRTC "Fact Sheet: A More Equitable Pricing Framework".
34. Ibid., p. 104.
35. H.N. Janisch, "Canadian Telecommunications: The Other Shoe Falls", Public Utilities Law (Forthcoming, Dec. 1994).
36. Ibid., p. 29.
37. Order in Council PC 1994-1659, October 8, 1994 and accompanying Press release, Oct. 11, 1994.

REGULATORY REFORM IN GERMANY: PRIVATIZING AND REGULATING DEUTSCHE BUNDESPOST TELEKOM

DR. JOACHIM SCHERER

LL.M. Partner, Baker & McKenzie. Döser Amereller Noack Bethmannstraße 50-54 60311 Frankfurt am Main, Germany. Tel.: (49-69) 29 90 81 89 Fax : (49-69) 29 90 81 08 ; Professor at Johann Wolfgang Goethe-University, Faculty of Law

Abstract : *Only four years after a major organizational reform of the German Federal Post Office (Deutsche Bundespost, "DBP") in 1989, the Federal Parliament has adopted, in July 1994, legislation privatizing the three DBP enterprises, namely DBP Postdienst, DBP Postbank and DBP TELEKOM. The privatization of DBP TELEKOM has not accompanied by a further liberalization of the German telecommunications market. Instead, the privatized Deutsche Telekom AG continues to exercise both the network monopoly and the voice telephone monopoly until December 31, 1997, when the European Union is expected to abolish the voice telephone monopoly.*

Key Words : *Deutsche Bundespost Telekom; Deutsche Telekom AG; Postal Reform I; Postal Reform II; privatization; monopoly services; mandatory services; optional services; network monopoly; telephone service monopoly; radio monopoly; mobile radio comunications services; radio paging; satellite licenses; licensing requirements; notification requirements; terminal equipment; multimedia.*

INTRODUCTION

Only four years after a major organizational reform of the German postal and telecommunications system in 1989 - the "Postal Reform I" [1], the German federal parliament has adopted, in July 1994, legislation privatizing the three German Federal Post Office (Deutsche Bundespost - "DBP"-) enterprises, namely DBP Postdienst (postal services), DBP Postbank (postal banking) and DBP Telekom (telecommunication services) - the "Postal Reform II" [2].

On January 1, 1995, the three DBP enterprises were transformed into stock corporations which are owned by the Federal government. In the case of DBP Telekom, now Deutsche Telekom AG, government ownership is foreseen for a period of five years.

The privatization of DBP Telekomwas not accompanied by a further liberalization of the German telecommunications markets. Instead, the privatized Deutsche Telekom AG continues to exercise both the voice telephone monopoly and the network monopoly until December 31, 1997, when the European Union is expected to abolish the voice telephone monopolies of its Member states.

I - POSTAL REFORM II: PRIVATIZATION OF DBP TELEKOM

Since the status of the DBP enterprises as public enterprises was guaranteed by the Federal constitution, their privatization was predicated upon a constitutional amendment which required a 2/3 majority in both houses of parliament.

Under Article 87 f of the constitution, as amended, the Federal government is empowered to discharge the regulatory functions in the postal and telecommunications sectors. The government is obligated to ensure, on the basis of a federal law, that nation-wide adequate and sufficient services shall be rendered in the postal and telecommunications sectors. The provision of such services is no longer a governmental task, but shall be open to competition between the former DBP enterprises and its competitors.

Based upon the constitutional amendment, the privatization was brought about by the following legislative acts [3]:

- an act establishing a Federal Entity for Posts and Telecommunications (Bundesanstalt für Post und Telekommunikation) which will hold the shares of the privatized enterprises [4];
- an act transforming the three DBP enterprises from public corporations into private stock corporations (the "Transformation Act") [5];
- legislation adapting the existing regulatory framework to the requirements of the privatized DBP enterprises [6].

The Federal Entity for Posts and Telecommunications is established at arm's length of the Federal government under the authority of the Federal Minister for Posts and Telecommunications. It will be governed by a president, to be appointed by the Federal President upon the recommendation of the Federal government, and by an administrative council consisting of

- a representative of the Federal Ministry of Posts and Telecommunications,
- a representative of the Federal Ministry of Finance,
- one representative of each of the privatized DBP enterprises,
- one representative of the personnel of each of the enterprises.

In addition to holding the stocks of the privatized enterprises (for a minimum of five years and until the adoption of a Federal law in the case of Deutsche Post AG), the Federal Entity has the following main tasks:

- to advise the corporations in cases of conflicting business plans, taking into account the infrastructural tasks of the enterprises;
- to advise the corporations with regard to their "corporate image";
- to advise the corporations, upon their application, on personnel management principles;
- to conclude, on behalf of the enterprises, outline ("skeleton") collective bargaining agreemens.
- to take appropriate measures to ensure the transition of the public officials who were employed by the DBP enterprises to the privatized enterprises.

On the basis of the Transformation Act, the three DBP enterprises were transformed into stock corporations, namely:

- Deutsche Post AG,
- Deutsche Postbank AG and
- Deutsche Telekom AG.

These stock corporations are the legal successors of DBP Postdienst, DBP Postbank and DBP Telekom, respectively. At the time of the transformation, the sole shareholder of all three enterprises was the Federal government, which will exercise its shareholder functions through the Federal Entity for Posts and Telecommunications. It is planned to sell the shares of Deutsche Telekom AG, following a capital increase.

The 1989 Postal Reform I had the dual objective

- to reorganise the DBP from a Federal administration to a public enterprise which was separated from the Ministry as its regulatory body and
- to establish a new, liberal regulatory framework for the provision of telecommunications networks, services and terminal equipment.

The 1994 Postal Reform II has limited itself to the privatization of the DBP enterprises and has refrained from further liberalizing the telecommunciations markets. Instead, the existing system of telecommunications regulation was merely amended to take into account the transformation of DBP Telekom into a stock corporation. Overall, the regulatory system appears to have been weakened through

- the establishment of an overly complex organizational framework for regulation,
- the introduction of cumbersome regulatory procedures and
- the lack of appropriate regulatory instruments aimed at ensuring fair and open competition between Deutsche Telekom AG and its private competitors.

The following analysis of German telecommunication laws and regulation is based on the situation prior to the privatization of DBP Telekom [7], since all of the administrative rules which were adopted subsequent to Postal Reform I remain in force until they are repealed repealed or replaced.

II - RE-REGULATION OF THE GERMAN TELECOMMUNICATIONS MARKETS SINCE 1989

The far-reaching liberalisation of the German telecommunications markets in 1989 has been brought about, in essence, by amendments to the Telecommunications Installations Act (Fernmeldeanlagengesetz, "FernmAG"). In its version of 1989 [8], the FernmAG establishes:

- an exclusive right of the Federal government to set up and operate networks and radio installations (network monopoly), subject to statutory and case-by-case exemptions;
- an exclusive right of the Federal government to provide voice telephony service (telephone service monopoly), subject to statutory and case-by-case exemptions;

- a new framework for the licensing of satellite systems;
- the freedom to provide telecommunications services for third parties over leased lines;
- the freedom to set up and to operate type-approved terminal equipment.

Almost five years after Postal Reform I and prior to Postal Reform II, the new regulatory framework - including new ordinances, administrative rules and licenceswas well established, and the market for telecommunications services has undergone considerable change. In principle, the Postal Reform II legislation has not affected this regulatory framework other than by transferring the network monopoly and the telephone service monopoly for a "transition period" to Deutsche Telekom AG.

Following the statutory reform of 1989, the Federal Ministry of Posts and Telecommunications (Bundesministerium für Post und Telekommunikation, -"BMPT"-) has gradually re-regulated the German telecommunications markets.

Since 1990, the BMPT has:

- adopted detailed administrative rules which delineate the network monopoly and the telephone service monopoly;
- issued a licence for the operation of a second nation-wide digital cellular mobile radio network based on the GSM standard (the "D 2" network), which competes with the D 1 network of TELEKOM;
- issued a licence for the operation of a nation-wide digital cellular radio system based on the DCS 1800 standard (the "E 1" network);
- issued a licence for the operation of a nation-wide mobile data network ;
- issued some 50 licences for satellite communications systems, including twelve licences which allow the provision of voice telephone service in and with the new (east German) Federal states;
- issued more than 40 trunk radio licences for regional trunk radio networks;
- received notifications from more than 380 telecommunications service enterprises which offer telecommunications services over leased lines provided by TELEKOM;

III - THE 1989 ORGANISATIONAL REFORM

As a result of the 1989 Postal Reform I, the BMPT bears the political and parliamentary responsibility for the three DBP enterprises. His tasks include the setting of medium- and long-term objectives for the three enterprises, ensuring their conduct is legal; the approval of important decisions of the bodies of the enterprises (e.g. their economic plan, annual financial statements, rates for monopoly services etc.), and the setting of national and international standards and certain regulatory functions, particularly in the telecommunications field.

After the privatization, the BMPT has remained the regulatory body for the telecommunications sector; however, Deutsche Telekom AG is no longer a public entity under the BMPT's authority and direction and can thus no longer be regulated merely by administrative rulings. Instead, regulation of Deutsche Telekom AG no requires the formal adoption of administrative orders, which are subject to court review.

Certain regulatory functions have been delegated to the Federal Office for Posts and Telecommunications (Bundesamt für Post und Telekommunikation, "BAPT") and the Federal Office for Telecommunications Approvals (Bundesamt für Zulassungen in der Telekommunikation, "BZT"). The BAPT, which has its seat in Mainz and 55 branches all over Germany, is in charge of the licensing of telecommunications installations, radio installations and high frequency apparatus; the adoption of technical standards; the adoption of approval standards and criteria for telecommunications installations and for persons who establish such installations; and frequency management. The BZT, which is located in Saarbrücken, with branches in Halle and near Berlin, is in charge of the approval of telecommunications equipment which is to be:

- used as terminal equipment on the public telecommunications network;
- connected to a leased line; or
- set up and operated as a radio installation.

One of the guiding principles of the 1989 reform was to introduce competition in the telecommunications sector while maintaining the infrastructural responsibility of DBP TELEKOM. Therefore, DBP TELEKOM was obliged not only to provide certain specific monopoly services, but also, on the basis of an ordinance having the force of law, to provide certain mandatory services. In order to ensure political influence on the development of the telecommunications infrastructure, a Council of Infrastructure (Infrastrukturrat) was been established, consisting of 11 representatives of the Federal Parliament (Bundestag) and representatives of the Federal Council (Bundesrat, i.e. the legislative chamber representing the Laender). This Council of Infrastructure participated, inter alia, in decision-making processes concerning monopoly tariffs, the definition of mandatory services and other decisions of infrastructural relevance.

IV - REGULATION AFTER POSTAL REFORM II

The Postal Reform II legislation has preserved and further expanded the political influence on telecommunications regulation in general and on the development of the telecommunications infrastructure in particular.

The 1994 Act on the Regulation of Telecommunications and Postal Matters (Gesetz über die Regulierung der Telekommunikation und des Postwesens, -"PTRegG"-) defines as the main objectives of regulation:

- to ensure a nation-wide and modern offering of telecommunication services at favourable prices;
- to ensure equality of chances between rural and urban areas;
- to ensure discrimination-free access of users to such service offerings;
- to ensure effective management of scarce resources, in particular with regard to frequencies and numbering;
- to take into account social requirements;
- to ensure effective consumer protection and data protection.

The BMPT is in charge of the supervision of the behaviour of Deutsche Telekom AG and all other providers of telecommunication services. Case-by-case regulation, in particular the supervision of compliance with licensing conditions and other legal obligations of telecommunication service providers, is subject to administrative orders to be issued by "adjudication bodies" (Beschlußkammern) which have been established within the BMPT. These adjudication bodies will issue their administrative orders following formal hearing procedures. The administrative orders, which must set forth the reasons why they have been adopted, are subject to an administrative appeals procedure and to court review.

Other regulatory decisions of the BMPT are subject to adoption by a Regulatory Council (Regulierungsrat), consisting of one representative of each Federal state and an even number of members of the Federal parliament, which will replace the Council of Infrastructure.

The regulatory powers to be exercised by the BMPT upon a decision of the Regulatory Council include:

- determination of mandatory services to be provided by Deutsche Telekom AG;
- determination of the legal framework for the provision of monopoly services and of mandatory services;
- decisions on the opening of telecommunications services markets to competition;
- adoption of rules on the scope of telecommunications licenses and on the licensing procedures;
- adoption of frequency administration measures;
- approval of tariffs and tariff-related business conditions for the provision of monopoly services;
- changes of the scope of the monopoly rights of Deutsche Telekom AG;
- decisions on the impermissibility of cross-subsidisation of competitive services by revenues generated from monopoly services.

In cases of disagreement between the BMPT and the Regulatory Council, the ultimate decision power lies with the Federal cabinet.

V - LIBERALISING THE TELECOMMUNICATIONS MARKETS

As part of the 1989 Postal Reform I, the Telecommunications Installations Act was amended with a view to liberalising the provision of telecommunications services.

The Telecommunications Installations Act, as amended in 1989, replaced the formerly sweeping monopoly rights of the German Federal Post Office (Deutsche Bundespost) in the telecommunications sector by narrowly defined monopoly rights of the Federal government, i.e. the network monopoly and the telephone service monopoly. These monopoly rights were transferred to DBP TELEKOM, on the basis of administrative rules, to the extent necessary to ensure fulfilment of DBP TELEKOM's statutory tasks.

For a "transition period" following the privatization of DBP TELEKOM, the monopoly rights have now been transferred upon Deutsche Telekom AG. This transfer of monopoly rights on the basis of a transitory constitutional provision was necessary, from a constitutional point of view, in order to justify the restriction of the freedom of competing telecommunications service providers to establish networks and to provide voice telephone service. Although the Federal constitution does not establish the exact duration of the "transition period", it is expected to end on 31 December 1997, when the European Union will abolish the remaining telephone monopolies of its Member states.

The network and voice telephone monopolies are subject to certain statutory exemptions and to ministerial licences. Telecommunication services other than voice telephone service can be provided freely by anybody over fixed or switched connections to be made available by Deutsche Telekom AG.

The provision of terminal equipment (Endeinrichtungen) has been fully liberalised: anybody has the right to install and operate type-approved terminal equipment if the conditions laid down to ensure the proper handling of telecommunications traffic are observed.

Under the German regulatory system (both, prior to and after the privatization of DBP TELEKOM) three categories of services to be provided by the dominant carrier may be distinguished:

- monopoly services;
- mandatory services; and
- optional services.

VI - MONOPOLY SERVICES

As the provider of monopoly services, TELEKOM is obliged to render such services without unfair discrimination. The provision of monopoly services is subject to stringent regulation, including regulation of tariffs. The scope of the monopoly services in the telecommunications field is defined, in general terms, in the Telecommunications Installations Act and has been specified by the BMPT in detailed administrative rules; the scope of the monopoly rights which were transferred upon the privatized Deutsche Telekom AG is identical to the monopoly rights prior to privatization.

The conditions under which TELEKOM must provide its monopoly services have been set forth in the Telecommunications Ordinance (Telekommunikationsverordnung, "TKV") of October 1992 [9]. This ordinance provides the legal framework for the general business conditions of TELEKOM with regard to monopoly services. The ordinance includes the following obligations of TELEKOM with regard to its monopoly services:

- to "unbundle" its monopoly services in accordance with market demand as far as reasonably possible;
- to provide its monopoly services in accordance with Community

law, in particular in accordance with the ONP directive;
- to provide fault repair service without delay and also at night and on Sundays and holidays;
- to adhere to specific liability rules with regard to damages suffered by customers in conjunction with their use of monopoly services.

6.1. Mandatory Services

Mandatory services (Pflichtleistungen) are those services which TELEKOM is obliged to provide in the public interest. The Postal Constitution Act of 1989 and, similary, the 1994 PTRegG provides that such mandatory services - which are not monopoly services - shall be imposed upon TELEKOM by Federal ordinance. In August 1992, an ordinance was adopted setting forth several narrowly defined mandatory services including the provision of directory information, provision of public telephones and an emergency call service [10].

6.2. Optional Services

Optional or competitive services (freie Leistungen) are those services which TELEKOM may provide freely within the framework of the generally applicable laws, in particular the laws on unfair competition.

Cross-subsidisation between monopoly services and optional services is permissible, but subject to fairly stringent procedural requirements and regulatory control.

6.3. The Network Monopoly

Deutsche Telekom's exclusive right to install and operate telecommunications transmission lines including the associated network terminations (Network Monopoly) is subject to regulation by the BMPT. The Network Monopoly as well as the voice telephone monopoly (see below) are subject to statutorily defined exemptions and to exemptions on the basis of licences.

6.4. Scope of the Monopoly

The precise scope of the Network Monopoly is not defined in the FernmAG. In view of the crucial importance to telecommunications service providers of a detailed definition of the Network Monopoly, the BMPT has adopted detailed administrative rules delineating the

Network Monopoly and DBP TELEKOM's obligations in exercising its monopoly rights [11]. The same rules apply after the privatization of DBP TELEKOM.

The Network Monopoly comprises the right to set up and to operate telecommunications transmission lines and their associated network terminations. Transmission lines are cable and radio communications installations together with their transmission equipment. They establish point-to-point or point-to-multipoint connections with a specific transmission capacity (bandwidth or bit rate). The transmission lines terminate at and, according to the Telecommunications Installations Act, also include the "termination installations". Termination installations are defined as the separation points between the network (reserved under the government's exclusive right) and the terminal and switching equipment (to be provided on a competitive basis).

The administrative rule governing the Network Monopoly restricts the monopoly to those functional units which fulfil two requirements. First, they must be necessary, from a technical viewpoint, for the service-neutral, transparent point-to-point or point-to-multipoint transmission of information within the transmission network. Second, terminal or switching equipment must not be capable of providing those functions instead.

Multiplex installations and similar installations shall be considered as part of a monopoly transmission path only to the extent that they are essential for the transmission of information over the respective transmission path between the termination installations and are thus included in the functions to determine the operability of the transmission path network.

6.5. Rights and Obligations under the Network Monopoly

Monopoly transmission paths must be made accessible to the user via service-neutral, freely reachable interfaces between termination installations of the transmission paths and the terminal equipment attached to them.

As regards the provision of monopoly transmission paths, TELEKOM is obliged to provide such transmission paths to users without discrimination. As far as is economically feasible, TELEKOM is obliged to provide such transmission paths in accordance both with market demand and with the state of technological development. The administrative rule specifies, in terms of bandwidth or bit rates, those transmission capabilities which TELEKOM is obliged to make available "as a starting offer". This offering must be adapted, from time to time, to technological developments. The Minister has reserved his right to demand from TELEKOM the additional offering, within a reasonable deadline, of additional types of transmission paths.

Furthermore, the administrative rule specifies the transmission quality for monopoly transmission paths and establishes a system of deadlines and reporting duties with regard to the timely provision of transmission paths to TELEKOM customers as well as a system of deadlines for the provision of repair services.

In order to guarantee equality of chances between TELEKOM and its private competitors, the administrative rule provides, in essence, that the "technical conditions" under which TELEKOM provides its network monopoly services to its customers shall not be less favourable than those under which TELEKOM uses them for its own competitive telecommunications services. Deviations from this basic rule are permissible only if they have been sufficiently justified to the Minister.

The administrative rule provides that the user has the right to use freely the monopoly transmission paths provided by TELEKOM, notwithstanding the voice telephone monopoly. In particular, transmission paths may be interconnected directly or indirectly. They may also be connected directly or indirectly with fixed lines (Festverbindungen) or switched lines (Wählverbindungen) provided by TELEKOM or other network operators.

6.6. The Radio Monopoly

Under the FernmAG of 1989, the Federal government had the exclusive right to set up and operate radio installations. Upon the privatization of DBP TELEKOM, this exclusive right has been transferred upon Deutsche Telekom AG for the duration of the "transition period". However, the Federal Minister of Posts and Telecommunications remains empowered to grant licences to set up and operate individual telecommunications installations, including radio installations.

Further, the BMPT has comprehensive frequency management powers, including the power to assign frequencies; in the case of frequencies which are to be used for broadcasting purposes, frequency planning and frequency assignment will in the future require the consent of the Federal states, which have the regulatory powers regarding broadcasting matters.

6.7. The Telephone Service Monopoly

German telecommunications law does not distinguish between basic and enhanced ("value-added") telecommunications services. Instead, the 1989 Telecommunications Installations Act reserves - in a somewhat meticulous language -

"the operation of telecommunication installations for the purpose of switching speech for third parties" as an exclusive right of Deutsche Telekom AG.

The purpose of this Telephone Service Monopoly (Telefondienstmonopol) is to make the telephone available to everyone on equal terms, thus providing for a basic means of communication. In addition, revenues from the Telephone Service Monopoly are to guarantee fulfilment of infrastructural tasks and of those tasks of special public interest with which TELEKOM has been or will be entrusted.

The statutory definition of voice telephony as the "switching" of speech for others designates activities which are more than just transmission. In other words: "Transmitting" speech without any "switching" is permissible under German telecommunications law and does not infringe upon the Telephone Service Monopoly.

Certain obligations of TELEKOM regarding the provision of voice telephony service and certain basic rights of the users have been set forth in the Telecommunications Ordinance: TELEKOM is obliged to provide telephone lines to its customers upon request without delay. The telephone lines must be furnished with a termination installation which is the termination of the network of TELEKOM. The customer is entitled to connect any type-approved piece of terminal equipment to such termination point. The customer has a right to use and to interconnect telecommunications lines freely as long as TELEKOM's exclusive right to "switch" speech for third parties is not violated.

In July 1992, the Federal Minister of Posts and Telecommunications has further delineated the scope of the Telephone Service Monopoly in administrative rules which were adopted following a public comment procedure [12]. The administrative rules, which also apply to the privatiedDeutsche TELEKOM AG, contain principles for defining technical and operational functions that may be discharged by TELEKOM under its monopoly as well as principles for fair and equal access to telephone network transmission lines. In particular, the administrative rules contain detailed definitions of the terms "switching", "speech" and "third parties", thus establishing the regulatory boundary lines between the Telephone Service Monopoly and the competitive sector of voice services which do not constitute "voice telephony".

VII - TELECOMMUNICATIONS INSTALLATIONS REQUIRING A LICENCE

The statutorily defined monopolies (i.e. the Network Monopoly, the monopoly to establish and to operate radio installations and the Telephone Service Monopoly) are not absolutes. Rather, the Federal Minister of Posts and Telecommunications may, at his discretion, grant a licence to install and to operate individual telecommunications installations insofar as exclusive rights exist. Such licences may be granted as individual licences (Einzelgenehmigung) or as general licences (Allgemeingenehmigung). Licences which have been issued under the amended Telecommunications Installations Act of 1989 include licences for the operation of:

- "Corporate networks";
- mobile radio communications systems;
- radio paging systems; and
- satellite networks.

Certain telecommunications installations are governed by statutory exemptions and therefore do not require a licence (see below).

7.1. Corporate Networks and Other Closed User Groups

According to the definition of voice telephony, under the German Telecommunications Act, as the "switching of speech for third parties", the provision of voice telephone services would be within the scope of the Telephone Service Monopoly even in those cases where the "third parties" are part of one single corporate network or of some other closed user group. This sweeping definition of voice telephony would not be in accordance with the European Commission's Directive of June 28, 1990 on competition in the markets for telecommunications services (90/388/EEC) [13], which defines voice telephony as the commercial provisison for the public of the direct transport and switching of speech in real time between public switched network termination points, enabling any user to use equipment connected to such a network termination point in order to communicate with another termination point.

In order to avoid violations of European telecommunications law, the Ministry adopted, in January 1993, a "Licencing Concept for Corporate Networks" consisting of

- a general licence for "corporate networks", i.e. telecommunications installations which serve the purpose to switch voice between consolidated enterprises

and

- administrative rules which restrict the Ministry's administrative discretion to deny licenses for the provision of voice telephone services for closed user groups [14].

Under the general licence, consolidated enterprises are defined as

- enterprises which are party to a control agreement within the meaning of the German Stock Corporation Act or to a corresponding contractual arrangement;
- enterprises integrated with each other within the meaning of the Stock Corporation Act or by virtue of a corresponding contractual agreement and
- enterprises one of which is majority-owned and the other holds a majority stake in it.

Such consolidated enterprises may establish "corporate networks" via leased lines or use the services of a corporate network service provider for their corporate voice telephony needs, without having to apply for an individual licence.

For the provision of voice telephone service to other closed user groups, an individual licence is required. The "Licencing Concept for Corporate Networks" sets forth the criteria for the granting of such licences. In an attempt to achieve conformity with the European Commission's Directive 90/388/EEC, the Licencing Concept establishes that a licence for the provision of voice telephony shall be granted for the switching of voice for third parties when it is done either on a non-commercial basis, i.e. not for payment, or not for the "public", i.e. for a closed user group. The Licencing Concept defines a closed user group as a group that is characterised by its members being linked by permanent company law or contractual relations either with one another or in each case with at least one and the same member of the group. The relationships between the members of the group may not serve exclusively or predominantly to switch voice for third parties. Whether or not this definition covers all of the situations which, under European telecommunications law, can be considered as voice telephony to groups other than the "public", is an open question.

7. 2. Mobile Radio Communications

As part of the 1989 reform, the provision of mobile radio communications services was liberalised. After a public bidding procedure, Mannesmann Mobilfunk GmbH was licensed in February 1990 to set up and to operate a nation-wide digital cellular mobile radiocommunications network - the "D 2 network". This D 2 network competes with the D 1 network which is operated by DeTeMobil GmbH, a subsidiary of TELEKOM. In order to ensure fair competition between the dominant TELEKOM and its competitor, the Ministry has adopted administrative rules and licensing conditions for DeTeMobil regarding the provision of leased lines of TELEKOM to the D 2 operator, establishing interconnection conditions and prohibiting cross-subsidisation of D 1 services by monopoly revenues [15].

Both TELEKOM and Mannesmann Mobilfunk GmbH are obliged to permit service providers to market D 1 and D 2 services, respectively, and to develop and to offer additional services. Service providers shall not be bound exclusively to one D network operator or for an unreasonable period of time [16].

The tariff structure and business conditions of the D 1 and D 2 operators vis-à-vis their service providers are subject to prior approval by the Ministry.

In 1993, following an open and competitive tender procedure, the BMPT granted a licence for the provision of a digital cellular mobile radio network based on the DCS 1800 standard (the "E 1 network") to E Plus GmbH. The E 1 operator has the right to set up and to operate the nation-wide E1 network. As with the D 2 operator, the E 1 licencee is obligated to achieve a certain coverage within five years of the award of the licence. The licencing conditions of the E 1 are laregly identical to those of the D 1 and D 2 licence.

In 1994, the BMPT granted a licence for a nation-wide mobile data network. As in the case of the E 1 licence, TELEKOM and its subsidiaries were excluded from the bidding procedure.

7. 3. Radio Paging

The provision of radio paging services over trunk radio networks (Bündelfunknetze) has been opened to competition on the basis of a newly established licensing scheme.

The Ministry has published guidelines for the granting of licences to set up and to operate trunk radio networks [17]. They provide, inter alia, that licences shall be granted for regional networks in regions selected by the Ministry (Type A licence), for regional networks in regions selected by an applicant (Type B licence), for trunk radio systems limited to local premises (Type C licence) and for nation-wide trunk radio systems (Type D licence).

Type A licences are granted upon public bidding procedures; Type B and Type C licences will be granted upon application provided that the licensing conditions are met. Type D licences will be granted on the basis of a public bidding process. TELEKOM is allowed to compete with all licensed private trunk radio system operators.

Following a public bidding procedure, the Ministry has granted, in April 1991, six trunk radio licences for the regions of Berlin, Chemnitz, Dresden, Frankfurt/Main and Leipzig. More than 34 additional regional and local licences have been granted and further licences are granted upon application, provided that the necessary frequencies are available.

VIII - SATELLITE LICENCES

With regard to satellite systems, the German regulatory regime is the most liberal in the European Community.

The 1989 Telecommunications Installations Act has established a dual regulatory scheme which distinguishes between satellite installations intended for the transmission of data at low bit rates and other satellite communications installations. Licences to set up and to operate low bit rate satellite communications installations must be granted, provided that the envisaged transmission does not affect radio communications. With regard to other satellite systems, the granting of a licence lies within the discretion of the Federal Ministry of Posts and Telecommunications. In a policy statement, the Minister has indicated that in the case of "other" satellite systems (i.e. systems which are designated for services other than data communication at low bit rates), the permits to be granted to private enterprises will be denied only if this is necessary to:

- ensure the proper handling of the radio communications services (radio engineering requirements, frequency management); or
- prevent the substitution of voice communication in the terrestrial network.

The purpose of the statutory distinction between the two types of satellite communications systems and the stricter licensing requirements for "other", i.e. high bit rates satellite systems, is to provide an additional, technological barrier for the provision of voice telephony service via private satellite links.

The - now somewhat obsolete - rationale for the distinction was that mass-produced speech and coding equipment was not available for bit rates below 16 kbit/s and that a legal right to obtain licences for satellite systems at lower bit rates would therefore not risk infringements upon the Federal government's voice telephone monopoly.

In order to simplify the procedure for the granting of permits to private operators of satellite earth stations and to define their rights and obligations, the Federal Ministry of Posts and Telecommunications has published "Key Elements Concerning the Handling of the Licensing Procedure", a "Standard Licence" and certain guidelines concerning the licensing procedure and the information to be submitted by applicants [18].

This regulatory framework provides two options for the granting of rights to private operators. A private operator may be granted the right to provide satellite services by way of a licence for the setting up and operation of a private satellite system and, subsequently, by individual authorisations for specific satellite earth stations within the framework of this licence.

Alternatively, a private operator may be granted an individual authorisation for the establishment and operation of satellite earth stations without a prior licence.

The purpose of the first option is to enable the development of private satellite networks in which satellite services may be provided either for private use or to the public. The licensee, who is considered as a telecommunications service provider in competition with TELEKOM, is free to use his satellite system partly or wholly for his own telecommunications purposes within the framework of the licence and its conditions. For the establishment and operation of satellite earth stations, the licensee needs an individual authorisation for each station. This authorisation, which is also issued by the BMPT, is designated to ensure the radio engineering compatibility with regard to the specific location of the individual earth station.

If a private operator does not intend to become a satellite service provider, but rather wishes to establish one-directional point-to-point connections or closed networks for his own telecommunications purposes, a simplified (and cheaper) authorisation procedure can be chosen. Instead of a licence and subsequent individual authorisations for satellite earth stations, the applicant is granted only individual authorisations for the establishment and operation of private satellite earth stations.

For the licensing procedure concerning private satellite networks, the Federal Ministry of Posts and Telecommunications has announced, inter alia, the following key elements:

- the number of licences to be granted is not limited, provided that there are no obstacles from the radio engineering standpoint;
- there are no filing deadlines for licence applications. Licence applications will be handled in the order of their filing;
- licences will be granted to those applicants complying with the conditions set by the licensor. There will be no competitive selection process among licence applicants;
- foreign and domestic enterprises are free to apply for licences on a non-discriminatory basis.

An applicant may apply for a licence:

- for satellite services for private use; and/or
- for satellite services for public use.

Each of these types of licence may be granted:

- for satellite networks for data communications at low bit rate; or
- for "other" satellite networks.

Consequently, four types of licence exist. They differ with regard to the licensing conditions. "Private uses" within the meaning of the standard licence are not only uses that are handled within the licensed enterprise. Rather, they also include such uses as are realised by enterprises which are connected with the licensee within the meaning of the Stock Corporation Act or by enterprises having a specific contractual relationship with the licensee (e.g. subcontracting agreements in the manufacturing industry). In the case of a licence to provide services for public use, the licensee is subject, inter alia, to certain "conduct rules", including the obligation to notify the Minister of the tariffs, to make the satellite available to all interested parties and to comply with certain notification requirements.

The distinction between satellite networks for data communication at low bit rates and "other" satellite networks is also reflected in the various types of licensing procedures and in the licensing conditions. The application procedure for a licence for the establishment of a satellite system providing data communications at low bit rates is less cumbersome than the application procedure with regard to "other" satellite services. In the latter case, the applicant has to describe in detail his proposed satellite network. The licence will, as a rule, be granted only for services that do not include voice telephone service, unless the applicant can establish that a service incorporating voice telephony, which he intends to offer, will not undermine the Telephone Service Monopoly of DBP TELEKOM.

IX - STATUTORY EXEMPTIONS

Certain types of telecommunications installations may be set up and operated without a licence even if they fall under the network and voice telephone monopolies.

These exempted telecommunications installations (genehmigungsfreie Fernmeldeanlagen, Sect. 3 FernmAG) include:

- telecommunications installations which are intended exclusively for the internal use of certain public authorities;
- telecommunications installations which are used by transport entities (such as the German Federal Railway) on their lines exclusively for their service operations; and
- telecommunications installations
- within the limits of the same premises,
- between several premises belonging to one and the same owner or united to form one undertaking, provided that the linear distance between the premises does not exceed 25 km and provided that these installations are exclusively intended for handling gratuitous traffic related to the use of the premises.

The Federal Ministry's administrative rules provide that the term "premise" will be construed to designate a specific real estate area recorded in the public land register, or, as the case may be, a piece of land which serves a specific economic purpose. This interpretation of Sect. 3 FernmAG is expected to enable private operators to establish and to operate networks in "smart buildings", airports, exhibition centres etc. without licensing requirements.

X - THE FREEDOM TO PROVIDE TELECOMMUNICATIONS SERVICES

The provision of telecommunications services other than voice telephony is permissible without a licence. Under the FernmAG, everyone has a statutory right to provide telecommunications services for third parties over fixed or switched connections to be made available by TELEKOM. Private telecommunications service providers are merely obliged to give notice in writing of the opening of operations as well as any modifications in or the cessation of such operation to the Federal Ministry of Posts and Telecommunications within one month. The Minister publishes the notifications in his Official Gazette.

Private telecommunications service providers as well as TELEKOM are subject to stringent data protection requirements which govern the collection, processing and use of personal data of customers and other users of their respective telecommunications services.

The applicable data protection ordinances [19] anticipate and - in many cases - go beyond the data protection requirements in the proposed European Community directives, which are presently under discussion [20].

The Data Protection Ordinance for Teleservice Enterprises (Teledienstunternehmen-Datenschutzverordnung) provides that telecommunications service providers shall collect, process and use "personal data" - i.e. data relating to an identified or identifiable natural or legal person - only if it is explicitly permitted by law or if the data subject explicitly consents thereto.

The ordinance specifies those data which may be stored and processed in order to:

- establish and maintain a customer relationship;
- provide a given telecommunications service;
- bill the customer for the provision of telecommunications services.

XI - TERMINAL EQUIPMENT

Under German telecommunications law, everyone has a legal right to set up and to operate type-approved terminal equipment provided that the orderly telecommunications traffic is not disturbed.

Under the 1989 Telecommunications Installations Act, type-approval is subject to the condition that neither the transmission lines of TELEKOM nor any terminal equipment or persons are harmed or endangered by the connection or operation of the equipment to be approved. Depending on the use of the individual equipment, the technical and operational conditions for the functioning of the telecommunications services concerned must be met and, particularly in the operation of radio installations, any avoidable interference to others or by others must be precluded. To the extent necessary for the avoidance of hazards to public telecommunications traffic, private terminal equipment may be set up, modified and maintained only by individuals who have been approved to render such services on account of their technical and professional expertise.

The administrative entity in charge of type-approval is the Federal Office for Telecommunications Approvals (Bundesamt für Zulassungen in der Telekommunikation, "BZT"). The type-approval procedure is governed by the Telecommunications Type Approval Ordinance (Telekommunikationszulassungsverordnung) of March 1991 [21]. This ordinance applies to telecommunications equipment, i.e. radio installations and telecommunications terminal equipment (Telekommunikationsendeinrichtungen). Terminal equipment is defined as equipment which is suitable to be connected to transmission lines, fixed lines or switched lines of TELEKOM. Provided that certain mandatory requirements are fulfilled, type approval must be granted. The mandatory requirements are:

- user safety;
- safety of the personnel of the network operator;
- electromagnetic compatibility;
- no harm to the network;
- economic usage of the frequency spectrum;
- interoperability of terminal equipment with transmission lines, fixed lines and switched lines of TELEKOM; and
- interoperability of terminal equipment with other pieces of terminal equipment in the case of the voice telephone service and mandatory services.

Type-approval may be granted in the form of general approval (Allgemeinzulassung) or individual approval (Einzelzulassung). General type-approval is granted for telecommunications equipment intended for use in the entire Federal Republic of Germany or in large parts thereof. Individual type-approval is granted for telecommunications equipment intended for use by one or several specified persons on one or several specified locations.

Type-approval shall be applied for to the BZT in writing. The application and supporting documents shall be submitted in German;

BZT may permit exceptions with regard to specific documents. BZT is required to use its best efforts to decide about an application for type-approval within six months of receiving the complete set of supporting documents. Denials of type-approval are subject to judicial review by the administrative courts.

XII - MULTIMEDIA SERVICES BETWEEN TELECOMMUNICATIONS AND MEDIA LAW

From a technological point of view, Germany is well prepared for the development of "multimedia" services, i.e. integrated and interactive telecommunications services to be provided via integrated telecommunications networks: TELEKOM's ISDN has almost achieved nation-wide coverage; the use of optical fibre in long-distance connections is more pervasive than in any other telecommunications network; "fibre to the home" is being studied in seven large-scale field trials. As from 1995, TELEKOM intends to offer fixed and switched broadband ISDN connections which will allow numerous multimedia services.

From a regulatory point of view, "multimedia" services will require a major modernisation of the German regulatory framework and, in particular, the definition of regulatory "interfaces" between media and broadcasting law, which is largely a matter of State (Laender) legislation and Federal telecommunications law.

CONCLUSION

German telecommunications law is increasingly influenced by - and has served, in several cases, as a model for - EC legislation in the telecommunications field. The legislative attempt to restrict the formerly far-reaching monopoly rights of Deutsche Bundespost will continue to require numerous regulatory measures in order to define and re-define the scope of monopoly rights and obligations and to ensure fair competition between the dominant TELEKOM and private enterprises, which are increasingly entering the telecommunications markets. For the years following TELEKOM's privatization, no further liberalisation is expected until on December 31, 1997, the European Union will abolish the telephone service monopoly.

REFERENCES

1 For a summary cf. Joachim Scherer, Telecommunications law in Germany, in: Joachim Scherer (ed.), Telecommunications Laws in Europe, Kluwer Publ. 1992, pp. 77.

2 For a discussion of political, economic and social aspects of Postal Reform II see the contributions in Brigitte Bauer/Karl-Heinz Neumann (eds.), Privatisierung der Telekom - Zielsetzung und Bedeutung für den Standort Deutschland, Wissenschaftliches Institut für Kommunikationsdienste 1994; for an assessment of regulatory and legal aspects see Joachim Scherer, Postreform II: Privatisierung ohne Liberalisierung, in: Computer und Recht 6 (1994), pp. 410 seq.

3 Gesetz zur Neuordnung des Postwesens und der Telekommunikation, to be published in the Federal Gazette.

4 Gesetz über die Errichtung einer Bundesanstalt für Post und Telekommunikation Deutsche Bundespost, BGBl. I 1994, w.o.

5 Gesetz zur Umwandlung der Unternehmen der Deutschen Bundespost in die Rechtsform der Aktiengesellschaft, w.o.

6 Gesetz zur Änderung des Gesetzes über Fernmeldeanlagen, w.o.; Gesetz über die Regulierung der Telekommunikation und des Postwesens, BGBl. I 1994.

7 Reference to "TELEKOM" indicates that a given legislative or regulatory provision will remain unaffected by the privatization.

8 BGBl. I 1989, p. 1455.

9 As amended by Ordinance of September 16, 1992; BGBl. I 1992, 1717.

10 Verordnung zur Regelung der Pflichtleistungen der Deutschen Bundespost TELEKOM vom 16.9.1992 BGBl. I 1992, 1614.

11 Verwaltungsvorschrift zur Konkretisierung der auf die Deutsche Bundespost TELEKOM weiterübertragenen Befugnis zur Ausübung des Netzmonopols vom 19.9.1991, Amtsblatt BMPT 1991, pp, 2267, last amended by Dritte Verwaltungsvorschrift zur Konkretisierung der auf die Deutsche Bundespost TELEKOM weiterübertragenen Befugnis zur Ausübung des Netzmonopols des Bundes vom 20.10.1993, Amtsblatt BMPT 1993, pp. 523.

12 Verwaltungsvorschrift zur Konkretisierung der auf die Deutsche Bundespost weiterübertragenen Befugnis des Bundes zur Ausübung des Telefondienstmonopols des Bundes vom 13.7.1992, Amtsblatt BMPT 16/1992, pp. 334 seq, amended in Amtsblatt BMPT 5/1993, pp. 99.

13 O.J.L. 192/10.

14 Genehmigungskonzept Corporate Networks, Amtsblatt BMPT 1/1993, pp. 3 seq; Verfügung 8/1993 zur Herstellungder Übereinstimmung zischen deutschem und europäischem Telekommunikationsrecht; Verfügung 9/1993 Allgemeingenehmigung für das Betreiben von Fernmeldeanlagen zum Zwecke der Vermittlung von Sprache für zusammengefaßte Unternehmen, Amtsblatt BMPT 2/1993, pp. 26 seq.

15 Vorläufige Lizenzen Mobilfunkdienste DeTeMobil, Amtsblatt BMPT 1/1994, pp.3 seq. with attachments.

16 For a summary of the regulatory framework regarding mobile telecommunications systems see Federal Ministry of Posts and Telecommunications (ed.), Licensing and regulation in mobile and satellite communications, Bonn, June 1992.

17 Regulierung zum Bündelfunk, Amtsblatt BMPT 60/1991, pp. 2224 seq.

18 For a summary see Federal Ministry of Posts and Telecommunications (ed.), Licensing and regulation in mobile and satellite communications, Bonn, June 1992, pp. 23 seq.

19 Verordnung über den Datenschutz für Unternehmen, die Telekommunikationsdienstleistungen erbringen vom 18. 12. 1991, BGBl. I 2337; Verordnung über den Datenschutz bei Dienstleistungen der Deutschen Bundespost TELEKOM vom 24. Juni 1991, BGBl. I 1991, 1390.

20 Amended proposal for a European Parliament and Council Directive concerning the protection of proposal data and privacy in the context of digital telecommunications networks, in particular the integrated services digital network (ISDN) and digital mobile networks, COM (94) 128 final - COD 288, 13.6.94.

21 BGBl. I 1991, S. 756.

TELECOMMUNICATIONS AND SPACE JOURNAL, Volume 2, 1995 ©

TOWARD FRENCH ELECTRONIC HIGHWAYS THE NEW LEGAL STATUS OF DATA TRANSMISSIONS IN FRANCE

LUCIEN RAPP

Professor, Director, Institut d'Etudes Internationales, Université de Toulouse I, Place Anatole France, F-31042 Toulouse, France. Tel. (33). 6163-3637 ; Fax. (33). 6175-9415. Serra Michaud & Associés, 2 rue de la Baume 75008 Paris., France. Tel. (33) 1 4421-9797 ; Fax : (33) 1 4289-5790.

Abstract *: Describes the new legal framework of data transmissions in France and in Europe which was adopted in 1993 in the context of the development of international electronic superhighways.*

Key Words *: Data transmissions, Policy, Legal Framework, ONP, Value Added Networks, Policy, European, Electronic Highways, Policy, SwitchedLeased Lines, Bearer Services, Policy.*

INTRODUCTION

During the course of 1993, France has gradually been equipped with a new legal system for data transmissions, characterized by a discernable relaxation in the rules which traditionally applied and by the opening-up of a market which was until now monopolistic, to competition. The concerns of the French forerunners in the field of telecommunications infrastructures are akin to those of numerous other governments which endeavour to promote the setting-up of vast data transmissions insfrastructures. We shall refer to them from now on by an expression made popular by the American Vice President Gore and President Clinton during the last American Presidential election campaign as "electronic highways".

In addition to the French and American governments, the European

Community itself has just decided to set up a trans-European network project in the field of telecommunications. Such network shall be used principally for data transmissions. It has already formed the object of a first release of European budgetary funds (67 billion ECUs). It is in line with the legal provisions of the entirely new Section XII of the Maastricht Treaty which recently entered into force.

It thus appears that all the elements of a real international data transmission network are in place. Despite the peculiarities of national regulations or of States' legal systems and in spite of their diverse industrial interests including those of their national operators, a consensus exists in favour of the development of infrastructures and of data transmission services which will have to be coordinated and interconnected one day in line with the pressure from users' needs.

France has thus recently embarked on this road.

The present article deals with the recent evolution of the rules applicable in France. We recall that despite the numerous improvements that we may still aspire to, France has been equipped over the last number of years with relatively modern telecommunication regulations and it was notably one of the first European countries to adapt European regulations to the entirety of its provisions. Its case may thus be used as a significant example of the new European regulatory system in force, with all its weaknesses and its strengths.

I - DATA TRANSMISSIONS AROUND THE WORLD : AN OVERVIEW OF THE MOST SIGNIFICANT RECENT EVOLUTIONS IN THE MARKET, INDUSTRY AND REGULATION

Data transmissions represent an increasing part of the world's telecommunications traffic. According to confirmed statistics [1] they represent today one fifth of the profits generated by the telecommunications market and their potential development may be very important in view of the fact that the returns which they generate are already growing at six times the rate of those emanating from simple telephonic communications. They thus contain considerable potential for growth in all their forms and they should absorb a major part of telecommunications traffic of most network operators before the end of this century.

It is also true that the telecommunications field has undergone a new revolution, the multimedia revolution, after the one which, a few of years ago, produced telematics, which is the merging of telecommunications and computer science. Multimedia is itself the product of the merging of telecommunications, computer science and audiovisual communication under the combined influence of the development of network digitilization techniques and data compression. It is therefore significant and without doubt premonitory, that over the last number of years, the number of computers connected to a network has increased in a most striking manner. In the United States, in 1989, they represented less than 10%, whereas in 1993 they represent over 60%. The increasingly important number of computers sold on the market every year, the correlative decrease in their sale price, their ongoing technical perfection must surely make up the rest...

It is thus not surprising today that the interest of all telecommunication or television network operators, from public operators to cable companies is focused on data transmissions and that it is the object of everyone's covetousness in the United States and in Europe.

1.1. In the United States of America

In the United States more than seventy agreements (repurchases, mergers, or joint-ventures) have been made in relation to the control of data transmission networks. This figure takes on a symbolic value if we add that it relates only to operations carried out within the last three years throughout the United States. It is incontestable that the multiplication of such agreements, which involve considerable sums, has given rise to a new industry whose existence was inconceivable only ten years ago.

Amongst said agreements, how can one fail to highlight the official announcement on 15 October 1993 of the merger of one of the most dynamic regional telephone companies, Bell Atlantic [2], with the leading American cable-operator TCI (Tele-Communications Inc.). Beyond the comments which it provoked, this operation [3] will remain without doubt the biggest plan of merger in the United States industrial history and probably in the world's industrial history. The figures involved are truly phenomenal : 33 billion dollars for the operation itself according to initial estimates, over 60 billion dollars of assets for the group stemming from the merger.

This is why among many others reasons, this merger recently failed. However these figures are in line with the very lively competition engaged in by the telephone companies and cable network operators in the United States regarding the control of tomorrow's television digitalized, interactive and "à la carte" [4].

A lot of emphasis was placed on the inadequacies of the regulatory framework of telecommunication activities in the United States which, because it deprives the RBOC's of access to the long distance communications market and restricts them to seeking the most profitable markets on American soil [4 bis]. This regulatory framework may be modified before long. Over the past months, the inadaptability of the provisins of the Cable Act to the present conditions of the American communications market has been frequently criticized. Such provisions have been relaxed by the courts themselves and again recently in August 1993 by a federal judge of Virginia for authorizing Baby Bell to offer their video services in their zones of activity [5]. The above will undoubtedly lead to a discernable change in the applicable rules in the United States which indeed has already begun. However, it is equally important to emphasise that this evolution in the rules can only be profound as it involves an important change in the regulations to adapt it to a new age. This new age is that of the transition from the classic and graded telephone and television networks over to that of data transmissions networks and services. Such networks will use "new communication machines" [6] which have probably nothing in common with the terminals of the preceding generation. They open the door to a new world which has often been described over the previous months, as the bearer of values and of a system of social relations until now unknown.

It is true that it may appear surprising to allude to the end of an age at a time when telecommunications and television enterprises are attaining the highest ever turnover in their history, in the region of billion of dollars for certain amongst them. However, such change is in line, in an apparently irremediable manner, with the logic of the previously described changes, which leads to the progressive substitution of the classic telecommunication and television networks by data transmission networks in high-performance computers which have become increasingly numerous and less costly and isolated [7] . With continious digitalization, as predicted four years ago by Pr Richard Solomon, of the well-known Media Lab of M.I.T. [7], the public telephone network PSTN, [itself] will transform into a vast processor".

The accelerated evolution over the last months, of the needs, the techniques, the enterprise' strategies and of the regulations themselves are not far from proving him right.

1.2. In Europe

In Europe, and particularly in France, which represents one of the most active videotex services markets in the world [8], such an evolution is equally consistent with the previous evolutions. It is particularly consistent with the logic of the development of the services market. With more than 23,000 (23,227 in 1993) access codes to the services in working order (an annual turnover in excess of 6 billion French francs (6,7 billion French Francs in 1993), total traffic in the region of 89 million consultation hours per year, it is difficult to find better. It is true that the videotex system is only offered through a fleet of terminals amongst the most important of which we can identify today, if we relate this fleet of terminals to the number of inhabitants, more than 6,5 million minitels. It is also based on one of the most widespread national data transmission networks, the Transpac network, by means of packet connection and on an original system of tariffing, the system known as "kiosk" [9].

The success of this system appears to be guaranteed for the years to come as it will undoubtedly be accentuated by that of the audiotex and will gradually be internationalised [10].

Will it be sufficient to meet the data transmissions needs which will inevitably manifest themselves more numerously and more importantly. How can we otherwise interpret the manoeuvres which are taking place in Europe and particularly in France in the field of mobile communications or for the control of leased networks. All that in the aftermath of the opening-up of data transmission services to competition in all States of the European Community [11] and despite the fact that the recommendations of the "leased lines" Directive [12] have already been adapted into their legal systems [13] .

Cofira, holding of the Compagnie Générale des Eaux is authorised to set up its own specialized lines for the transmission of the voice and especially data on its radio-communication networks [14]. It decided to immediately launch the construction of a specialized lines infrastructure for an investment of 400 million francs [15]. The public operator, France Télécom, announced at the same time a significant decrease in the price of its leased lines and for the services of its subsidiary, Transpac, effective from 1 January 1994 [16].

The Cigref, a French association which consists of 70 major data processing and telecommunications users, made public its propositions in a White Paper, dealing with "International Connections" [17]. The most dynamic French cable-operators, notably, Lyonnaise Communication, a subsidiary of Lyonnaise des Eaux, are awaiting authorisation to offer their networks, which are in the process of digitalization, to telecommunications traffic, particularly in Paris [18].

We notice immediately, that with the emergence of alternative solutions to those offered by France Telecom, the evolution in the rules gradually opens up to French users a new perspective in the choice of the technical support for their data transmissions. However, this choice is not neutral from the point of view of their development strategies. We notice also that beyond France, on the scale of European Union in particular, that veritable electronic highways are being set up similar to those that are underway in the United States in accordance with a procedure and pace which is specifically European. It does not involve competition between continents but merely arises from the same necessity which manifests itself simultaneously on both sides of the Atlantic. This necessity corresponds to the satisfaction of an identical need in accordance with methods and procedures which are in conformity with the legal, administrative, economic and commercial traditions of each State or of each group of States.

We wish to clarify such choice through the example of the recent evolution in the rules applicable to data transmissions in France, in order to describe the various elements and also in an attempt to draw up an initial appraisal.

II - DATA TRANSMISSION IN FRANCE : A NEW REGULATORY FRAMEWORK FOR THE SWITCHED LEASED LINES

Data transmissions were for a long time in France a network issue. It is only starting to become a service supplied to users. It really is a radical transformation which corresponds not only to a complete change in the approach to the problems raised by data transmissions, but also to a discernable change in the behaviour of the suppliers as well as of their clients. The network was quasi-monopolistic, while the services are now supplied in a market which is more and more open and competitive. The network was unique, while now the services are by their nature diversified. The diversification of competitive services leads in turn to an increase in networks which appears inevitable today.

2.1. Former Article L33 of the Post and Telecommunications Code

In the beginning there was a network: Essentially, the PSTN, entirely controlled by the Directorate General of Telecommunications, (DGT) i.e. by a directorate general of the minister in charge of telecommunications. The provisions of the Post and Telecommunications Code are in this regard instructive : since January 1, 1991, Article L33 which dates from the July Monarchy provides as a general principle that "no telecommunications installation may be established nor used for transmission of correspondence other than through the Minister of Post and Telecommunications or upon its authorisation". It is moreover on the legal basis of a concession lines from the State network, that Transpac, the packet connection network, operated through a subsidiary of DGT in which it initially owned 97% of its assets, came into operation in 1978. Data transmissions are carried out partly on the Transpac network and partly by way of specialized lines, i.e. lines detached from the PSTN which are made available to one or more users pursuant to a leased-maintenance system. Several specialized lines joined from end to end may form a specialized network. However, the most important characteristic of the specialized lines system is the limitation of their use "for the exchange of conversations or signals relating to the particular interests of the lessee(s)" (Article D371 of the Posts and Telecommunications Code). If we exclude the existence of private lines, which notably allow for the possibility of vocal transmissions or signals on a private network and which do not take in whole or in part from the infrastructure which constitutes the general administrative network of the P.T.T., users have the choice between two types of data transmissions networks : an unadapted network, the public network designed for the slow speed of the speaking voice (55 bps/bits/second) and specific networks which are probably more adapted but detached from the public network and which as a result are largely dependent. Such specific networks are opened to them either in the form of a concession (Transpac network) or in the form of a lease (specialized networks). The dependency of such networks vis à vis the public network is evidenced by a relatively high tariffing system and by a level of services which is singularly limited to taking account, for example, of a system of quasi-irresponsibility of the telecommunication administration for damages caused by a defective transmission of private correspondence on the public network. The offer made to users takes the form of an administrative contract, unilateral and heavily biased in favour of the administration to the extent that it contains numerous conditions over and above those required by law.

2.2 Decree n°87-775 of 24 September 1987 on Value-Added Networks

9. Decree n°87-775 of 24 September 1987 introduced a first gap in this legal system which had placed data transmissions under the quasi-exclusive control of the telecommunication administration. It is true that previously the Law of 30 September 1986 had made changes by making express reference to the existence of networks open to third parties (Article 10). The Decree of 1987 completes said provisions by introducing two important modifications to the specialized lines system :

- Firstly, a new definition, which creates the possibility of an exchange of signals on users private network with a third party designated in the leasing agreement. Article D369 provides that "A specialized line is a leased line made exclusively available to the signatory of the leasing agreement for its own needs or for the exchange of signals with a third party designated in the agreement". The designation of a third party in the leasing agreement enables the administration to limit the risks of traffic diversion, and since being designated in the leasing contract made with the administration, the third party is known thereby and may only be replaced with its agreement. This rule of designating the user appears both simple and objective, and it represents important progress when compared to the previous system of specialized lines.
- The second change involves the possibility of connecting several specialized lines in such a manner as to constitute closed private networks. Such networks may be connected at one of their extremities to a subscription line, i.e. to the public network, subject to the obligations of the user to comply with the norms of transmission and traffic applicable to the subscription line. They may be established and operational in the premises of each user upon completion of the declaration for the authorities. They may form aborescent and multipoint networks allowing for the permanent interconnection between a principal terminal installation and secondary terminal installations, the reception of all categories of messages and particularly of data for public use or directed to particular categories of corporate clients and the extension of a subscription line to an exceptional link up switch.

However, the conditions of the concession set out in a Circular of 13 August 1994 (BO PTT Document 206 T 38) by reason of the restrictions which they create, do not permit the contemplation of the exponential development of these type of networks which the Decree of 1987 prudently calls "telematic networks open to third parties" and which the professionnals in the telecommunications and computer sector baptised "value added networks", an expression in the air at that time. This is nevertheless the logic of the provisions of the Decree of 1987. From the reform of the specialized lines system, it is a matter of admitting for the first time the possibility of data transmission networks distinct from those of the administration, whether it consists of a network which it operates directly, of that which it concedes to Transpac or of the networks leased to subscribers for their exclusive use, even though such new networks still use its physical infrastructures. Article D385-2 defines them as "all computer and communication resources managed by the same legal entity, which offers third parties services associated with transmission and data processing for the most part. It results therefrom that these networks are authorised provided that they are not used exclusively for data transmissions and that transmitted data processing represents a larger than majority share of services billed to users, fixed at 85%. The networks so constituted are grouped into two categories according to their power : networks of 3.5 megabits/s and over are subject to authorization and constitute type 1 networks, the other networks whose power and capacities are inferior belong to a declaratory system and constitute type 2 networks.

The assessment of the conditions of application of the Decree of 1987 remain modest. Over 100 declaration requests have been filed since the beginning of 1988, in relation to the regularization of existing networks which were until then illegal. Four authorisations were delivered during 1990 to Sligos, SITB, Reuter and Axone.

2.3. The "L.R.T. of 29 December 1990

10. The Law of 29 December 1990, completed the changes instigated by the Decree of 1987. It is true that meanwhile the DGT first changed its name by adopting a more commercial name, "France Telecom", and it later changed its status, to become from 1 January 1991 a legally independent public operator. It is also true that the Council of Ministers of the European Community ratified the proposals of the Commission's Paper which promoted the liberalization of telecommunications networks services system, in particular, those relating to data transmissions.

By fixing new telecommunication regulations, the law of 1990 overturned the previous rules. The telematic networks open to third parties did not appear as such in the telecommunications "scene" of which it traces the contours. The law retains in effect only two types of networks : those open to third parties (Article L33-1 of the Posts & Telecommunications Code) and those reserved for a closed group of users (G.F.U.) known as "independent networks" (Article L33-2). On the other hand, it largely opens up the field of telecommunications services, amongst which one can find on the one hand bearer services (Article L34-2) and on the other hand a catch-all category (Article L34-5) in which are included services using leased lines facilities.

Finally, a discernable evolution of the rules applicable to data transmissions in France, has emerged under the guise of an apparently limited reform of the provisions of the French Code of Posts and Telecommunications :

- Firstly, because the existence of independent networks, in the sense of physical infrastructures as distinct from those of the public network are recognized, even if such networks are not by definition open to the public and remain reserved to a closed group of users. However, as this concept was not defined by the law of 1990, it lends itself to a case-by-case interpretation which is all the more flexible since no general definition exists. The"3RP" or more recently "2RC" networks and more importantly, the "VSAT" networks open up the possibility of channels of communication, particularly in relation to data transmissions, alternative to those of the public network. It is undoubtedly not by chance that a decision of 18 November 1993 authorised, subject to certain conditions, the occasional connection of telephonic communications between a PSTN subscriber's set and terminal equipment of an independent radio-electrical network for private use of the terrestrial mobile service.
- But especially because data transmissions ceased to be exclusively the preserve of networks and became an entirely separate service which had as an immediate consequence the placing of the legal system on competitive and no longer monopolistic grounds.

2.4. "Switched Leased Lines" and "Bearer Services"

On the basis of the provisions of the new Articles of the Post and Telecommunications Code cited above, two texts were adopted which placed data transmissions on the road to liberalization.

(i). The Decree n°92-286 of 27 March 1992 in relation to telecommunication services which emerged from Article L34-5 of the Post and Telecommunications Code, introduced four important changes in the system of value-added services :

- a simplication of applications submitted to the DRG within the framework of authorization or declaration procedures which are henceforth open not only to legal entities but also to individuals ;
- a relaxation of the authorization procedure which may be tacit, following upon the silence of the administration during a period of four months after the date of filing of the application. Moreover, the threshold beyond which such procedure is required has been increased by a Ministerial Order of 27 March 1992, from 3.5 megabits to 5 megabits ;

- the abolition of the quantitive threshold initially provided for by Article D385-4 of the Post and Telecommunications Code and which required that 85% of the global annual turnover of the services proposed to the public should come from the operation of value-added services and that the transmission service could represent at most 15% of such global annual turnover ;
- the disappearance of two previous restrictions which could be inconvenient in the operation of value-added services : that of joining a specialized line to both ends of the public switched network and that of overcharging for a leased line used by a person other than the co-contracting party.

As an illustration of the opening effected by the Decree of 27 March 1992, we will compare the definition quoted above and contained in Article D369 of the Post and Telecommunications Code of what we called "Switched Leased Lines" to that which the new Article R11-1 henceforth proposes for what we call "leased lines". "Leased lines mean the making available by the public operator within the framework of a leasing contract, of a transmission capacity between two fixed termination points of the public network for the benefit of a user, to the exclusion of all controlled commutation by the said user". There is no longer any reference to the exclusive needs of the title holder, nor to a third party designated in the contract. The generic term, users, which is undoubtedly more suitable, covers one and all; a single precaution relating to commutation which remains the key to all networks open to the public competing with the national operator, France Telecom.

(ii). The Decree of 30 December 1992 provides for the legal rules governing bearer-services referred to in Article L34-2 of the Post and Telecommunications Code, effective from 1 January 1993, the date from which such services were no longer exclusively provided directly by France Telecom or through its subsidiaries, particularly Transpac, but may be provided to the public by other service providers, subject to an authorisation request to DRG and within the limits set forth in the schedule of conditions annexed to each authorization decision. The result thereof is that the simple resale of leased line capacities to the public operator is no longer forbidden, which represents, were it not the application of a European recommendation, one of the most unexpected changes in the French regulatory framework only ten years ago. It is noteworthy that the bearer-services may themselves be interconnected either upon the initiative of the services supplier or upon that of potential clients, as a result of the Ministerial Order of 30 December 1992, which authorizes the construction of veritable infrastructures for data transmissions, parallel to those used by the companies of the France Telecom Group. The Decree of 28 July 1993 which incorporated into French law the recommendations of the European Council's Directive relating to the supply of an open network (ONP) to leased lines has extended this regulatory change by providing the conditions in which the French public operator may supply its leased lines, in the manner of its European counterparts : the obligation to make public, subject to the control of the guardianship authority, information concerning its offer of leased lines and consequently to keep to it, the strict application of the principles of objectivity, transparency and non-discrimination, the limits relating to restrictions of access and use, the necessary respect for equality of treatment of the public operator which provides services in the competitive market on such leased lines and for other users, providers of the same bearer or other similar services, recommendation of a cost-oriented tariff policy.

This Decree, which was swiftly completed by three orders of the same date, is henceforth in force in France in all its provisions and through said provisions are also in force the provisions of the above cited Directive of 5 June 1992 which it completely incorporated into French law.

2.5. Legal consequences upon contracts and agreements providing services to consumers

Such regulatory changes have undoubtedly caused a parallel evolution which is certainly slower but nonetheless already noticeable, in the contractual conditions concerning the offer of leased lines. We reason more in terms of service-orientated offers, which represent a veritable revolution of contractual ways which were until now dominated by a certain lack of knowledge, more or less voluntary, of the users needs. It is worth recalling on this subject the conditions in which the offer of leased lines are currently proposed. The agreements define the obligations of the client and are more circumspect in relation to those of the suppliers of infrastructure. These obligations are in line, in all cases, within the framework of an obligation of means which does not have any real significance unless we consider that national suppliers of infrastructure cannot make obligations except in relation to the semi-circuits they control. The events of force majeure are of course defined in the agreement in wider terms than those admitted by the court decisions or even by law, which leads or may lead to complete exoneration from liabilily. The practice of mandates granted to the public operator is frequent ; it is in reality a simple management facility given without guarantee of any results. The system of contractual penalties consists of discounting prices and such penalities are only applicable where the rate of availability is not reached ; provided that such rate is correctly provided for in the agreement. So many elements which still attest today to the limited number of guarantees given to the client and to a certain extent to the little importance attached to its needs. The changes which are taking place are fortunately orientated henceforth in the opposite direction.

2.6. The Future : Toward French Electronic Highways

The question which is posed today in relation to this first evolutionary phase of French legal texts is to know whether the next stage is not the complete relaxation of the value-added services system and simultaneously the pure and simple suppression of the bearer-services system.

The success of the Teletel services on the PSTN is due to a large part to its legal system. This system is dominated by the principle of free supply of Teletel services. It confirms the possibility of a real development in value-added services under a liberal system with which they may be endowed. It also demonstrates the difficulty of maintaining for too long within the same category of services legal barriers which today have become more and more artificial.

With respect to bearer services, it is clear that the law of 29 December 1990 introduced into French telecommunications law this concept without succeeding in defining it in an operational manner. Article L32.9° of the Post and Telecommunications Code provides that bearer services mean, "the commercial use of simple data transmissions, i.e. services whose object is either to transmit or to transmit and convey signals between the termination points of a telecommunications network, without submitting these signals to treatments other than those necessary for their transmission, their conveyance and control of their functions". It is probably because it does not cover a sufficiently homogeneous category of services. The bearer services correspond particularly to the offer of specialized lines whose regulation is henceforth fixed by Article L34-5° of the Post and Telecommunication Code and by the regulatory provisions cited above although not exclusively. Again we may talk of the " catch-all category", as far as it is concerned. Has the moment not arrived to take note of the consequences of an experience which, it must be acknowledged, has not produced the results one expected. How many bearer services have they authorized other than those proposed by companies from the France Telecom Group ? A single one was recently granted. How much longer will this system of authorization last ? May one believe today that the abolition of the Decree of 30 December 1992 and the implementation of real competition for this type of services could serve as the beginning of a positive response to this observation, particularly correct, of the Dandelot Report of the Summer of 1993 : "The profound upheavals connected with the evolution of technology and with the pressure on the market towards competition have had until now very little practical effects on the French telecommunications services market" ?

These questions are all the more important today when the French telecommunications market is enjoying a confirmed success after a modest start, involving communications from mobile terminals, i.e. by means of radio-electric networks. However, such networks constitute useful infrastructures for data transmissions or are likely to become so in the near future.

They may even, if we are not careful, favour the emergence of alternative networks to those of France Telecom, which will become operational when vocal traffic is completely liberalized in 1997.

The recent authorization given to COFIRA to establish its own network, at first by way of an experiment, and later very likely in a definitive manner in 1994, led to the establishment of a private network completely distinct from that of France Telecom.

This network combines the physical and technical aspects of a cable network and a Centrex type commutator, which allows for the dispatch of communications and consequently data transmissions without using the infrastructure of France Telecom's public network. It is probably one of the most remarkable decisions made over the past years which the most liberal States, particularly the United States and the Great Britain, have not wished to make until now. It is unusual that such a decision was made in a country where legal rules on telecommunications were for a very long time rigid. It remains however, that despite many very remarkable advances over the past years like the above, the system applicable in France still suffers from many inadequacies, which the government and notably the new DGPT (Directorate General of Posts and Telecommunications) shall no doubt wish to overcome. They have launched over the past months a number of important advisory sessions. These advisory sessions attest to their willingness to inform themselves completely of the needs of network operators, service suppliers and their clients before adopting the necessary reforms. They inaugurate a very new method in the French administration. And they have enjoyed an important success if we judge by the number and especially the importance of the contributions which were submitted to the government in response to such advisory sessions. It will produce, certainly within the next couple of years, new rules which will be adopted. And these rules shall very likely lead to a new evolution of the regulations applicable to data transmissions. It is at this price that we shall soon be able to speak in France and Europe, as well as the United States of "electronic highways" for data transmissions.

REFERENCES

1. Quoted by George Gilder in "Adieu au téléphone et au téléviseur". "Farewell to the Telephone and Television" Courrier International 16-22 September 1993, n°150, p. 56.
2. In accordance with consolidated business statistics, Bell Atlantic is the third regional American company (RBOC) to emerge from the dismantling in 1984 of ATT.
3. See in particular Business Week October 25, 1993, page 32, "Bell-ringer, How Bell Atlantic and TCI hooked-up and what it means for the information age".
4. For another illustration of this phenomenon we recall that Bell Soutah invested in the capital of Prime Management, the Texan cable-operator, at the level of 22.5%, Nynex invested 1.2 billion dollars to buy into the capital of Viacom, Southwestern Bell repurchased two cable-operators in Virginia, US West acquired 25,5% of the capital of Time Warner Entertainment, which is the number two cable television network in the United States, for a global investment of 2.5 billion dollars.

4bis. See in particular the point of view of Mchaël Schrage published in the Los Angeles Times of 19 August 1993 under the title "Free the Baby Bell 7 ! Let them compete - With conditions".

5. For an analysis of the regulation of television cable services in the United States, see particularly the note of Jean-François Leprince of the American Trade Commission with the French Embassy in Washington n°77, February 1993.
6. To use the expression of George Gilder, in "Adieu au téléphone et au téléviseur", article quoted above.
7. In the United States, between 1989 and 1993 the percentage of computers connected to the network changed from less than 10% to over 60% (source : George Gilder, article quoted above).

7bis. Quoted by George Gilder in "Adieu au téléphone et au téléviseur", article quoted above.

8. According to Martine Georges and Alain Vallée. "Loi du 29 décembre 1990. Analyse et commentaires. Perspectives pour les télécommunications". (Law of 29 December 1990. Analysis and Commentary. Telecommunications perspective.") IDATE. La Documentation Française. Paris 1992 p. 123.
9. The Kiosk system is a means of tariffing the videotex services which notably allows for a discernable reduction in the formation cost of the information services market. It is based on a triparite convention between the data transmittor, the supplier of services and the dealer.
10. There exists specific cross channels which allow German, Italian, Luxembourgeois, Portugese, Danish and even Finish subscribers to connect up to services available in France. The same possibilities are today offered to American or French speaking African users.
11. Since the 1st of January 1993.
12. Directive 92/44/CEE of 5 June 1992 of the Counsel regarding the application of the network supply opened to the leased lines JOCE n°L165/27 19 June 1992. In relation to this Directive, see particularly Geneviève Castanet et Jean-Paul Combenègre "Télécommunication : fourniture d'un réseau ouvert aux lignes louées". Ed. Techniques Europe, Octobre 1992. Adde : Lucien Rapp Lamy Droit de L'Informatique et des Télécommunications. Editions Lamy Paris 1994, n°1684 et s. ; Klaus Kraatz "Establishing Value-Added Network and Providing Value-Added Services in Europe "IBA Conference, Committee Cm. Cannes 1992.
13. In France the recommendations of the "Leased lines" Directive were adapted by Decree n°93-961 of 28 July 1993 and by three decisions dated the same day and which were taken for its application (JORF 29 July 1993 p.10666).
14. The French Minister of Telecommunications, Mr Gérard Longuet announced at the end of September 1993 that the Société Française de Radiotéléphone, a subsidiary of Cofira, may affect its

own specialised connections enabling it to broadcast the voice and data throughout radio-communications network instead of using those leased by France Télécom. See Arrêté of 29 March 1994. JORF 12 April 1994, p. 5387. Most recently see Arrêté of 9 May 1995 (JORF 11 May 1995 p. 7555) antlivising the french railway operator, SNCF, to open its independant network to third parties, operating telecommunications networks for providing opened to the pulic services such as the PSTN.

15. The Tribune 20 October 1993.
16. The decrease envisaged is up to 36.6% for a connection of 300 km at 64 kbits/s. The Tribune 20 October 1993.
17. White paper produced by the working group CIGREF "Liaisons Internationales" "International Connections". Cigref 1992. 56 pages.
18. The Fig-Eco 23 October 1993.

- V -

EMERGING MARKETS

LEGAL ISSUES IN TELECOMMUNICATIONS LAW IN POLAND AND THE OTHER COUNTRIES OF CENTRAL EUROPE RECENT DEVELOPMENTS

MICHAEL DAVIES,

Partner, Nabarro Nathanson, ul. Senatorska 12, 00-082 Warsaw, Poland [1].

INTRODUCTION

There is general acknowledgment that the condition of telecommunications in the four countries of Central Europe, the Czech Republic, Hungary, Poland, and Slovakia, remains a serious impediment to business in general and to exports in particular. It is an area where there is a need for a large amount of investment; some estimates put this as high as $94 billion [2]. This article will deal principally with Poland but will also refer to developments in the other countries of Central Europe, namely, the Czech Republic, Slovakia and Hungary.

A report of the OECD produced in 1991 described the telecommunications position in the countries of Central Europe and referred to the need for substantial investment. A later report produced in 1993 [3] shows that since the political changes of 1989/90 the telecommunications systems of these countries remain in a generally poor condition. This report states that "[o]n average, the density of mainlines is less than a third of and revenue per capita is barely 5 per cent of OECD levels".

According to the 1993 OECD report, in the three (now four) countries under consideration the number of mainlines per hundred of population is the following (statistical information as of 1 January 1992):

- Poland: 9.3 (population 38 million);
- Czechoslovakia (as it then was): 15.7 (population 15.7 million);
- Hungary: 10.9 (population 10.3 million).

The report then shows how many lines would have to be introduced to achieve 40 lines per hundred of population by the year 2005 [4] : 11.7 million in the case of Poland; 3.8 million in the case of the Czech Republic and Slovakia; and 3 million in the case of Hungary. The growth rate to reach this target is 20%, 12.4% and 17.6% respectively.

As is plain from these figures Poland has the biggest task ahead of it. Progress is, however, already being made. The number of mainlines per 100 people increased to 10.3 during 1992. Outside the cities, however, the position is much worse. In addition to the mainline density, the other issue of concern referred to in the OECD report, is the low level of revenue generated - less than 5% of OECD levels.

The primary reason why the telecommunication sector was not developed was due to ideological reasons; the technology of these countries therefore lagged much behind that of Western European nations.

It is recognised that a large amount of investment must take place and that the development of the sector is a high priority in the process of promoting business activities whether foreign or domestic. For this reason, all the countries under consideration have enacted legislation permitting private sector activity in the area of telecommunications.

I - POLAND

1.1. The Legal Framework

a). Introduction

The Polish Parliament adopted the Post and Telecommunications Act on November 23, 1990 [5] (the "Act"). The Act governs the operational rules in the area of post and telecommunications [6].

The Act provides that activities in the area of telecommunications can only be carried out by: a) the Polish state-owned telecommunication entity (Telekomunikacja Polska S.A.), b) para-statal agencies under the authority of the Ministry of National Defence or Internal Affairs, and c) entities granted telecommunication licences.

b). Licences

A licence is generally required for the installation and operation of telecommunication equipment, lines or networks. The Ministry of Post and Telecommunications (the "Ministry") can, however, determine that certain telecommunication equipment, lines and networks do not require a licence.

There are some restrictions contained in the Act with respect to the issue of licences. A licence may not be granted in the following circumstances:

i) for the provision of international "public service" type services;
ii) for the use of radio-telecommunication equipment designated for the transmission of radio or television programmes, by foreign entities or joint venture companies with foreign equity participation where the foreign capital exceeds 33%;
iii) for the use of telecommunication equipment, lines and networks linked directly by cable, radio or through international satellite arrangements with lines, equipment or networks outside Poland by a foreign entity or a joint venture company with foreign capital participation; or
iv) for the use of long-distance telecommunication lines and networks and for the provision of long-distance telecommunication services to a foreign entity or a joint venture company with foreign capital participation where the foreign capital exceeds 49%.

The licence will contain the following details:

i) the person authorised, plus domicile or place of residence;
ii) the object, scope and area of operations;
iii) the conditions of performance, and in particular the form in which the services are to be provided, the technical specifications, the type and size of telecommunication network, the method for interface with "public service" telecommunication networks and the main technical parameters, the place of installation of the equipment, and the allotted frequencies;
iv) the date for commencement of operations and duration of the permit.

The Ministry may refuse to issue a licence when:

i) the issue would jeopardise national security;
ii) the issue would contravene international treaties to which Poland is a party;

iii) the use of the equipment would interfere with already operating equipment or there are no technical possibilities for linking up to the "public service" telecommunication network;
iv) the applicant cannot be assigned frequencies requested by the applicant because they are essential for other operations;
v) the applicant had a licence which was rescinded within the last 5 years.

The Ministry may rescind a licence when:
i) the operations stipulated in the licence were carried out in a manner contrary to the Act;
ii) the entity holding a licence fails to eliminate the violation of the Act or of licence terms defined in an administrative decision;
iii) the entity failed to start operations within the time specified in the licence.

Licenced entities must pay annual fees for the use of telecommunication lines, equipment or networks. The Ministry, together with the Ministry of Finance, will define the amount and manner of paying the fees.

The fees are collected by the State Telecommunications Inspector (Panstwowa Inspekcja Telekomunikacji - "PIT") and the State Radio-telecommunication Agency (Panstwowa Agencja Radiokomunikacji - "PAR"). Fees collected constitute revenue for the State Treasury.

c). Homologation Certificates

Telecommunication equipment and systems installed and operated on the territory of the Republic of Poland, including telecommunication cables and lines (except for equipment or systems used by the Ministry of National Defence or Internal Affairs), must have a homologation certificate authorising their use (Swiadectwo Homologiczne). This certificate is issued by the Ministry.

A Homologation Certificate may be issued as a general or specific certificate. A general certificate is issued for a system or type of telecommunication equipment supplied by a given producer. A specific certificate is issued to a producer or operator of a specific piece of equipment or system.

The equipment or system may be installed and operated only during the period of validity of the certificate. This period will vary depending on the type and age of equipment or system used. The Ministry may rescind a certificate if there is a technical change in the telecommunication equipment or system.

d). Allocation of Frequencies

The Ministry allocates the frequencies and identification signals to entities who have applied for a licence to operate telecommunication equipment. When it would not be possible to allocate frequencies to all the applying entities, the Ministry, together with the Ministry of Culture and Arts, may determine the terms on which given frequencies should be used and may organise a tender for their allocation.

The Ministry may refuse to allocate frequencies when: (a) there are no spare frequencies available in the area in which the applying party intends to operate; or (b) it is unable to protect other frequencies from interference.

e). Control Agencies

There are certain agencies which may be authorised by the Ministry to perform functions of control and supervision. These agencies will then be empowered to: (a) supervise and examine telecommunication lines, equipment and networks; (b) observe the use of allocated frequencies, call signs and identification signals; and (c) observe their conformance with regulations and with terms and conditions specified in the licence.

Such agencies include the PIT which supervises telecommunication cable networks, lines and equipment and the PAR which supervises radiocommunication networks, lines and equipment and the use of allocated frequencies.

Personnel of these agencies have the right, upon presentation of official identification and written authorisation, to:

i) enter the premises of the supervised unit as well as buildings in which the telecommunication networks, lines and equipment are placed or are being installed;
ii) inspect, test and measure the networks, lines and equipment.

If the operation of telecommunication lines or equipment fails to conform with the regulations in force or terms of the licence or interfere with operation of other telecommunication lines or equipment, the agency can issue an administrative decision in which it sets out the nature of the violation and the deadline for its elimination.

In addition to the above, PAR also performs other functions such as:

i) the securing of frequency allocations for Poland to enable the development of domestic radiocommunication services; securing access to given positions on geostationary orbits and satellites; and international coordination in this area;
ii) control and suppression of harmful radio-electrical interference leading to reduced efficiency in use of the frequency spectrum.

f). Public service telecommunications

An entity providing telecommunication services of a "public service" type is required to set out the scope and conditions for the provision of such services in a published document [7]. Such entity may not refuse to provide public services, unless the party demanding the service has failed to comply with the terms of the document.

The owner or user of real property has the duty to enable entities rendering public telecommunication services to place on the property, within certain limits, telecommunication equipment.

The operator of a "public service" telecommunication network cannot refuse to connect another telecommunication network to its network. The terms and charges for these connections are to be defined in a contract concluded between owners of the networks.

The fees for telecommunication services are set by the network operator. The terms of settling the fee amounts due for services provided must be defined in a contract concluded between the network operators.

The Ministry may issue ceilings for fees charged for telecommunication services of a "public service" nature. The Ministry sets the rates for international telecommunication services.

g). Registration

Telecommunication networks, lines and equipment operated on the territory of Poland, with the exception of those operated by agencies under the authority of the Ministry of Defence or Internal Affairs, are required to be registered.

h). Other functions of the Post and Telecommunication Ministry

The Ministry is empowered to determine: i) the standards for the installation and operation of telecommunication systems; ii) technical and operational requirements for telecommunication equipment, lines and networks installed and operated on the territory of the Republic of Poland; and iii) conditions for interface of the telecommunication equipment, lines and networks.

The Ministry, along with other appropriate Ministries, is further empowered to set down conditions for installation of telecommunication lines along public roads, canals and waterways and in the vicinity of airfields.

1.2. Privatisation and foreign investment.

Under the communist system Poland was isolated from technological development and the authorities had no interest in providing a modern telephone system. Following the fall of the communist system, one of the aims of the Government in the area of telecommunications was to avoid domination by imported equipment. Hence, in 1991 it ordered that telephone equipment sold to the public system be 50% Polish. This led the way to the privatisation of the telephone equipment sector and the need for major foreign companies in the field to buy into existing state-owned companies.

Poland had the lowest number of telephone lines in Europe and therefore the market was, and remains, extremely attractive for foreign companies; provided that some form of market protection was provided. The Ministry of Post and Telecommunications decided to limit the number of long-term suppliers to three. The purchase of shares in privatised telephone equipment suppliers therefore became the only way to gain access to the Polish market.

During 1992 and 1993 all five major companies in the Polish telecommunications equipment supply sector were involved in privatisation whereby the majority of their shares were sold to foreign strategic investors. These transactions ensure that there is access to capital and modern technology for telecommunications equipment production and development.

The transactions in question were the following:

- the acquisition in November 1992 by AT&T of the USA of 80% of the share capital of Telfa SA [8] ; the price paid was $28 million;
- the acquisition in March 1993 by Alcatel Sesa of 80% of the share capital of PZT SA and Teletra SA; the price paid was $37 million;
- the acquisition in September 1993 by Siemens of Germany of 80% of the share capital of ZWUT SA and Elwro SA [9]. The price paid for

> both was $38.3 million. Liabilities of $30 million remained in the companies. Siemens agreed to invest $57 million and to increase the share capital of the two companies by an amount of $35 million. Siemens also provided employment guarantees for 100% and 75% respectively of the staff of the two companies for a period of 18 months.

All of these companies were joint stock companies wholly owned by the State Treasury following their transformation from state enterprises pursuant to the Privatisation Law of 13 July 1990. As a result of these transactions, three international manufacturers and suppliers of telecommunications equipment now dominate the market in Poland.

There are a number of advantages to the Polish government in addition to the proceeds of the sales themselves. The companies have been required to meet certain local content criteria laid down by the Ministry of Post and Telecommunications. No market share guarantees were given. This effectively means that the sales of the suppliers to Telekomunikacjja Polska SA will depend upon the terms of the offers submitted, e.g., with respect to price and technical specifications. The suppliers are given protection against imported goods for a period of six years.

The combined effect of the three transactions was as follows:

- 5,600 jobs kept for eighteen months
- investment commitments of $162 million over a six year period (penalties are payable if the commitments are not met)
- capital increase commitments of $70 million of cash
- $75 million earned by the State
- debts of $45 million assumed by the private sector.

As will have been clear from section II. A. above there is no longer a monopoly of telecommunications services in Poland. Telekomunikacja Polska Spólka Akcyjna is obviously the largest provider of such services but licences have been issued to private sector entities. For example, licences have been issued for local services to companies with both domestic and foreign participation. The Law does not set down any procedure for the granting of licences. In practice, there has been a tender where foreign companies have been involved.

A cellular telephone system is now operating in Poland. It started in Warsaw but is being expanded to most major cities with a view of covering the whole country within five years. The system is operated by Polska Telefonia Komórkowa Sp. z o.o., which is a joint venture between

Telekomunikacja Polska S.A. (51%), Ameritech (24.5%), and France Telekom (24.5%).

A joint venture is to be formed between Sprint, a US company, and RP Telekom, a Polish company, to develop a broad band fibre optic network. Other Polish investors may become involved and the International Finance Corporation is also considering becoming an equity investor.

Satellite connection is also possible. Komertel was set up in Poland in 1992 and is a joint venture between Telekomunikacja Polska S.A. and AT&T Network Systems International. Other private telephone systems include a digital network established by Telbank SA, a joint venture between a number of Polish banks.

As at the end of June 1994 Poland had 12.11 telephone subscribers for every 1000 persons. At the end of 1993 the figure was 11.47. Similar increases can be seen in the other forms of telecommunications, eg, Komertel international subscribers increased by 7.8% in the period from the end of 1993 to the end of June 1994; and the cellular network showed the greatest percentage increase, 65.4%, during the same period (a total of 22,400 subscribers).

II - THE CZECH REPUBLIC

2.1. The legal framework

a). Introduction

On March 12, 1992 the Czech and Slovak Federal Republic adopted amendments to Act No 110 of 1964 on Telecommunications (the "Act").

The Act defines what comprises the Unified Telecommunications Network and it provides for the role of the State in the protection of telecommunications. The Act also regulates the extent of using a third person's property for public interest and provides for confidentiality of any information gathered while providing telecommunication services.

b). The Unified Telecommunications Network

The Act establishes the Unified Telecommunications Network. It consists of long distance and local networks. The network includes transmission methods to transport all types of information by means of telecommunication facilities or equipment and equipment to provide telephone services. Excluded from the Network are radio and television.

c). Regulating Bodies

The Act permits licences to be granted "by way of exception" for operation of telecommunication services, including transmission and telephone facilities, by the appropriate state administrative telecommunication body ("Communication Organisation").

The Ministry of Telecommunications establishes the general rules for the operation and the technological capacity of telecommunication equipment. It considers the security and reliability of telecommunication equipment, ascertains whether all provisions of the Act and regulations issued to implement the Act are being followed, and whether the conditions set for the Communication Organisation are satisfied.

One of the Ministry's functions is to protect the availability, cost, speed and reliability of telephone and telegraph connections as well as the concealment of communicated data. It does this by the introduction of new technology (automation), and the increase of the number of subscriber stations and other telecommunication equipment (e.g., public call offices, public phones). The Ministry also determines the conditions for use of the Unified Telecommunication Network for telecommunication connections, creation and use of telephone and teleprinting subscriber stations, and the conditions for enabling other telecommunication equipment to be used by subscribers through the Uniform Telecommunication Network.

The Ministry also controls the use of frequencies. It namely:

i) assigns frequencies and frequency bands in consultation with the Ministry of Defense and the Ministry of Interior; ii) exercises state supervision over the assigned frequencies; iii) cooperates with the Federal Council for Radio and TV transmission. The Ministry also coordinates negotiations on international telecommunication matters and enters into international agreements.

d). Protection of Telecommunication

The Act provides for the protection of equipment and operation of the United Telecommunication Network. For example, all telecommunication equipment outside of the United Telecommunication Network (e.g., radio and television) must be established and operated without harmful interference, threat or damage to the equipment or activities of the United Network.

If the acts of a natural or legal person cause damage to or endangers or interferes with equipment of the Unified Telecommunications Network, the natural or legal person must compensate for the damage and pay a fine. An agency authorised by the Ministry investigates the sources of threat, damage or interference with the operation of the United Telecommunication Network or other telecommunication equipment.

e). The Use of Real Property for Telecommunication

For public interest reasons the Communication Organisation is authorised, at its own expense, to: (a) build and operate on a third person's property (above and underground) telecommunication lines including its supporting components; (b) enter (as reasonably necessary) other property when designing, constructing, operating, repairing, modifying or removing telecommunication equipment; (c) perform necessary lay-outs of land and plant growth, including the removal of trees, within certain limits (e.g. protected parts of nature, health factors), which are in the way of the telecommunication lines.

The Communication Organisation is obliged, prior to the execution of the above, to notify the owners or users of the property concerned. The owner or lessee of the property is entitled to adequate compensation if any of the acts mentioned above limits the use of or damages the property.

Once the Communication Organisation has completed its work, it must put the property back into its original or appropriate state. The owner or user of the property must exercise reasonable care when using and working on the property so that the telecommunication equipment is not jeopardised or damaged.

f). Telecommunication Secrecy

Natural and legal persons (as well as their employees) must keep confidential any information gathered while providing telecommunication services. Courts, prosecuting bodies and other state authorities may obtain any messages or transmitted documentation or can use the telecommunication equipment for special circumstances as specified by law.

2.2. Privatisation and foreign investment.

Plans are currently underway for the sale of approximately 26% in the national telecommunications company, SPT Telecom. This 26% will be disposed of through the distribution of vouchers. At present the proposal is to deal with the balance as follows: 4% will be kept in a restitution investment fund, and 70% may eventually be sold to foreign investors. At present 27% is to be offered to a strategic investor through a tender expected to take place in early 1995. Foreign investors who are to take part in the tender include a consortium of Deustche Telekom and Ameritech, France Telecom and Bell Atlantic, the Dutch company KPN and the Italian company Stet, and Teledenmark and British Telecom. Regulations require that the state retain 51% for a period of 5 years.

A cellular telephone system is now operational, having begun in 1991. It is operated by Eurotel, a joint venture of SPT Telekom, US West and Bell Atlantic.

III - HUNGARY

1.1. Legal framework

a). Introduction

The Hungarian Parliament adopted a Telecommunications Law on 23 November 1992 (the "Act"). Its stated purpose is to develop effective, economical, and up-to-date telecommunication services as well as to promote competition in accordance with international rules and regulations.

The Act establishes the mode and conditions of providing telecommunication services, the rights and obligations of participants (providers, operators, owners, users, and subscribers) of telecommunication services, the conditions pertinent to the planning, implementation and operation of a telecommunication network and to the use of telecommunication devices and equipment. In addition, some provisions of the Act are related to the protection of telecommunications and to the role of the State in telecommunications. The Act also provides for the freedom for anyone to establish a telecommunication network for its own purposes or for special purposes.

b). Licence for Telecommunication Services

Telecommunication services may be provided by either a company with a concession or by an economic association established by the State or with State majority equity ("telecommunication organisation") operating in accordance with a telecommunications concession. This telecommunication organisation may provide the following telecommunication services: (a) public telephone service; (b) public mobile radio telephone service; (c) nationwide public paging service; (d) nationwide and regional distribution and broadcasting of public service radio and television programmes.

In addition, the telecommunication organisation may carry out the following activities: (a) other telecommunication services; (b) marketing, sale, maintenance and repair of products for telecommunications; (c) computer application services; (d) measuring and safety services; (e) general engineering development activities; (f) telecommunications research and development; (g) organisation, maintenance of building, investment contracts; and (h) telecommunication training courses and other professional education activities.

Under this Act tenders may be invited by the Ministry of Transport, Telecommunications and Water Management ("Ministry") for concessions in telecommunications. A local government can also initiate an invitation for a concession tender concerning public telephone service for the area under its authority if it is justified by the development of the telecommunications network or by the need for the improvement of the standard of service.

Contracts for concessions must include the following:

i) quality, quantity and technical data, and the terms of co-operation with other telecommunication service providers;
ii) terms related to the concession fee;
iii) the way of establishing the service tariffs and the conditions of modifying them provided this is not stipulated by regulation;
iv) the area where the activity is carried out ;
v) the Ministry's right from time to time to review and to modify the contract under conditions as stipulated in the contract in the interest of users, or to meet the development requirements with respect to international obligations assumed in the meantime;
vi) the indication of all legal rules which determine the terms of fulfillment of the contract.

The concession contract may only be extended once by half of its original duration without the need for an invitation to tender.

c). Telecommunication Service Contracts

The Act provides for various forms of telecommunication service contracts. A network contract is one whereby the provider of services agrees to connect telecommunication networks to the public network.

Under a subscription contract the provider of services must provide the subscriber, within 30 days, a fixed or mobile network service access point. The provider of services may refuse to conclude such contract only if the subscriber owes payments to the service provider for previous telecommunication services, or if the previous subscription contract was terminated by fault of the subscriber.

A provider of service may cancel the subscription agreement upon notice to the subscriber if, among others, the subscriber uses the equipment in a manner not in compliance to what was stipulated in the contract or hinders the checking of his/her equipment. The subscriber may also cancel the subscription contract upon notice to the provider of service.

d). Authority Functions

Authority functions related to telecommunications (e.g., licensing, interconnection, modification and dismantling of telecommunication links and networks) are performed by a state agency working under the control and supervision of the Minister, the Telecommunications High Authority as its central organ, and local organs acting as regional telecommunication authorities. The duties, jurisdiction and competence of the agency is established by the Government.

e). Protection of Data and Secrets

The Act generally provides for the obligation of a provider of services to keep secret messages of a private nature he/she acquired while providing service in fulfillment of his/her duties. The obligation of data and secret protection also applies to the employees, members, or agents of the provider of service.

f). Rights Related to the Use of Real Property

The owner or user of real estate is under the obligation to allow, upon prior notification, the duly authorised representative of the public telecommunications service provider to enter the land for maintenance and fault elimination purposes. The provider may also place certain telecommunication equipment, lines, or antennas for the purpose of public telecommunication services on the owner or user's property if there are no alternatives as to its placement. The authority may establish, for public interest, an easement or other right to use a real estate.

g). Protection of Telecommunications

The Act provides for the general protection of telecommunications. This includes the obligation of the operator of an equipment (e.g., apparatus, line, vehicle, or other installation) causing interference in the operation of telecommunications apparatus to prevent or to eliminate the interference. The authority may demand the owner of the equipment causing interference to eliminate, reduce the interference or relocate the equipment. The Act further provides for the party who caused the deterioration in the quality of telecommunication services to bear the costs incurred for the reinstatement of the original service quality.

h). Functions of the State Relating to Telecommunications

Under the Telecommunications Act, the State retains the following functions:

- development of the national telecommunications policy and the conditions necessary for its implementation, the enforcement and control of governmental, security (private and public);
- control of the telecommunication market, uniformity of concession contracts, assurance of equal opportunities for new market entrants;
- provision of public telecommunication services for hearing-impaired and disabled persons;
- exercising ownership rights related to State property used by telecommunication networks;
- fulfillment of authority duties concerning telecommunications;
- economical and technical regulation and legislation with the goal of ensuring a unified national telecommunications network;
- supervision of the telecommunication service providers and of operators of telecommunication networks;
- taking the necessary actions in case of termination for unforeseeable reasons or breach of the concession contract;
- participation in international telecommunication organisations and the conclusion of telecommunication agreements, conventions and contracts;
- operation of the Telecommunication Fund;
- assurance of the operation condition for the Telecommunication Interests Reconciliatory Forum and the Telecommunication Engineering Qualifying Committee.

The above functions are to be performed by the Government and the appropriate ministries and authorities as set out in the Act. The coordination of State functions related to telecommunications rests with the Ministry of Transport, Telecommunications and Water Management.

i). Telecommunication Fund

A Telecommunications Fund ("Fund") is established for purposes of:

- covering necessary expenses deriving from the unforeseeable termination of the concession contract before its expiration or from the violation of the contract by the concession company;
- financing the operation of the Telecommunication Interests Reconciliatory Forum and the Telecommunication Engineering Qualifying Committee; and
- creating an intervention and development fund.

Sources of the Fund include: (a) concession fees due to the state budget deriving from concession contracts on telecommunication services; (b) moneys transferred for this purpose either with or without consideration to the state budget, state or private organisations, or private persons; (c) a certain fixed sum by the Government deriving from the privatisation of the state share in telecommunication; and (d) fines received for installing network, operating or selling equipment without a licence.

3.2. Privatisation and foreign investment.

A consortium of Deutsche Bundespost Telekom from Germany and Ameritech International of the U.S. (MagyarCom) won a tender for the purchase of 30.2% of the shares of the Hungarian state-owned telecommunications company, Matav, and for an operating concession. The purchase price was $875 million for both the shares and the concession. The consortium has also agreed to increase the share capital of Matav by $400 million. Both EBRD and the IFC have a small share holding in the company but the consortium has an option over the shares. The Hungarian state retains a majority interest and the majority of the board members will be Hungarian. The company has a 25 year concession for domestic, long distance and international public telephone services throughout Hungary. This is the first sale of shares in a state telecommunication company in the region.

Concessions are being awarded to other consortiums with foreign partners to run local networks. For example, concessions have been granted in Budapest to VTS, a partnership of Alcatel of France and US Telecom East. In Szeged concessions have been awarded to Compagnie Generale des Eaux. However, Matav retains its monopoly on trunk and international calls. The local concessions give the operators exclusive rights for eight years. In March 1994 awards were made following tenders for 25 regional concessions. For the grant of the concessions the state received $79.4 million in total. $193.6 million of further investment is expected to follow. Matav won 10 of the concessions.

CONCLUSION

In the countries under consideration there is general recognition that the telecommunications sector requires modernisation and investment, and that this can only be achieved through privatisation and foreign investment. Most major international telecommunication companies have targeted these countries and others in Central and Eastern Europe. According to a recent survey the ten major investors are Deutsche Bundespost Telekom of Germany, Ameritech of the USA, AT&T of the USA, Siemens AG of Germany, Alcatel-Asthom of France, Cable & Wireless plc of the UK, US West Inc of the USA, Telecom Finland, France Telecom, and Telia of Sweden [10].

Private investment is being invited into the national telecommunications companies, with the exception of Poland. All countries are encouraging foreign investment in new systems and the market position of the previous state monopolies is therefore being steadily eroded. In Poland, however, there are no immediate plans to sell shares in Telekomunikacja Polska S.A. and the company under current legislation will retain its international monopoly.

In 1993 investments into Central and Eastern Europe from the private sector in the West were between $3 to $4 billion. Commitments for further investments are substantially higher. There has been progress, both in terms of modern equipment manufacture and supply, and in the operation of new systems. Cellular systems in particular are seeing substantial growth. At the end of 1993 the figure for cellular phones in Eastern Europe stood at 95,000. Hungary now claims 150,000 cellular subscribers. However, the targets remain daunting: extra lines required for Poland to reach 17 per 100 of population - 2,700,000; for the Czech Republic and Slovakia to reach 24 per 100 of population - 1,200,000; and for Hungary to each 23 per 100 of population - 1,100,000. The investment costs are estimated to be $4 billion for Poland; $1.8 billion for the Czech Republic and Slovakia; and $1.6 billion for Hungary. These are figures produced by the International Telecommunications Union with a target date of 2000. To bring the countries up to the number of lines as exist in Western Europe would require even higher investment.

REFERENCES

1. The author acknowledges the valuable assistance he has received in the preparation of this article from Karen Buschardt-Pisarczyk of Nabarro Nathanson, Warsaw.
2. Central European Review, Winter 1994, Telecommunications Survey.
3. Communications Outlook, 1993, OECD, Paris.
4. The Western European average is 43.
5. Dziennik Ustaw (Journal Of Laws) No 86, item 504.
6. Provisions of international treaties to which Poland is a party will prevail over provisions in the Act.
7. Article 2.1(2) of the Act defines telecommunication public services as being those used in the carrying out of telephone or telegraph connections for profit; Article 2.2 gives the Ministry power to define other telecommunications as being of a public service nature.
8. Telfa SA makes telephone exchanges and components.
9. ZWUT stands for Zaklady Wytórcze Urzadzen Telekomunikacyjnych. It is the largest manufacturer of telecommunications equipment. Elwro makes components and circuits for the electronics industry.
10. Central European Economic Review, Winter 1994

REGULATION OF TELECOMMUNICATIONS IN URUGUAY

ANDRES M. CERISOLA

Partner, Estudio Ferrère-Lamaison, Acevedo Díaz 996, Montevideo 11200, Uruguay, Tel. (598-2) 424243, Fax (598-2) 424241 ; Director of the Institute of Law, ORT School of Business and International Affairs, Bvar. España 2633, Montevideo, 11300, Uruguay, Tel. (598-2) 771806, Fax (5989-2) 788810.

INTRODUCTION

The regulation of telecommunications is particularly complex in Uruguay and business possibilities in the industry largely depend on issues on which no consensus yet exists among the primary parties involved. That complexity derives in part from the lack of comprehensive bodies of rules and from the fact that regulation of the most important issues depends on the interpretation given to rules drafted on successive occasions without a clear view of the subject at hand. Added to this is the impact of the popular referendum of 1992, which derogated only some of the articles of a law that had established the basis for a competitive telecommunications market.

Regulation of the telecommunications market is currently shared by the Executive, the National Communications Bureau, and the government-owned telecommunications company, ANTEL. The attitude and interest of these bodies has, however, varied. In fact, since 1992 ANTEL has had a strong policy geared explicitly to broadening its sphere of power and defending its monopolistic presence in as many markets as possible.

This article proposes an interpretation that differs from the official one as to the impact of the 1992 referendum on regulation of the telecommunications market. Strong legal arguments support the fact that the 1991 privatization law put an end to the telecommunications monopoly, and that derogation of five articles of that law by the 1992 referendum does not change the situation. Beyond the discussion as to the effect of the derogation (by the 1992 referendum) of the derogation of the ANTEL monopoly (by the 1991 privatization law) on which attention has focused thus far, other provisions of the 1991 privatization law, which empowered the Executive to authorize and oversee the installation of new telecommunications services, are still fully in effect.

The impact of this interpretation is extremely important in connection with three aspects: (i) If there is no legal monopoly the Executive can authorize new telecommunications services in all areas including traditional telecommunications and long-distance services; (ii) If the requirement for such installation is an authorization, under the traditional interpretation given by administrative law to the concept of authorization (as opposed to that of a concession), authorization cannot be denied for reasons of convenience, but instead only for reasons of legality; and (iii) It does not make sense to regulate in a convoluted way, within the sphere of ANTEL, telecommunications companies that clearly exist in Uruguay, when the Executive could grant them authorization and subject them to a fair, objective and reasonable legal system.

In terms of regulatory policy recommendations, the Executive, after assuming its responsibility under law on the subject of telecommunications, would do well to establish a practical authorization procedure for telecommunications services, particularly services having a limited scope and those for own use, which should not generate major discussions as to regulatory policy. This would eliminate significant stumbling blocks faced today by commercial companies because of ANTEL's monopolistic aspirations.

I - TYPICALLY URUGUAYAN REGULATIONS

Uruguay has in many respects been an atypical country of Latin America. After enjoying impressive economic development and becoming one of the world's ten most prosperous countries during the second half of the 19th century, and after a relatively successful experience with social-democracy at the outset of the 20th century, it closed its economy during the decades of 1950 and 1960, preventing its benefitting from the great surge of growth during the second post-war period. Uruguay nevertheless maintains many indicators in line with developed countries, and in this context it is not strange that it has the best and most modern telecommunications network in Latin America.

Telecommunications regulation, however, has not escaped the typically Uruguayan characteristic of tortuous evolution, often against the tide of history, to which the population is accustomed. As will be seen later in this article, the last attempt at deregulation and privatization in 1992 paradoxically complicated regulation and in fact reinforced government power in many areas where it had attempted to limit itself.

Generally speaking, an historical introduction would seem to be appropriate to place this kind of subject in context (and in the Uruguayan case, history is crucial to the interpretation of the provisions applicable to the most sensitive telecommunications issues). Nevertheless, the aim not to turn this article into a catalogue of laws and decrees has led the author to abstain from such a recounting.

II - PRIMARY ACTORS

The Uruguayan telecommunications market is currently heavily influenced by the presence of a government-owned telecommunications company (ANTEL) which claims, though highly debatably, to hold a monopoly on all the country's telecommunications. Its power is further strengthened by its dual capacity as provider and regulator of telecommunications services.

Additionally, there is a National Communications Bureau, located for historical reasons within the Ministry of National Defense. Among other functions, it oversees the radioelectric spectrum, authorizes paging service providers, and has other major responsibilities regarding television and radio. The overlapping of responsibilities and long-standing competition between the Bureau and ANTEL were traditionally significant, but in recent times ANTEL has managed to prevail. This "conquest" was symbolized by the 1992 appointment of the influential ANTEL General Counsel, José Camelo, as director of the National Communications Bureau.

The third body with responsibilities in the area of telecommunications is the Executive, whose attributes were notably broadened by the 1991 law. Since the 1992 failure to privatize ANTEL, however, the Executive's attitude has been passive, and like the National Communications Bureau, to date it has allowed ANTEL to expand its power, without encountering resistance, to practically all issues linked to telecommunications.

The private telecommunications market has attempted to develop in the scant spaces left by regulation, and particularly by the interpretations expoused by the regulatory agencies. There is currently a private cellular company (a consortium led by BellSouth and Motorola) which operates under a sui generis contract which formally is not a license but instead a service contract with ANTEL. As of 1994, moreover, this company has been subjected to a curious situation of competition with a cellular company operated by ANTEL itself, who since then has become, in addition to the regulator and user of the private company's services, its competitor. It is a legally complex situation that would probably horrify any observer looking at it with a criterion of regulatory reasonability or legality. The parties are trying to manage the situation by means of contract checks and balances, but the possibility of this type of alternative being appropriate and sufficient to fairly regulate such an atypical situation with so many risks of abuse by the regulator remains to be seen.

There are also three paging companies subject to regulation by the National Telecommunications Bureau. These companies are operated freely despite the fact that according to ANTEL's interpretation paging services are clearly included within the scope of its monopoly. As of 1995, however, the novelty for these companies is that they will have to compete with an ANTEL-operated paging company in a regulatory and commercial scenario that is still unclear for everyone involved.

In addition, almost all the international telecommunications companies in one way or another have a presence in the Uruguayan market, but given the characteristics of the sector's evolution to date, in most cases they have tended to define themselves as suppliers of ANTEL, rather than prospective telecommunications operators.

Other international companies have attempted to break into the Uruguayan telecommunications market with a variety of services, but ANTEL has thus far managed to play its cards so as to prevent or substantially limit the entry of any type of direct or indirect competition. The absence of antitrust regulations limiting possible abuses of monopolistic power has aided in keeping out or unimaginably hampering the operations of value-added service providers seen as particularly undesirable by the government-owned company.

III - CHIEF CONCEPTUAL ISSUES

Telecommunications regulation in Uruguay has become a complicated web where it is easy to get lost, even for those who deal with it on a daily basis. Some of the chief issues currently under debate are as follows:

What telecommunications regulations prevail following the popular referendum of 1992 that prevented privatization of ANTEL and partially derogated the new framework created for the post-privatization stage?

Who is responsible for regulating telecommunications? Is there a government monopoly on telecommunications?

What regulations should apply to the prosperous cellular market and future competitive developments (PCS, for example)?

What regulations should apply to other radio-based telecommunications markets, such as paging?

IV - REGULATION OF TELECOMMUNICATIONS

Under the legal framework predating the 1991 privatization law, ANTEL was a monopoly provider of telecommunications services via cable (most specialists agree on the restriction to cable services, despite the broader-reaching interpretation put forth at the time by ANTEL). It was, however, a curious monopoly, since until the mid-1970s a telegram system operated by Western Union competed with ANTEL. Reference was even made to it by the law establishing the monopoly, which, contrary to the idea of a monopoly market, made ANTEL responsible for overseeing "companies authorized to operate telecommunications services."

On the other hand, radio communication services (then much less technologically and commercially developed than now) constituted from the conceptual standpoint a free private activity subject only to a system of authorization by the National Communications Bureau. This refers both to ham radio services and to paging systems, and eventually also cellular telephony.

ANTEL, however, had defined control of the radioelectric telecommunications market as a vital objective, both for rural and cellular telephony. When the launching of cellular telephony in Uruguay was proposed in 1989, ANTEL developed a theory which, implicitly and without much conceptual rigor, implied treating it like a value-added service deriving from fixed telephony and thus, according to ANTEL's extensive interpretation, subject to its monopoly. This, together with the effort to make an otherwise difficult concession politically viable, led ANTEL to call for bids for cellular telephony under the guise of a contract for provision of services to ANTEL. ANTEL would formally be the service provider, with the awardee operating the service entirely for ANTEL.

This circuitous structure was contrary to the interpretation we considered legally correct in the sense that cellular telephony is a form of radiocommunication that as such should be subject to authorization by the National Communications Bureau, without ANTEL involvement beyond authorization of the interconnection with the fixed telephony network. Despite this, and the fact that the issue was at that time taken up by the main parties involved, the bid was held and, after a long and extremely complex negotiating process, a contract was signed at the end of 1990 for operation of a cellular telephony service by a private company "on behalf of ANTEL." From the substantive point of view, however, the private company makes all of the investment, fully operates the system, takes all commercial risks, and receives all the profit deriving from operation.

As regards paging activity, consistent with the interpretation we consider appropriate for the regulations in effect until the time of the 1991 privatization law, companies were authorized and overseen by the National Communications Bureau. No one ever raised the doubt that such activity, as user exclusively of the radioelectric spectrum, should fall under the ANTEL monopoly as "telecommunications." However, the legal reality of the industry after the 1991 law would be, in line with the text of the legislation, that said companies should be considered providers of telecommunications services and, as such, require a license from the Executive. The problem that reappears more strongly now that ANTEL is proposing to become involved in this activity, is that if ANTEL is subject to its own interpretation regarding the effectiveness and scope of its monopoly, it will have no other choice but to conclude that paging is a telecommunications activity included in same.

The 1991 privatization law and its subsequent partial repeal by the popular referendum in 1992 contributed further elements to the industry's already confused situation. The privatization law covered a broad spectrum of activities (including utilities, airlines, concessions, etc.), and only part of it referred to privatization of ANTEL and to the new telecommunications framework.

As necessary elements deriving from ANTEL privatization, the law provided for (i) derogation of the telecommunications "monopoly," (ii) establishment of a system of Executive authorization for new companies, and (iii) elimination of ANTEL's regulatory authorities, which were basically transferred to the Executive. With the primary objective of defining the scope of the Executive's regulatory power on the subject, the new law defined telecommunications as broadly as possible, following the well-known definition set forth in the Nairobi Convention. Telecommunications are thus comprehensively defined for the purposes of the principal laws on the subject as communications via cable and radioelectric spectrum. It never occurred to any of the authors of the privatization law that this definition could come to be used as an argument to broaden the scope of the monopoly that was supposedly "reborn" after the unexpected popular referendum of 1992.

The law provided a deregulatory package which, as we said, faced the unexpected opposition of a popular referendum that repealed five of the thirty-two articles of the 1991 law, and which was basically geared to blocking privatization of ANTEL. The articles derogated by the referendum included the repeal of the telecommunications monopoly (repeal of the repeal) and the redefinition of the purposes of ANTEL, which had eliminated its regulatory authority. Nevertheless, the referendum did not repeal the new and broader definition of telecommunications for the purposes of these laws, or (and this is critical for the argument raised by this article regarding the current situation of telecommunications) the articles empowering the Executive to regulate the market and to authorize new telecommunications companies.

The regulation of telecommunications is, following the aforesaid referendum, a true legal nightmare. Without getting into abstract theoretical discussions, it is important to underscore that based on the current legislation it is not possible to affirm, as ANTEL does, the existence of a legal telecommunications monopoly in Uruguay.

To clarify this it is not necessary to answer the hotly debated question of whether derogation of the article derogating the monopoly can give rise to the "rebirth of the monopoly." In fact, beyond that discussion, there are articles of the 1991 law not affected by the popular referendum which clearly make the Executive responsible for authorizing and overseeing the installation of new telecommunications services, and that authority is totally incompatible with effectiveness of a legal monopoly.

The legal reality, in our view, is that the Executive has the power to authorize such new services, and only insofar as it does not do so will it in fact be maintaining the monopoly privilege in favor of the state-owned company. If in the future the Executive should decide, in the exercise of the powers it undoubtedly has, to authorize new services, same would be able to compete with ANTEL in all telecommunications areas, be they by wire or radiocommunication, local or long-distance. The issue is even more interesting since pursuant to the interpretation provided by doctrine to the concept of "authorization," the Executive would not have any discretion in accepting or denying requests for licenses, but instead would be obligated to authorize all services meeting objective parameters determined generically.

Doubt could arise, within that framework, regarding which agency would be responsible for regulating such telecommunications services, since the referendum repealed the article that eliminated ANTEL's regulatory authority. Without getting into overly technical discussions, however, it seems clear to us that even after the referendum ANTEL's regulatory authority has at least been limited and subordinated to that of the Executive, as a result of the subsequent nature of the law entrusting the Executive with such authority.

The question of the scope of the monopoly is of enormous commercial importance which is not limited to the significant market of telecommunications companies. ANTEL has informally, but very inflexibly, defended an interpretation which in our opinion is as theoretically absurd as it is legally unsustainable, to the effect that a company cannot even establish a radio link between two computers if ANTEL is capable of providing such a link using the same or another technical system. The de facto subordination of the National Communications Bureau, which is the agency that legally should authorize the concession of the frequency band, vis-a-vis ANTEL, however, hampers a discussion of this subject in legal terms when the time frames for implementation of projects are important.

V - CELLULAR TELECOMMUNICATIONS

Uruguay's cellular market regulations are what a medical doctor would call a case of multiple and divided personality. The private company operating in the market since 1991 does not formally have a license, but instead provides ANTEL with a complete cellular telephony service, which ANTEL, in turn, sells to the public. In substance, however, the relationship between the cellular company and the public is direct and does not involve ANTEL. Since the price for the service to ANTEL coincides exactly with the payments the public makes for cellular service and ANTEL has authorized the private company to negotiate the service vis-a-vis clients on its behalf, in fact few users notice that they are buying a service from ANTEL, and not from the private company.

Moreover, since 1994, ANTEL has started up its own cellular operation that competes with the earlier one. The risks of abuse of regulatory power, of improper use of the private company's confidential information for competitive purposes, of the advantages of a larger authorized area of coverage, the availability of an indefinite term for amortization of the investment by the new ANTEL-owned company, and above all the possibility of indiscriminate cross-over subsidies in an operation that has more political than economic importance for ANTEL, are so significant that it seems difficult that a structure of this sort could be maintained. That, however, is what is being attempted in a very "Uruguayan" solution.

In this context one could ask if this cellular market regulation has any resemblance to the legally correct regulation of this type of telecommunications under the new regulatory framework appearing in the 1990s. Based on the analysis set forth in the foregoing point, there is no doubt that the Executive is competent both to authorize and to oversee telecommunications services. Telecommunications via radioelectric means were, moreover, until then, and to some extent afterwards, to be regulated by the National Communications Bureau and not by ANTEL.

The reasonable route, from the standpoint of legality as well as regulatory policy, should therefore be to grant the private company a license to provide telecommunications services and, on equal footing with ANTEL, subject both companies to oversight by the Executive and, insofar as applicable, the National Communications Bureau. It does not seem reasonable to continue fostering the illusion of a contract for provision of services between the private company and ANTEL, and the non-existence of any private telecommunication company per se. Moreover, it is extraordinarily clear that regulation of a telecommunications company by its competitor does not constitute a reasonable or advisable policy.

VI - PAGING

Paging is currently subject to regulation by the National Communications Bureau. This derives from its nature as a form of radioelectric telecommunication and, for reasons impossible to justify except in historical terms, ANTEL has excepted this service from its ultra-broad definition of telecommunications subject to its monopoly.

The formal argument ANTEL has used to "tolerate" the existence of similar services in the past is that to the extent that ANTEL does not provide the service, it is not interested in exercising its monopoly and therefore permits the existence of private operators on a "temporary basis." Even this argument (unsustainable from the legal point of view) is going to disappear over the coming months, since ANTEL has called for bids for acquisition of a turnkey paging service, and there is great anticipation in the industry to see how ANTEL is going to reconcile, if that is possible, its interpretation of the effectiveness of its monopoly with the existence of three private paging companies in competition with ANTEL, who it is difficult to believe can be forced to shut down.

* The author and his firm are advisers to several telecommunications companies. The opinions set forth in this article, however, are his personal opinion as to the subjects in question and in no way commit his firm or his clients.

OPPORTUNITIES IN THE ARGENTINE TELECOMMUNICATIONS' INDUSTRY

JUAN JAVIER NEGRI AND FEDERICO BUSSO

Negri, Teijeiro & Incera, Abogados, Avde, Corrientes 330, 4° p. 1378 Buenos Aires, Argentina.

Abstract *: Since July 1989, after the current administration took office, Argentina made a 180 degree change in its approach towards telecommunication services: from the monopolic and obsolete state-owned Entel to a situation where most telecommunication services are open to competition. Furthermore, services which still remain free from competition and the providers of which enjoy exclusive licenses will not remain in such situation for a long time. This article describes the licencing process in Argentina to obtain a license to provide telecommunication services on a non-exclusive basis.*

INTRODUCTION.

Decree 731/89 (September 12, 1989), as amended by Decree 59/90 (January 5, 1990) provides that telecommunication services in Argentina may be provided by any operator (except for those considered to constitute "basic telephone services" or upon which an exclusive license or concession has been granted). For this purpose a license must be obtained from the Comisión Nacional de Telecomunicaciones (Federal Telecommunications Committee), the relevant regulatory agency of the Argentine Government ("CNT"), pursuant to Decree 62/90 (January 5, 1990) and Decree 1185/90 (June 22, 1990), as amended from time to time, and their implementing regulations.

The term "basic telephone services" -which are the only telecommunication services the operation of which has been granted in exclusivity- is defined by Decree 62/90 as the supply of fixed telecommunication connections which are part of, or are connected to, the public telephone network, thereby providing both local and long distance (domestic and international) live voice services.

Domestic and long distance basic telephone services are provided by Telefónica de Argentina S.A. (in the southern provinces of Argentina) and Telecom Argentina Stet-France Telecom S.A. (in the northern provinces of the country), while international basic telephone services are provided by Telecomunicaciones Internacionales de Argentina TELINTAR S.A., formed by the two companies mentioned above.

All three companies enjoy an exclusive license granted for a seven-year term counted as from the date in which the operations of the former state-owned telephone company were transferred (November 8, 1990). Provided the companies meet certain mandatory operational objectives set forth in the bidding rules under which Entel was privatized, a three-year extension period may be granted. The mandatory objectives set number of new lines to be added and public telephones to be installed, call and operator efficiency levels, failures allowed, repair time delays and waiting time for installation of new lines.

The above licenses were granted with no time limits. Once exclusivity terminates, basic service operators may continue to act as operators in a competitive market.

Recently, the above described exclusivity has been limited by a judicial decision allowing the use of "call back" services. By using these services frequent long-distance telephone users can cut their costs an average 50%.

I. LICENCING PROCESS FOR THE PROVISION OF TELECOMMUNICATION SERVICES ON A COMPETITIVE BASIS

As mentioned above, operators interested in the provision of telecommunication services not deemed to constitute a "basic telephone service" must obtain, upon compliance with certain requirements, a prior license from the CNT.

Exhibit I and Exhibit II of CNT Resolution 477/93 (February 17, 1993), set forth the requirements to be complied with for the obtention of a license for the provision of telecommunication services on a non-exclusive basis.

1.1. The request :

Petitioners must file an application in writing indicating their intention to operate telecommunication services, specifying the type of service for which the license is requested.

1.2. Contents of the application:

License applications must include:

a) Background and legal, corporate, economic and technical capacity of the applicant (and, if a corporate body, of its members).
b) Experience in the production of goods and services with documentation evidencing the capacity of the applicant or the capacity of its members or shareholders (in the case of companies) for the provision of the service.
c) The forms required under CNT Resolution 4744/92 for the registration of providers of telecommunication services.
d) General and technological description of the project. Applicants must present a pre-project containing:

    i) General description of the service to be provided;
    ii) Description of the operating characteristics of the system;
    iii) Description of the equipment and most important elements that will comprise the system;
    iv) Coverage Area; and
    v) Deployment schedule indicating date of commencement of the service and expansion of the technical project of the system.

e) Legal entities must:

    i) file their articles of incorporation (acta constitutiva) and by-laws (estatutos) duly registered at the Public Registry of Commerce;
    ii) provide information regarding the identity and expertise of the members of the board of directors of the applicant in the event their technical background is invoked:
    iii) legal representatives must provide a certified copy of the instruments relative to their appointment;
    iv) copies of the instruments in which it is evidenced the capital and net-worth of the copmany and copies of the last two balance sheets. In the case of a new company or if the CNT deems it necessary, the company shall attach the balance sheet of its members or shareholders. The CNT may request additional financial information.

f) Individuals must:

   i) provide their personal identification;
   ii) be a merchant registered as such before the Public Registry of Commerce providing evidence thereof;
   iii) present a statement about their financial capacity certified by a CPA;
   iv) the signature of the applicant must be certified by a notary public.

1.3. The resolution:

Once the application is received by the CNT and provided the system is compatible with other existing non-exclusive systems it shall issue a license. Otherwise, the CNT must justify the denial.

1.4. Other conditions:

The CNT may determine the term for the start up of the systems. In the event of default, the authorization shall automatically terminate. Further, if there are a limited offer of certain services exists (or limited ways -such as reduced number of available channels- to satisfy the demand for such services), the CNT may call for bids to operate those services (i.e., cellular telephony services). Licensees must use CNT-registered equipment or in compliance with security and interconnection regulations.

II. TERMS AND CONDITIONS OF THE LICENSE

In addition to certain requirements for the obtention of the license, Exhibit I of CNT Resolution 477/93 sets forth the licensing mechanism for the provision of telecommunication services on a non-exclusive basis.

2.1. Territory:

Licences may be granted to provide services within a limited geographical area or throughout the territory of the Republic of Argentina.

2.2. Term:

Non-exclusive licenses are granted with no time limitations.

2.3. Corporate structure :

Licensees may be either a physical person or a legal entity. There are no restrictions to foreign individuals or companies to be granted a license.

2.4. Conditions of the license:

Licenses are only granted for the provision of the services for which they have been applied for and within the authorized geographic area.

2.5. Interconnection agreements:

During the exclusivity period granted to Telefónica de Argentina S.A. and Telecom Argentina Stet-France Telecom S.A. (and any extension thereof) any interconnection agreement between them and a licensee shall be freely negotiated between the parties. If the parties fail to reach an agreement the CNT shall decide the matter.

2.6. Assignability:

Licenses cannot be assigned (in whole or in part) without CNT's prior authorization.

2.7. Costs - Payment of fees:

There are no costs associated with the obtention of the license. However, Decree 1185/90 provides that suppliers of telecommunication services must pay a control, supervision and inspection fee equal to 0.50% of the gross income obtained from the services provided, net of taxes and charges imposed on such services, except for the above mentioned 0.50% fee itself. This fee is payable to the Federal Telecommunications Fund created under the jurisdiction of the CNT. Fees are payable pursuant to CNT Resolution 4744/92 (December 4, 1992), as amended by CNT Resolution 155/93 (January 26, 1993).

CNT Resolution 4744/92 provides that payments made to other suppliers of interconnected telecommunication services may also be deducted from gross income up to the amount of those payments. Finally, credits due to discounts, reimbursements or bonuses shall also be deducted from the tax basis.

Fees must be liquidated and paid by means of a monthly sworn statement presented to the CNT within the first 10 calendar days of the month following the liquidated period, by filing the corresponding form and evidence of the deposit. Omissions, errors or forgeries of the return are subject to penalties and compensatory and punitive interests.

2.8. Penalties:

Article 38, Decree 1185/90 provides for the following penalties in case of violation of the applicable regulations or the terms of the respective licenses (except for licensees of the "basic telephone services"): admonishments, fines, total or partial revocation of the exclusivity regime (if applicable) or revocation of the license or permit. The CNT may order the publication of the sanction if the licensee repeats the same violation [paragraph (a)].

1. Penalties shall be applied taking into account: (i) the magnitude of the violation; (ii) the impact that the violation may have caused on the service, the clients and third parties; (iii) their impact on the public interest; (iv) degree of compliance with mandatory and not mandatory objectives and other conditions set forth in the license (if applicable).
2. If the event of sub-paragraph (ii) above, the licensee shall not be punished if (i) the violations are caused due to force majeure or reasons not attributable to the licensee or (ii) the licensee corrects or ceases the violation upon notice by the CNT. However, in this latter case the CNT may impose sanctions without further notice if the damages are serious and irreparable or have a great public repercussion. Additionally, notwithstanding the lack of sanctions in the above mentioned scenarios, the licensee shall have the obligation to cease the violatory conduct and repair the damages it may have caused.
3. Applicability of sanctions shall be independent from the obligation to reimburse or offset fees unduly collected from the clients (plus applicable interests), or to indemnify the government, clients or third parties for the violation.

4. Sanctions shall be imposed independently of the negligence or willful misconduct of the licensee or its employees.
5. Sanctions not subject to appeal shall be considered a negative precedent in case of further violations.
6. The application of sanctions will not prevent the CNT from bringing legal actions seeking compliance with the terms of the license.
7. The affected licensee shall enjoy a 10 business day term to appeal sanctions.
8. Fines shall not exceed the equivalent of 3,000,000 telephone beats per violation (the value of a beat is now US$ 0.04190). However, the CNT may increase the fine to the equivalent of 12,500,000 beats if the licensee continues with its violatory conduct or the violation has a great public impact. Every fine must be paid within 30 days once the resolution which imposed cannot be subject to further appeals.

If the CNT resolves the revocation of a license the same licensee cannot obtain another license for a 5 year-term counted as from the date in which the resolution could not be subject to further appeals. Furthermore, a license cannot be granted to a company in which the controlling company of the former licensee is a participant.

2.9. Means:

The granting of the license is independent from the existence and allocation of the necessary means for the provision of the services. The licensee must obtain authorization for the installation and operation of means, telecommunication systems or links.

2.10. Accounting:

If the CNT so determines the licensee shall organize its accounting systems so as to separate the accounts of the different services it may provide.

2.11. Number of licenses:

Any individual or legal entity may obtain several licenses for the provision of different telecommunication services or for the same service in different areas if the license is not valid to operate throughout the

Republic of Argentina or even for the same area if certain conditions are met.

2.12. Review of the terms of the license:

The CNT shall, every five years, review the terms of each license. Changes or amendments in the conditions for the provision of the services may be requested.

2.13. Prices:

The general price structure shall be published in the Official Gazzete and in a newspaper of the area where the licensee provides its service within 30 days after the license becomes effective. The publication shall set forth the value of the unit used to calculate prices. Any change to the value of the unit shall also be published.

Prices shall be valid as from the day following the publication, without need of prior approval. However, prices must be notified to the CNT.

III. ASSIGNMENT OF FREQUENCIES

As mentioned above, granting of a license is independent from the existence and allocation of the necessary means for the provision of the service. The authorization for the installation and operation of facilities, links or telecommunication services must also be obtained from the CNT.

IV. REGULATORY FRAMEWORK - CNT POWERS

The CNT is the regulatory agency of the telecommunication industry. It was created pursuant to Decree 1185/90 (as amended). It is under the jurisdiction of the Argentine Federal Executive through the Ministry of Economy, Works and Public Services. Its principal office is in the city of Buenos Aires.

4.1. Functions:

The CNT provides administrative and technical regulations, control, supervision and verification of the regulatory framework of the telecommunication business.

4.2. Powers and obligations:

The following are among the CNT's most important powers and obligations:

1. application, interpretation and enforcement of any law, decree or regulation regarding the telecommunication business;
2. administration of the radioelectric spectrum. Authorization of use and installation of satelite systems;
3. authorization of equipment and material for exclusive use in the telecommunication industry;
4. prevention of non-competitive behaviours;
5. determination of technical and service standards and supervision of compliance with them;
6. resolution of disputes between companies providing services and their customers and between companies and other suppliers of telecommunications services;
7. collection of service fees;
8. application of sanctions; and
9. review of interconnection agreements.

4.3.Management:

The CNT has a board of directors composed of five members appointed by the Federal Executive. They are elected for a five-year term.

V. REGULATORY APPROVAL REQUIREMENTS FOR FOREIGN COMPANIES

No Argentine approval is necessary for the operation of a foreign company by itself or through a local subsidiary.

There are no restrictions or exchange controls in Argentina for the payment of principal, interests or dividends to a foreign lender, bondholder or shareholder.

- *VI* -

INTERNATIONAL ORGANISATIONS

GLOBAL PARTNERSHIPS FOR GLOBAL COMMUNICATIONS : CHALLENGES FACING THE NEW ITU *

DONALD J. MACLEAN

Head, Strategic Planning Unit ; International Telecommunication Union ; Place des Nations 1211 Geneva 20, Switzerland.

INTRODUCTION

The International Telecommunication Union (ITU) is a remarkable institution. Its roots can be traced to the International Telegraph Union which was founded at the Paris Conference in 1865, and to other conferences that took place in central and western Europe in the preceding decades.

Throughout its history, the ITU has adapted to successive waves of technological and political change. The question to be examined in this paper is whether it can continue to do so as the world leaves the age of government controlled "plain old telephone service" and enters the age of market-driven, global, multimedia communications.

In responding to this challenge, the ITU's long history gives it obvious advantages and disadvantages. However, the ITU has one characteristic, unusual for an international organization, which is clearly of great strategic value in the changing telecommunications environment. Although the ITU was established as an intergovernmental organization, provision was made for private sector participation at the ITU's second conference in Vienna in 1868. Private members were first admitted to the Union at its third conference in Rome in 1871-2.

Today, the ITU has 184 "Big-M" members and about 325 "small-m" members. The latter number includes all of the major telecommunication service providers, equipment makers, and international satellite organizations. It also includes representatives of "convergent" industries such as broadcasting and computing; major user groups; financial and development institutions; and consultancies. Several thousand experts from these companies take part in the work of the ITU, alongside government members.

Partnership between government and the private sector is universally agreed to be an important element in the development of global, multimedia communication systems. The ITU is therefore clearly positioned to continue to play a leading role in the future evolution of world telecommunications. But will it? Is it realistic to think that an organization with roots in the 19th century can play a central role in the global information society of the 21st century? In response to this challenge, the members of the ITU have reorganized its structure, modernized its working methods, redefined its mission, and mapped out a strategy designed to ensure that the ITU plays a leading role in the emerging global information economy and society.

I - THE NEW ITU: STRUCTURAL CHANGE AND FUNCTIONAL INNOVATION

At an extraordinary Additional Plenipotentiary Conference (APP) which took place in Geneva in December 1992, the ITU was reorganized for the first time since becoming a part of the United Nations system in 1947. As a result of this reorganization, the ITU now has three main functions - standardization, regulation, and development. These functions are largely although not completely reflected in the ITU's current organizational structure [1].

1.1 Standardization

The ITU's original function was to standardize the technical, operational, and tariff aspects of telegraphy in different countries, in order to facilitate communications between them. The early telegraph conventions covered such matters as the technical systems which should be used for international telegraphy, the languages in which international telegrams should be composed, the physical links that should be established between different countries, and the prices that should be charged for different services.

Today, the ITU performs analogous functions for a very wide range of wired and wireless telecommunication services. The ITU's approach to standardization is minimalist. It aims to standardize only those elements necessary to ensure that telecommunication networks can interoperate and interconnect on a global basis. The resulting ability to communicate through public telecommunication networks from anywhere in the world to anywhere in the world is one of the most striking characteristics of the telecommunication industry. The universality and transparency which telecommunications have achieved, largely through the work of the ITU, stand in sharp contrast to the divisions that have marked computing and broadcasting.

As the pace of innovation in telecommunications increases and as the different information industries converge into multimedia networks, the ITU's standardization function faces a number of critical challenges. Some have already been successfully met. For example, by re-engineering the ITU standardization process, the average time it takes to produce a standard has dropped from four years to eighteen months. Productivity has been increased to the extent that the same amount of work was done in the 1988-92 period as in the previous twenty years. These results were achieved purely by process re-engineering, without any increase in resources.

With these productivity gains in place, the ITU is re-focusing its standardization activities on the priorities of the global marketplace. In this respect, the technologies underlying the intelligent, multimedia, personal communication networks of the future are of central concern. Customer numbers are becoming an increasingly scarce and valuable resource as telecommunication markets become more competitive, differentiated and personalized. Numbering plans have therefore emerged as a key priority for the ITU. The principles and practices underlying international telecommunication tariffs continue to be among the most important items on the ITU's agenda, as they have been from the beginning. In this latter respect, the position of the ITU is clea: tariffs should reflect costs. Implementation of this policy is, of course, the responsibility of ITU members.

1.2 Regulation

The ITU's second main function is to regulate radiocommunications and international telecommunications. This is done through treaties which have the force of international law and which oblige ITU member states to apply the provisions of the Radio Regulations and the International Telecommunication Regulations in their national law.

Thus described, this function probably sounds more intimidating - and more constraining of technological progress and market development -

than it actually is. In fact, the aim of ITU regulations is to provide the minimum treaty-level agreements required to permit the existing global telecommunication system to function, and to facilitate the development and application of new technologies.

a) The Radio Regulations

Because radio waves cannot be contained within national boundaries, it is essential that governments agree on which radiocommunication services will use which frequencies, as well as on rules and procedures to avoid harmful interference between terrestrial radio stations and earth orbiting satellites. The ITU Radio Regulations embody these agreements.

The agreements reached at the last major ITU radiocommunication conference at Torremolinos in 1992 opened the way for the introduction of personal communication services provided by mobile satellite and terrestrial networks, as well as for high definition television and digital audio broadcasting. The next World Radiocommunication Conference, which will be held in Geneva in October-November 1995, will consider proposals to re-write the Radio Regulations, in order to make them simpler and easier to use. It will also examine how to coordinate the introduction of these new services with existing ones.

b) The International Telecommunication Regulations

The ITU also overhauled the International Telecommunication Regulations (ITRs) at the World Administrative Telegraph and Telephone Conference (WATTC) which took place in Melbourne in 1988 [2]. These regulations represent a somewhat uneasy compromise between the old régime of monopoly service providers operating according to universal, uniform, symmetrical principles, and the new realities of a competitive, differentiated and asymmetrical marketplace.

On the one hand, the International Telecommunication Regulations state that ITU member governments should apply ITU recommendations regarding technical, operational and tariff standards and that they should also encourage companies under their jurisdiction to do the same. By their very nature, these standards are intended to be universal.

On the other hand, however, the ITRs permit ITU member administrations and companies operating under their jurisdiction to enter into "special arrangements" which do not concern ITU members in general. These special arrangements can involve matters such as the apportionment of accounting rates, the provision of resale services, and general conditions of market access.

The ITU and the World Trade Organization (WTO) are currently studying the fit between the provisions of the International Telecommunication Regulations, the principles underlying the General Agreement on Trade in Services (GATS), and the negotiations now underway in the WTO on trade in basic telecommunications services.

1.3 Development

The ITU's third main function is to assist developing countries in developing their telecommunication networks. This is a relatively new function in the long history of the ITU, but one that has acquired increasing importance and visibility in recent years.

The mathematics of the development challenge have been well known since the 1984 report of the Maitland Commission drew attention to the fact that there were more telephones in Tokyo than in all of Africa. In the last ten years, the gap between the telecommunication haves and have-nots has narrowed only slightly, if measured in terms of access to basic telephone service. It has actually widened if certain other yardsticks are used.

The ITU has been active in telecommunications development for forty years. For most of this period it has concentrated on providing technical advice and assistance, using its own resources or those of the United Nations Development Program, at a level of effort on the order of twenty to thirty million dollars a year.

While important to ITU members, this effort is clearly marginal in the greater scheme of things. Meeting the telecommunication needs of developing countries requires the investment of hundreds or thousands of millions of dollars per year. There is consensus that these funds are available, largely from private sources, if market conditions are right. On the advice of his World Telecommunications Advisory Council (WTAC) and on the basis of a feasibility study conducted by McKinsey & Co., the Secretary-General of the ITU recently endorsed the establishment of a privately-owned, specialized investment fund for telecommunications development. This fund will be known as WorldTel. It will be established at arms-length from the ITU, but will carry the ITU imprimatur.

At the first World Telecommunication Development Conference (WTDC) in Buenos Aires in March 1994, the ITU re-vamped its development strategy to bring it in line with market realities. Without abandoning its role in technical cooperation, the ITU will henceforth study and advise its member countries on the nuts and bolts of sector restructuring. The ITU will seek to do this in partnership with the private sector, as well as with other international and regional telecommunications, financing and development organizations. ITU members will concentrate their efforts on the Least Developed Countries (LDCs) and on rural and remote areas of the world, where market solutions are often problematic.

II - THE NEW ITU: GOALS, MISSIONS AND STRATEGIES FOR THE GLOBAL INFORMATION SOCIETY

One of the unusual features of the ITU is that it is a federal organization composed of different Sectors, each of which enjoys a significant degree of autonomy. The important reforms which have taken place in the areas of standardization, radiocommunication and development were the product of sectoral conferences. In organizational terms, they came from "the bottom up". In the ITU, "top down" direction is provided by Plenipotentiary Conferences (PPs). These conferences now meet every four years to establish policies, strategies and expenditure levels for the Union as a whole.

As previously mentioned, the Additional Plenipotentiary Conference which took place in Geneva in 1992 reorganized the ITU. The main task facing the ordinary Plenipotentiary Conference which took place in Kyoto, from 19 September to 14 October, 1994, was to decide from a "top down" perspective what the new organization should do.

For the first time, the Kyoto conference adopted a strategic plan for the ITU. This plan will serve as a guide and a benchmark for ITU members and managers for the next four years.

The goal of the strategic plan is ambitious. It is to establish the Union as the international focal point for all matters relating to telecommunications in the global information economy and society of the twenty-first century. To achieve this goal, the plan recognizes that the ITU must become more "client-oriented", and that its activities must respond to "the changing roles, needs and functions of Member administrations, the associated changes in the wider communications industry; and the changing needs of all ITU participants".

The plan analyses the needs of the ITU's major client groups in the changing telecommunications environment and lays out a three-point strategy for the Union as a whole.

2.1 Strengthening the ITU's Foundations

The first element of the ITU strategy is to strengthen the foundations of the Union. This is to be done by increasing synergy among the ITU's major functions and, most importantly, by enhancing private sector participation in ITU activities.

Strong private sector participation is vital to the ITU's future. In recognition of this fact, as well as approving the general guidelines included in the strategic plan the Kyoto conference adopted two specific resolutions on private sector participation. One of these resolutions formally recognized the rights currently enjoyed in practice by the private sector, including the right to be included in "any decision-finding process aimed at facilitating the achievement of consensus" in ITU study groups. The other resolution instructed the ITU Secretary-General to establish a Review Committee to recommend how the rights of the private sector can be enhanced over the next four years, by constitutional amendment and otherwise.

2.2 Broadening the Scope of ITU Activities

The second element in the ITU strategy is to broaden the scope of the Union's activities in response to the identified needs of government and industry members.

Insofar as ITU government members are concerned, the main new requirement they have identified is for a forum where they can discuss their telecommunications policies and strategies.

In the past, ITU member administrations were primarily concerned with operational matters of the kind that are dealt with in the Radiocommunication, Standardization and Development Sectors. However, many countries have now separated telecommunications operations from government administration. Instead of being operators, the ITU's "Big-M" members increasingly are becoming policy-makers and regulators. At the same time as the role of "Big M" members is changing, so are the ground rules of telecommunications policy. In many ITU member countries, market-oriented approaches to telecommunications policy and regulation are taking their place beside or beginning to displace traditional, public utility-based approaches.

In light of these changes, the Kyoto conference decided to establish a new forum, to be called the World Telecommunication Policy Forum, where all ITU members, "Big-M" and "small-m" alike,will be able to exchange information and share experience on emerging policy issues arising from the changing telecommunications environment. One or two such events are to be held in the next four years, in conjunction with other ITU conferences and meetings.

Insofar as the ITU private sector members are concerned, the main requirement they have identified is for improved electronic access to ITU documents and publications. Many of the ITU's most active "small-m" members do a significant amount of their work in "cyberspace". The Kyoto conference recognized that in order to retain the allegiance of these members and to facilitate the speedy development and implementation of ITU standards and regulations on a world-wide basis, the ITU must increasingly become a networked organization.

2.3 Increasing the ITU's Leverage

With stronger foundations and a broader portfolio of products and services to offer, the ITU will be well-placed to implement the third element of its strategy - to increase its "policy leverage" through partnerships and strategic alliances with other international organizations.

By itself, the ITU cannot hope to achieve its strategic goal of becoming the international focal point for all matters relating to telecommunications in the global information economy and society. However, with vision, leadership, and good working relationships, the ITU and its sister organizations may be able to achieve together what none could hope to achieve separately. Key partnerships would include established institutions such as the World Bank, the World Trade Organization, UNESCO and the OECD, as well as new organizations such as the Internet Society and other groups active in promoting the development of the "Global Information Infrastructure".

III - THE NEW ITU : GLOBAL CHALLENGES FOR A GLOBAL FUTURE

Do the results of the 1992 Geneva Additional Plenipotentiary Conference and the 1994 Kyoto Plenipotentiary Conference mean that a truly "new ITU" has be born, that the Union has succeeded once again in re-inventing itself, and that its place is secure as the world approaches the global information society of the 21st century?

Time will ultimately provide answers to these questions. For the moment, the reforms introduced by the 1992 APP appear to be working to the satisfaction of the ITU membership. There are also encouraging signs that the Kyoto Plenipotentiary Conference marked a new departure for the ITU [3]. However, the final verdict on the ITU's new structures and strategies will not be available for several years, and will be conditional on actions yet to be taken and decisions yet to be made.

What are the longer-term "success criteria" for measuring the effectiveness of the structures put in place at the 1992 Geneva APP and the strategies adopted at the 1994 Kyoto PP? In the next five years, a number of issues are likely to be of particular strategic importance for the ITU.

- The ITU will have to move quickly to establish its credibility with respect to emerging telecommunications policy and regulatory issues. Becoming the focal point for discussions about the "Global Information Infrastructure" (or whatever version of this concept ultimately prevails) may be the greatest strategic challenge and opportunity.
- To establish its policy credibility, the ITU will have to attract greater participation by decision-makers from government and industry, instead of relying as it now does almost exclusively on technical experts. This requirement is recognized in a number of key Kyoto resolutions which instruct the Secretary-General to encourage "an appropriately high level of representation" at ITU meetings dealing with strategic tasks.
- Technological convergence means that the ITU will also have to attract a broader range of participants than it has in the past. Although technical representatives of the broadcasting and information processing industries and major user groups take part in ITU work, policy discussions in the multi-media telecommunications environment will require the presence of corporate planners, strategists, marketers and content developers as well.
- While expanding its horizons along the lines proposed in the strategic plan, the ITU must of course continue to meet the basic needs of its core constituencies.

The ITU's fate is not entirely in its own hands. As well as depending on the quality of its own efforts, the Union's success in meeting these criteria will depend on developments in key Member countries and in the international environment more generally. In many cases, the ITU will have little or no control or influence over these events.

In this connection, one of the key strategic questions facing the ITU and its members is whether telecommunications will in future continue to be a distinct domain of public policy.

An increasing number of countries are adopting market- and trade-oriented approaches to telecommunications policy. Trade organizations are beginning to exert a strong influence on telecommunications - domestically, regionally and internationally. In a number of important ITU Member States, formerly distinct telecommunication administrations have been incorporated into trade and industry administrative structures.

In many countries, telecommunication regulatory bodies are being established at arm's-length from government administration. However, at the same time a debate is beginning in a number of other countries where arrangements of this kind are already in place about whether there will be a requirement for such bodies in the longer term. As telecommunications ceases to be a public utility and becomes a media business, can it be regulated under "generally applicable statutes and regulations", such as laws governing competition, freedom of speech and protection of privacy?

As different communications technologies converge, as competition increases not only within traditional communications industry sectors but across media boundaries, and as "networks of networks" begin to take shape, the only thing that is clear is that none of the traditional policy and regulatory paradigms provide an adequate basis for facilitating the development of the emerging "Global Information Infrastructure".

Striking the right balance between all the interests now at play in the telecommunications field and finding new policy and regulatory paradigms are perhaps the greatest challenges facing the ITU in the years ahead. The 1992 Geneva APP and the 1994 Kyoto PP were steps in the right direction. Their decisions have positioned the ITU to play a leading role in shaping the global, multimedia communication networks of the future. It is now up to the ITU members, both "Big M" and "small m", to decide how best to use the new ITU to further global cooperation and development in the world's information economy and society.

REFERENCES

This paper is based on a presentation originally given to the Royal Institute of International Affairs Conference on "Global Policy and Multimedia Telecommunications", London, 7&8 November, 1994.

1. The new structure adopted by the APP established ITU Sectors for Radiocommunication, Standardization and Development. Each Sector includes conference, study group and advisory board meetings involving ITU members, as well as a permanent bureau at ITU headquarters in Geneva. Although World Conferences on International Telecommunications (WCITs) are also part of the structure, they do not have a bureaucratic counterpart.
2. Under the new ITU Constitution adopted at the APP, the WATTC has been replaced by the WCIT.
3. For an extended analysis of the results of the Kyoto PP, please see the author's paper "The ITU Kyoto Plenipotentiary Conference : A View from the Inside" in the April 1995 issue of Telecommunications Policy.

PRIORITY ORDER FORM

VOLUME 2, 1995
VOLUME 1, 1994

How to Order

• Post your completed order form to : Rachel Mayne, SERDI Publishing Company, 29 rue de Saint Petersbourg, 75008 Paris, France.
• Telephone your order on (33) 61.75.79.12.
• Fax it on (33) 61.75.94.15.

Please Send Me :

[ ] copy/ies of "Telecommunications & Space Journal", Volume 2, 1995. (FF 450 ; £45 ; US $ 72,00).
[ ] copy/ies of "Telecommunications & Space Journal", Volume 1, 1994. (FF 450 ; £45 ; US $ 72,00).

Please add postage to your order.
For single copy orders please add FF 20 within France, US $ 5.00 outside France. Postage free for two or more copies.

Name : ....
Address : ....
Telephone : .... Fax : ....

Payment by :

☐ Chèque ☐ American Express Card ☐ Master card

☐ Visa Card

Expiry date of card : ....
Number : ....
Signature : ....Date : ....

EC countries only - please complete details below :

I am registered for VAT, my VAT registration number is :

I am exempted from VAT and enclose proof.

ESA'S SPIDER

RALPH P. KRÖNER [1]

Partner at Trenité Van Doorne, Advocates and Notaries, Weena 666, 3000 AD Amsterdam

Abstract : *Continuing growth of outer space activities militates for the establishment of an international authority to investigate cause, liability and extent of damage in space accidents. ESA, together with the responsible institutions of its Member States, should create an independent investigative body to meet the demands of space travel industry in quickly and reliably settling disputes involving man-made space objects.*

Key words : *ESA, space investigation, dispute settlement, SPIDER.*

INTRODUCTION

Developments in the space field and the ever-increasing exploitation of outer space give rise to some legal concern. Not only in Europe, the United States and Russia (C.I.S.), but also in China and Japan, commercial space industry activities are increasing daily. Likewise, the interests and number of parties involved are increasing correspondingly. This article highlights briefly the European Space Agency's work, together with the consequences of the combined efforts in Europe and worldwide to explore and exploit outer space, specifically with regard to space accidents. Finally, a proposal is outlined for an independent investigative deal with the inevitable disputes which will arise along with the incidence of space accidents.

I - THE SPACE INDUSTRY IN EUROPE AND WORLDWIDE COOPERATION

The last European Space Agency ("ESA") Ministers' Conference held in Granada, Spain on November 9th and 10th, 1992, resulted in a revised strategy among ESA members concerning Mid- and Long-Term Planning. In the eighties, ESA successfully realised the launch of a series of Ariane rockets and a number of telecom and meteorologic satellites [2]. These gave Europe a standing in space matters that is politically important as well as industrially rewarding to its Member States.

In response to political and financial constraints of some Member States (due to the worldwide recession), ESA has reoriented and divided its programmes to enhance its capability to react flexibly to financial and technological developments [3]. More importantly, ESA has resolved to continue focussing more then ever on international cooperation, especially with the C.I.S. and the United States [4]. Those countries are also confronted with financing difficulties for their space programmes. Thus it has become apparent that for financial, technical, and political reasons, further exploration and exploitation of space will be achieved on a multinational level. Therefore, international cooperation is necessary to create a broader platform of concerted efforts on a political level [5].

Coupled with the international tendency of ESA Member States towards cooperation, international syndicates may soon be formed through joint ventures and mergers, restructuring the space industry. Only four or five syndicates worldwide will remain at the end of the nineties, of which two will be European [6].

The tendency towards vertical integration of the space market participants will afford users a better entry into the space industry [7]. During the Granada Conference, ESA geared its projects to the users' activities, such as establishing a more clear relationship with users and operational organisations like EUTELSAT [8], EUMETSAT [9] and Arianespace [10]. ESA's work results in industrial developments and operational products like the Ariane rockets and application satellites such as ECS and METEOSAT, all managed by commercial companies [11].

ESA's political ambition - evident during the 1987 The Hague Conference - to become fully autonomous with regard to the entry into space has since shifted towards far-reaching cooperation with the United States and the C.I.S.. Despite (or perhaps because of) this shift the use of outer space will almost certainly not decrease.

II - INCREASING LAUNCH ACTIVITIES AND ACCOMPANYING ACCIDENTS

Notwithstanding the rapidly growing use of outer space, especially the increase of satellite launches, little attention has been given at European Ministers' level (ESA/EU) or at the national level to handling possible disputes arising out of this growth in use of outer space. Since the first launch of Ariane in 1979, seven out of ninety launches failed, with a total loss of over one billion Dutch guilders. A steady growth of the launch market for commercial satellites is expected to continue for at least the near future. At this moment, about 110 satellites have been launched and this total will increase by the turn of the millennium at a rate of 20 to 25 satellites a year [12].

About a year ago, Japan launched its first fully autonomously developed and constructed rocket H-II, taking its first step toward acquiring a market share in the commercial space industry [13]. However, on January 15, 1995, a German/Japanese experimental capsule carried by a M3S rocket was destroyed after losing radio contact with the Japanese space center.

The Arianespace commercial launch consortium suffered its most recent launch failure on December 1st, 1994, destroying PanAmSat's newest satellite, the PAS-3 [14]. After reaching an altitude of about 200 kilometres (120 miles), the third stage fell back into the Earth's atmosphere, carrying the PAS-3 with it. The stage and the satellite likely broke up under the stress of reentry, with the smaller segments burning up in the atmosphere. The remaining larger pieces of metal fell into the Atlantic Ocean over a large area between 1,000 and 1,500 kilometers (620-930 miles) off the west coast of Africa [15].

Arianespace had planned to launch Brazil's Brazilsat B2 and the 45-nation Eutelsat organisation's Hot Bird 1 on December 29, 1994. Eutelsat's flight was a priority mission in compensation for Eutelsat's loss of its Eutelsat 2F5 spacecraft in the January 1994 Ariane failure [16].

III - ESA AND THE NETHERLANDS' AUTHORITIES

In March 1995, the ESA Council [17] will commit ESA to (some of) the United States and Russian space programmes [18]. Due to the French Presidential elections in May 1995, ESA's next Ministers' Conference - initially scheduled for February 1995 - was shifted to late 1995 [19]. This means that revised strategy and policy plans will be decided upon on at a later date, giving the Member States more time to determine their own policy positions as to ESA's present and future role in the space industry in Europe and worldwide.

For the Netherlands, the consequence of this delay is that the Interdepartmental Space Commission (ICR) [20], cooperating with the Netherlands Institute for Aviation and Space (NIVR) [21] in preparing Dutch space policy, will further gear the mutual interests of Dutch delegations to ESA. As ESA's role involves the coordination of its own work with the national programmes of its Member States, the latter can be progressively integrated with the European programmes. All ICR's proposals with regard to the Dutch Government's space policy are sent to the Scientific Technology and Information Council (Raad voor het Wetenschaps-, Technologie- en Informatiebeleid) and the Council of Ministers (Ministerraad). The ICR is authorized to create working groups for specific subjects. For the reasons set out in the following paragraph, this article could contribute to the installation of such a working group.

IV - DISPUTE SETTLEMENT PROCEDURE IS NOT ENOUGH

Apart from the financial implications for the insurance industry [22], the legal implications of disastrous launch failures can be enormous. Not only the dispute settlement procedure [23], but also the technical investigations, identifying the cause of the accidents, should in this respect be addressed. The Paris Symposium "Aeronautical and Spatial Accidents: Investigations, Liability, Damages Settlement", organised by the French Aeronautical and Space Association (AAAF) discussed the creation of an independent investigative body [24]. Such an investigative body should have the powers to adequately determine cause, liability and extent of damage. There the idea to form a "Spider [25]" was born.

The December 1st, 1994, Ariane launch failure was immediately investigated by the Arianespace commercial launch consortium itself. Difficulties will rise, however, when catastrophes occur at, or just above the launch site, or when debris or large pieces of metal fall on the territory of another State. The parties involved in these space accidents may include :

1. the State on which territory the damage occurs;
2. the (State) party responsible for the launch (site);
3. the owner(s) of the satellite;
4. the inflicted third party;
5. insurance companies;
6. suppliers of components of launch sites, rockets or satellites.

All these parties will justifiably demand a thorough and fair investigation [26]. The interests at stake are always high. There is often no time to wait for the arbitrators or judges to appoint a team of investigation experts. Furthermore, dealing with space cases demands not only high professional standards of the adjudicating authority itself but also a solid technical knowledge of the launched object. All these points militate for the establishment of a standing independent body in charge of space accident investigations.

Use of arbitration in space law disputes suggests itself as the most preferable method of dispute settlement [27]. More than in any other field of international trade, however, the ad hoc appointment of arbitrators will not suffice to handle most space cases. Taking into account the need to start as quickly as possible the investigation of the cause and extent of the damage, an independent international investigative body should be included to any kind of dispute settlement procedure.

In this respect, a discussion now on European as well as on national levels would be helpful in order to address the problems which we will inevitably face in the (near) future. The integration of national and European space programmes should also focus on the field of implementation of space investigative rulings. Just as space insurance differs from other areas of insurance, space transactions differ from ordinary business in other fields.

CONCLUSION

In their article, Böckstiegel/Stoffel [28] conclude with the remark that arbitration might be the better dispute settlement procedure. That may under estimate the possible difficulties which accompany the enormous interests at stake, the complexity of the matters, and the usual time restraints. Although all or a great number of arbitration acts contain provisions concerning the appointment of experts, proposals need to be made on (inter)national political levels to establish one independent European investigative body. A commission should thus be founded within ESA to explore the possibilities of a permanent European investigative body with cross-border powers to immediately and effectively perform its investigatory duties.

REFERENCES

1. The work of Koen J.T.M. Hehenkamp, my assistant, for this publication is gratefully acknowledged.
2. ESA's capabilities span the fields of science, Earth observation, telecommunications, space segment

technologies, ground infrastructures and space transport systems, as well as microgravity research.
3. All Member States fund the science programmes in ratio of their Gross National Product.
4. ESA seeks the possibilities of cooperation with Russia in defining a manned in-orbit infrastructure and the definition of ESA's contribution to the Russian MIR-2 station. Equally, cooperation with NASA could be reinforced by a joint definition of an Automated Transfer Vehicle (ATV). Under the latest planning, ESA proposes to contribute the Columbus Attached Laboratory to the International Space Station. In general, before the year 2000, ESA will develop and launch several advanced telecommunications satellites as well as payloads flown on non-ESA satellites.
5. Letter from the Minister of Economic Affairs, J.E. Andriessen, to the Chairman of the Dutch Parliament, Doc. 22 800 XIII nr. 8.
6. The first one will be commanded by DASA, Alenia and Aerospatiale, the second one by Matra and Marconi Space.
7. High Key Communications report, "Ruimte voor de gebruikers", 1991, by order of the Netherlands Industrial Space Organisation (NISO).
8. European Telecommunications Satellite Organisation. The European Communications Satellite (ECS) gave impetus to this intergovernmental organisation.
9. European Organization for the Exploitation of Meteorological Satellites. ESA plans a joint venture with EUMETSAT which is illustrated in the proposed METOP (meteorological operations) satellite series.
10. Arianespace was set up to market the Ariane launchers. EUTELSAT and EUMETSAT have grown from original ESA satellite programmes.
11. Flights with the Ariane 4 vehicle are fully booked for 1996. Two Ariane 5 demonstration launches are scheduled for Spring 1996. ESA spends the bulk of its budget on contracts awarded to industry in its member countries. Its policy assures each Member State in return for its investments, both a financial return and a share in the technological spin-off. The Granada Conference decided to continue the advanced development of the Ariane 4 and 5 because the project promises to be commercial self-sufficient, earning enough through commercial satellite launches to recoup the investment.
12. About 70% of all satellites perform telecommunication services for television, radio, telephone, data-exchange and contact with ships and aircraft.
13. Japan's rocket was developed by the National Space Development Agency of Japan (NASDA) and launched on February 3, 1994. If the test launches all prove successful, NASDA plans to launch commercial satellites, making it a direct competitor of Arianespace. Arianespace currently has a 60% market share. The rest of the space launch market industry is divided between the United States, Russia and China.
14. On January 24, 1994, the Ariane 4-4LP carrying two communications satellites, Turksat 1 and Eutelsat 2F5, both built by Aerospatiale, fell into the Atlantic Ocean.
15. Space News, December 1994, p. 21.
16. See footnote 13.
17. The Council is constituted of representatives of the Member States. It meets four times a year and takes general policy decisions and rules on scientific, technical, administrative and financial matters.
18. See footnote 3.
19. At this late 1995 Conference, for example, a detailed program for working with the Russians on studies of space plane and space stations will be discussed. The agenda will also include a plan of action for the United States-led international space station projects.
20. Interdepartementale Commissie Ruimtevaart (ICR). Within the ICR, representatives of several Ministries have a seat, as well as representatives of the Netherlands Institute for Aviation and Space (Nederlands Instituut voor Vliegtuigontwikkeling en Ruimtevaart [NIVR]), the Dutch Space Research Institution (Stichting Ruimte-onderzoek Nederland) and the Policy Commission Remote Sensing (Beleidscommissie Remote Sensing).
21. The NIVR is the important intermediary between ESA and Dutch industry (and Dutch scientific institutions). The NIVR meets six to seven times a year.
22. The net loss (less all premiums paid) to the insurance industry for the failures in 1994 (including the loss of AT&T's Telstar 402 satellite which expired shortly after its launch in September 1994) will amount to about $ 220 million.
23. In the field of space law, arbitration could be the most suitable way of settling disputes.
24. Colloque: "Les accidents aéronautiques et spatiaux: investigations, responsabilité, réparation des dommages", Mai 17 et 18, 1994, Paris, "Legal aspects of the setting-up of inquiries and technical assessments in the space field" (Ralph P. Kröner).

25. SPIDER: Space Investigation and Damage Establishment Ring ((c) Ralph P. Kröner).
26. It suffices to just mention the Cosmos 954 case.
27. Telecommunications & Space Journal, Volume 1, 1994, p. 327-337, K-H. Böckstiegel/W. Stoffel, "Private outer space activities and dispute settlement".
28. Supra, p. 335.

TELECOMMUNICATION AND ITU: SUGGESTIONS FOR CHANGES AND IMPROVEMENTS

DR. J. GAVAN

Center for Technological Education Holon affiliated with Tel-Aviv University ; 52 Golomb Street ; POB 05 ; Holon 58102 ISREAEL

Abstract *: The present tremendous development in telecommunication necessitates efficient international coordination and cooperation in order to avoid regression and chaos. The most significant growth in telecommunication users, services and systems occur in the radio mobile field characterized by an exponential expansion. The increasing numbers of mobile users, equipments and systems necessitate mandatory efficient radio standardization and regulation policies as well as thorough ElectroMagnetic Compatibility (EMC) considerations. The International Telecommunication Union (ITU) UN specialized agency may be qualified to support telecommunication optimized expansion for the benefit of humanity. However, as it is almost impossible to attain perfection, significant changes and improvements are suggested. As an ex-ITU expert with an active experience of about 40 years in the radio communication and transmission fields, I try to deduce telecommunication and ITU main challenges. Suggested changes and improvements are analyzed by the following order of importance : Formation and training, standardization, library and documentation centers, technical publications, computers and quality management centers. Funding for the required functions is also discussed. However, the subject of computers and quality management centers remain open for further discussions.*

INTRODUCTION

Telecommunication enables communication from distances.

Telecommunication systems are actually the most mature application of Electronics and one of the most important factors in our expanding information era. They can provide instantaneous and simultaneous data, voice and picture informations to numerous places in our world. The continuation of the latest tremendous developments of Telecommunication, essential to human welfare, depends on worldwide cooperation, national and global harmonization and standardization, and on the efficient support and coordination of a specialized international agency [1 ; 2 ; 3].

The International Telecommunication Union (ITU) is the most veteran organization concerned with Telecommunication. The ITU was founded in 1865 by a few nations in Europe to regulate telegraphy and telephony traffic across their frontiers. The first main ITU activities were : language translations, service tariffs and the beginning of activity concerning compatibility of equipments. Nowadays, the ITU is a specialized agency of the United Nations consisting of about 180 member nations, usually PTT administrations, and cooperates with numerous industrial firms, scientific and specialized organizations all related to the field of telecommunication [2 ; 4]. The ITU main activities have to be dedicated to the development and improvement of Telecommunication and have recently undertook several significant changes [5 ; 6]. The most obvious are the merging of the CCIR and major part of the IFRB in the ITU-R Radiocommunication Bureau, and the CCITT has changed its name to the ITU-TS Telecommunication Standardization Bureau emphasizing the increasing importance of standardization to the development of Telecommunication. However, standardization is no less important to the ITU-R than to ITU-TS. Parts of the cooperation activities will be handled by the development department [7 ; 8].

An efficient organization has to change constantly and improve in order to approach optimum efficiency as absolute perfection is impossible.

It is a very difficult task to prepare serious improvement suggestions for such complex organization as the ITU, especially because of the recent tremendous technological, economical and political changes [6 ; 9 ; 10]. This was a challenge when the ITU Secretary General, Dr. P. Tarjanne, proposed that I write suggestions for enhancing ITU role in global improvements of Telecom services.

But I have hesitated a lot, in spite of my past and present long experience in radio communication and electronics, including 7 years as an ITU expert in the field for development, planning, management and teaching tasks.

However, considering that a person active in the radio

communication field for about 40 years but not involved directly in the ITU activities for more than 20 years, owns a special perspective that can be useful to the ITU and the telecommunication community.

Therefore I dedicated my efforts and was able to answer to the challenge. More recently a member of the Secretary General's staff encouraged me to propose part of my confined proposals as a paper for a dedicated Telecommunication Journal.

I - TELECOMMUNICATION AND ITU ROLES AND CHALLENGES

The main achievements in the last half of this century are in the fields of Transportation and Telecommunication, as stated in several publications. However, the achievements in telecommunica-tions are significantly more important as quoted by Arthur Clarke and other serious forecasters [1 ; 10 ; 11]. Telecommunication developments are very important for the welfare of humanity in the social, economic, scientific security and other aspects of human life. Therefore a better world for mankind and a brighter future for coming generations require optimum improvements in telecommunication services. The optimum improvements require enormous efforts [12]. The ITU main contributions are in the coordination and regulation of worldwide telecommunication services and systems [5 ; 6]. Therefore the ITU is one of few candidates to achieve the huge challenges in telecommunication improvements on condition that it will enhance education, quality and efficiency wisely and benefit from adequate assistance and budgets. The principal challenges in telecommunication developments are as follows:

1. Enabling access to wired telephone or Radio wireless systems to most people in the world on a national scale in order to improve life quality and increase the probability of saving life in emergencies. Nowadays the global number of telephones is approaching 1000 millions related to a market value of more than $500B per year. However less than 20% of the world population account for more than 90% of all telephones in use. This and other inequalities necessitates radical changes [10 ; 13].

2. Optimizing long range communication services between countries on the regional and international scale in order that people become aware of what is happening all over the world, even in the most remote countries and regions. This target may enhance efficient international cooperation assistance in case of disasters, reduce

regional and international conflicts and improve human conditions of life [7 ; 11].

3. Enhancing communication links between computers and scientific centers on national, regional and international scales to improve research efficiency, reduce duplication and enhance quality in medicine, education, nutrition, agriculture, space research and other domains important for human welfare and development [3 ; 10].

The present expansion trends will enhance merging between Telecommunication, computers and broadcasting to a powerful ISDN or multimedia global network [6 ; 12]. These ambitious challenges are associated with the vision of a global peaceful and prosperous (village) world served by optimal telecommunication networks [1 ; 10].

To achieve optimum improvements in Telecommunication, ITU efforts have to be extended especially for the future development of telecommunication systems and networks and improvements of current services and systems. Telecommunication of tomorrow will be based on the present and future research efforts and directed by students trained by us today in our institutes. Therefore the principal improvements have to deal with the following subject : Telecommunication education and training, standardization and planning teams, library and documentation centers, technical publications, computer centers, and quality centers. Nowadays, telecommunication education and training projects are only side-line activities of the ITU; and most of the training projects budgets come from external organizations such as the PNUD, the World bank and bilateral contracts allocations [2 ; 4]. However, implementation of efficient telecommunication education and training systems are mandatory for the development of telecommunication but require tremendous financial and qualified manpower investments [14].

Therefore a serious proposal for education and training plans is presented due to its crucial importance towards the efficient development of Telecommunication.

II - EDUCATION AND TRAINING

Proper design, construction, operation and maintenance of telecommunication systems, networks and services requires sound training projects. The optimal training in the telecommunication speciality is unique and requires a thorough formation in several subjects of science, engineering, and management [2].

Efficient formation requires considerable and long term investments as the fruits of education are not obtained immediately or in the near future, but can contribute tremendously to the future development of telecommunication, technology, science and the welfare of human society. But at present ITU does not have the means. However, its leaders can significantly contribute by sensitizing the world opinion and leaders to the importance of contributing and enhancing the education in telecommunication.

From pedagogic principles, the educator usually has to be one or two steps higher than the students levels [15]. Therefore high level formation centers are mandatory for forming leaders, experts in telecommunication R & D personnels, and professors to teach in these and lower level centers.

The ideal instructional pyramid top will be an international post graduate institute with English as the working language. The teaching personnel will be composed of permanent staff and other qualified scientists and engineers for mid term contracts recruited for their knowledge, skill, research records, teaching ability and integrity. The post graduate students up to the Ph.D. level will be chosen according to records and achievements. Every two to three years the main teaching efforts will concentrate on different required Telecommunication specialities, such as switching, transmission, Radio systems, etc. In parallel, several short frontly, or better, video teleconference courses will be organized for top level PTT employees and specialists during the academic year to enhance permanent education. The teaching programs and laboratory equipments will be selected in cooperation with the leading institutes in the world offering Telecommunication programs.

Projects for the development of high level international telecommunication institutes in the USA and Japan have failed as well as several international cooperation projects which were successful for several years, such as the Philips International Institute in the Nederland, but ended because of financial difficulties. However, a few national programs are very successful, efficient and productive such as the University of Colorado at Boulder Interdisciplinary Telecom. Program (USA) or the France Institute of Telecommunication now belonging to the France Telecom [14].

The future education projects management board has to be international, including high-quality members from ITU and other high standard organizations, university and institutes in the telecommunication field [15 ; 16]. The links with other UN international organizations will be limited to pedagogic and management subjects

Figure 1

SUGGESTED FORMATION AND TRAINING PYRAMID IN TELECOMMUNICATION

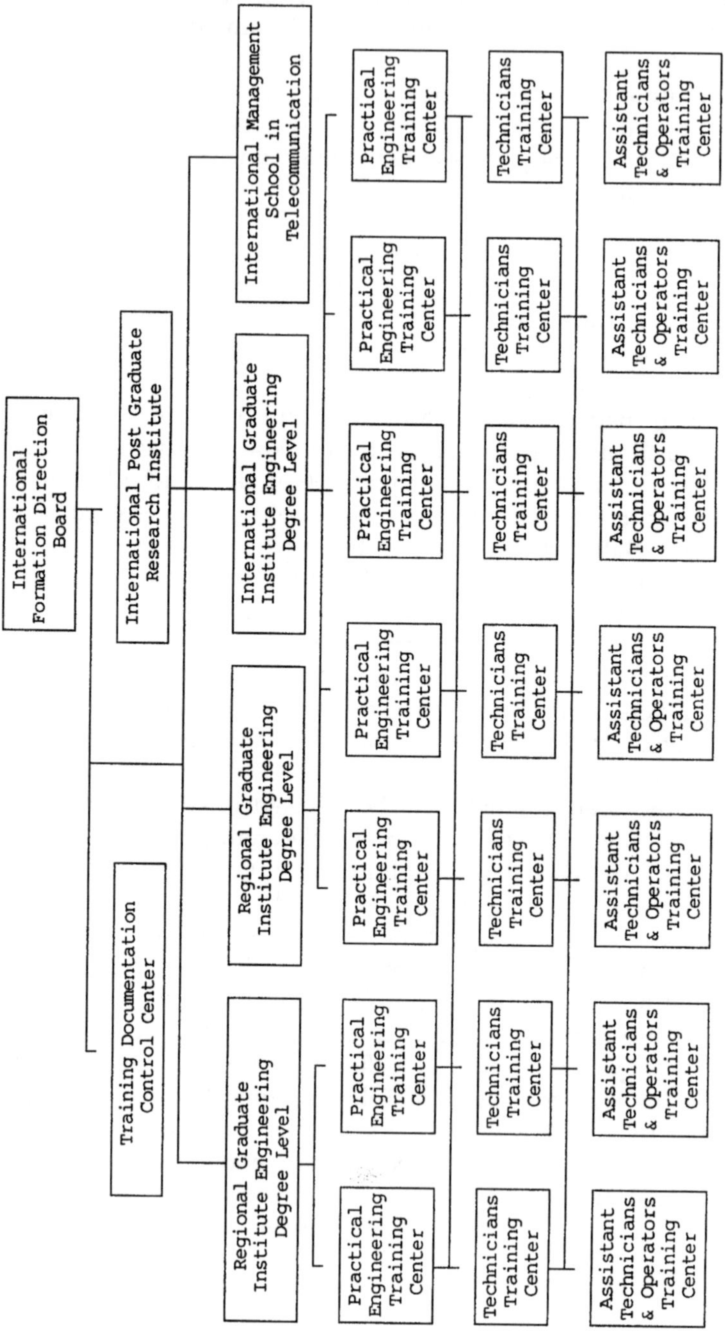

with UNESCO and to some technical subjects with ICAO, OMM, Intelsat and other organizations involved in important activities in the telecommunication field.

Using similar criteria, ITU may contribute to the establishment and operation of a few regional telecommunication institutes on the graduate engineering level, especially for developing countries. The institutes will be linked via a common video and computer networks, but the working language of part of them may be other than English. The main product of the regional telecommunication institutes is the formation of able engineers, top and middle level personnel for the PTT administration and companies and skilful instructors for the undergraduate institute and other telecommunication technical training projects.

ITU may collaborate with international, regional and national organizations in the operation of several undergraduate regional institutes all over the world for the training of superior technicians and middle level staff for telecommunication organizations.

A significant number of instructors and graduates will hold leading positions in telecommunication administrations, universities or companies throughout the world, and therefore may enhance positive international cooperation required for optimal improvements in telecommunication operation standards and developments.

The base of the training pyramid shown in Figure 1 is a significant number of national or regional centers for the formation of technicians and assistant technicians. These centers will offer specialized training in the different telecommunication branches and occasionally will be advised by higher level institute experts to enhance quality and efficiency.

ITU may participate in the establishment and direction of an international special model training center in telecommunication. In this center, the most efficient relevant formation programs, curriculums, printed literature, audio visual or video, modular training and laboratory planning documentations will be found [16]. The staff of this center will provide relevant advices, consultations and assistance at all levels in most fields of telecommunication to training centers or other organizations interested in this specialized training.

III - STANDARDIZATION

Efficient standardization achievements are also a necessity and first priority. Otherwise, the actual ultra fast evolution in the fields of telecommunication may stop completely, resulting in regression and chaos.

Main development trends in telecommunication are as follows : A huge increase in the number of telephone sets and users in the developing countries and also in mobile radio communication users, equipments and systems, including the emerging Personnal Communication Services (PCS) in developed countries. Important shift from analog to more efficient digital modulation methods create new regulation problems as well as massive utilization of computers and software. These trends enable progress toward the Integrated Services Digital Network (ISDN) target in spite of their immense complexity. The success of all these developments heavily depends on effective standardization [2 ; 17 ; 18].

The tremendous exponential increase in telecommunication services, users and equipments, especially in mobile radio, have enhanced the importance of the Electro-Magnetic Compatibility (EMC) field [17 ; 19].

EMC is also connected to the efficient utilization of the limited frequency spectrum and satellite geostationary orbit resources. Development of novel methods to mitigate Electro-Magnetic Interference (EMI) sources and effects for increasing the quality of telecommunication systems, are required to support an efficient spectrum management policy [20 ; 21]. Optimal developments of High Definition Television (HDTV), ISDN synchronization, Asynchronous Transfer Mode (ATM) and Local and Wide Area Network (LAN) and (WAN) formats are also dependent on efficient standardization measures [4 ; 9 ; 22].

The increasing importance of standardization and regulations for telecommunication development result in the establishment of numerous organizations in the EMC field on the international, regional and national scales. The most important EMC organizations are presented in the synoptic diagram of Figure 2 and the abbreviation list of Table 1 [2 ; 4 ; 17 ; 20 ; 22].

The ITU awareness is reflected by the recent change in the name of its two main operating departments the telephony Standardization ITU-TS and the radio communication bureau ITU-R [7], which is not the best choice as both are heavily involved in standardization.

In the operation of the radio communication study groups alone there are involved more than 1000 specialists from all over the world including numerous volunteers, plus a permanent bureau staff ITU-R especially dealing with standardization and EMC problems. The ITU-TS department is heavily involved in performances and standardization of telecommunication systems and in the interconnection with public networks [23 ; 24].

Figure 2

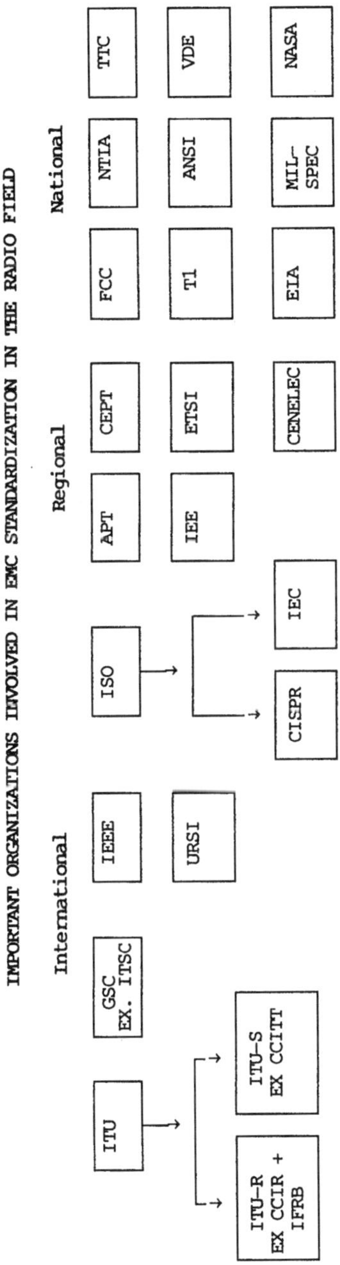
IMPORTANT ORGANIZATIONS INVOLVED IN EMC STANDARDIZATION IN THE RADIO FIELD
International
Regional
National
ITU
ITU-R EX CCIR + IFRB
ITU-S EX CCITT
GSC EX. ITSC
IEEE
URSI
ISO
CISPR
IEC
APT
IEE
CEPT
ETSI
CENELEC
FCC
T1
EIA
NTIA
ANSI
MIL-SPEC
TTC
VDE
NASA

Table 1

Abbreviations for Figure 2

ANSI	American National Standard Institute (U.S.A.)
APT	Asia Pacific Telecommunity
CCIR	International Radio Consultative Committee, now ITU-R
CCITT	International Telegraph and Telephone Consultative Committee, now ITU-TS
CENELE	European Committee for Electrotechnical Standardization
CISPR	International Special Committee on Radio Interferences
Com.	Communication
EIA	Electronic Industries Ass. (U.S.A.)
ETSI	European Telecom. Standard Institute
FCC	Federal Communications Commission (U.S.A.)
GSC	Global Standards Collaboration Group
IEC	International Electrotechnical Commission
IEE	Institute of Electrical Engineering (Europe - G.B.)
IEEE	Institute of Electrical and Electronic Engineers (International - USA)
IFRB	International Frequency Registration Board
ISO	International Standardization Organization
ITSC	Interregional Telecommunications Standard Conferences
ITU	International Telecommunication Union, R (Radio) TS (Standardization), D (Development)
NASA	National Aeronautic and Space Agency
NTIA	National Telecommunication and Information Administration (U.S.A.)
T1	Committee and non profit organization to draw up standards for Telecom. (U.S.A.)
Telecom	Telecommunication
TTC	Telecommunication Technology Committee (Japan)
URSI	International Union of Scientific Radio
VDE	Communication and Standard Commission (German)

Standardization and ITU operations have to cope with the recenttrends of privatization and liberation in telecommunication services. This can be reflected by significantly increasing the participation of private industrial and scientific organizations in the ITU study groups and other activities. Privatization reduces government control and heavy protected monopoly of supply and services to a more flexible and competitive participation. These new trends may enhance productivity and innovation [18]. However, too much privatization may cause excessive redundancy, waste, chaos and degradation because of insufficient control and inadequate standardization cooperation. An efficient international telecommunication organization has to indicate the golden path. The ITU has to consolidate its role in the application of an efficient standardization policy and contribute to the human development and welfare by protecting and optimizing the evolution of telecommunication services and systems using the following steps :

1. *Information and warnings*

To convince to the authorities and the great public that development of international standards and coordinating their implementation are crucial measures for the establishmentof a global efficient and useful worldwide telecommunication network. Insufficient standardization and EMC measures may stop telecommunication development and cause regression and chaos. The potential degradation of telecommunication services quality can be worse than the state of High Frequency (HF) long distance communications in the sixties before the successful development of Geostationnary satellite systems [3 ; 17 ; 25].

2. *Coordination and Cooperation*

Coordination and harmonization of the standardization and EMC activities inside each ITU department by increasing the number of permanent skilled experts. An efficient coordination is also required between the radio communication and the standardization departments. The coordination and fruitful cooperation with UN, IEEE, and other international organizations, regional and national public and private organizations, dealing with telecommunication systems and standardization are very important ; especially in the radiocommunication, broadcasting and other radio systems [22 ; 24]. Common efforts are crucial especially in the management of the frequency spectrum limited resources and the development of mitigation methods to reduce mutual radio intra and intersystem interference [17 ; 25 ; 27].

3. *Standardization Activities and Assistance*

Development of international telecommunication standards at an increased rate so that they will not fall too far behind technological advance. Increasing the quality, usefulness and numbers of publications, reports, handbook and computer software concerning standardization of telecommunication equipments, services and systems using the cooperation of other organizations [4 ; 22]. Assistance to needed administrations and organizations for the regional and national development and implementation of telecommunication standards coping with the international regulations [24].

4. *Operation of a Unique Test Center*

To design, build and operate an innovative model laboratory for most standardization and EMC measurements in the field of telecommunication [20]. This unique test center directed by skilled experts can significantly contribute to the R & D in global telecommunication and improvement of international standards and EMC techniques in cooperation with national and regional centers [2 ; 5]. This test center can become financially self supporting by operating impartial arbitrage tests between competitive telecommunication industries and other consultation and training activities.

5. *Organization of Seminars, Workshops and Exhibitions*

To organize, assist and participate in efficient seminars, workshops, conferences and exhibitions concerning telecommunication subjects. The ITU World Telecommunication exhibition and conference every second year in Geneva is an example of successful contribution to the development of the telecommunication in the world and to interest the public in the role of ITU and the importance of investments in telecommunication for the benefit of humanity [4 ; 6 ; 9].

6. *Education and Training*

The implementation of efficient and innovative standardization and EMC policy necessitate the participation of highly qualified and skilled telecommunication specialists in technology, management and legislation. The number of qualified personnel is insufficient, especially in developing countries. Therefore efficient and high quality education and training programs are required as presented in Section 3 of this paper.

Priority in development and standardization have to be allocated to the improvement of wireless life rescue projects contributing to the saving of human life, coordinating international aids in case of emergencies and reducing conflicts between people [7;10;11;29].

IV - LIBRARY AS DOCUMENTATION CENTER

In order to enhance ITU expertise, the ITU library has to be enlarged as a model documentation center, in order to include the most important documentations in the fields of telecommunication. The documentations will include books, audio visual, compact disks and videotapes. The library administration can be assisted by several technical experts on part time jobs and most of its content will be efficiently computerized and micro filmed. The purpose is to furnish promptly the most precise and relevant information or statistics to ITU staff or outside administrations and customers even on a payment basis in order to support itself financially and to reach the highest possible level.

ITU staff will be involved partially in reviewing books, journal and audio video packages in their fields of speciality. The target is that the books or video package editors will be interest in supplying free of charge all samples in telecommunication subjects to the ITU library for evaluation by impartial experts in their fields.

V - TECHNICAL PUBLICATIONS

The ex and the new Radio Communication and Standardization and other ITU technical publications and handbooks are useful to technicians, engineers and scientists in Telecommunication. However, they can be significantly improved.

The actual technical classification is too administrative and the ITU regulations could be completed by the edition of new handbooks, design manuals and application notes on different theoretical and practical levels. Also compact disk techniques could be useful. The aims of the publication effort are to help more efficiently technical staff from different professional levels in conception, design, standardization, maintenance, exploitation and other tasks in the telecommunication domain.

The ITU editing and publishing efforts could be reduced by

cooperation with several important publishing houses, organization like the IEEE, IEE and URSI and several other organizations also involved in telecommunication publications. Joint editions on special subjects important to telecommunication may enhance quality and diminish expenses. More participation of ITU staff in the editing of the above publications can serve ITU interests in the promotion and development of telecommunication.

The ITU Journal of Telecommunication published in three languages is unique and could be very useful if it continues. More contact between the readers and the editor and more publicity could be helpful. Circulation could also be increased by publishing more papers on telecommunication at various levels. Most papers should be popular, practically oriented but with sound scientific and technical basis at the level of the IEEE Spectrum or the Communication Magazine, and a few articles very theoretical but useful with practical foundations on the level of the IEEE Transactions on Communication or URSI publications. The result should be a significant increase in the number of interested readers.

VI - COMPUTERS AND QUALITY MANAGEMENT CENTERS

The merging of telecommunication and computers significantly enhanced the use of computers in : managing networks and systems ; simulating circuits, equipments and systems design ; and automatic measurements of equipments and systems performances.

The recent Total Quality Management "TQM" fashion can be promoted by the ITU for useful applications in the manufacture of Telecommunication equipments and may contribute to enhance the reliability of the services and systems. However, my knowledge on computers and TQM is too limited to provide expert advice.

VII - FUNDING

Achievements of efficient standardization, formation and other projects for the development of telecommunication require enormous funds, human resources and materials. The authorities, public and private organizations have to be convinced that even for their own sake and profit, contributions and investments in telecommunication will be very beneficial in long term planning [3 ; 6 ; 9 ; 10].

Providing ITU or other international organizations with the proper financial resources is not an easy task and necessitates very dedicated personnel. However, it should contribute to a harmonized and optimal development of telecommunication crucial to progress and human welfare.

CONCLUSIONS

The main conclusions on Telecommunication and ITU : Suggestions for changes and improvements are as follows :

1. Telecommunication services and systems tremendous growth, beneficial to human welfare and progress crucially depend on efficient training programs, standardization and considerable funding [8].
2. Efficient international experts cooperation is mandatory to assure telecommunication development, especially due to the recent privatization trends [18]. Otherwise regression and chaos may occur.
3. Education and training in telecommunication can be significantly improved [14]. At present, training is only a side-line activity of the ITU [9]. However, allocation of sufficient budgets and expert personnel may permit ITU or other international organizations to contribute in long term education programs crucial for an efficient development of the telecommunication, progress and human welfare especially in developing countries. If the ITU is unable or not interested in leading the presented global education and training projects in telecommunication [12], other organizations, such as the IEEE may be designed for these important tasks.
4. The ITU has to significantly improve in order to successfully assume the coordination functions required for optimally contributing to a stable and important progress in the telecommunication field. Recently (December 1992) an international high level committee has been nominated to review the structure and functioning of the ITU [6 ; 13] and it is possible that this committee will be efficient and successful.
5. In spite of the significant changes in the International Telecommunication Union (ITU), many improvements are still required in several fields. For instance : a). the frequency management allocation policy using 3 different geographical zones is obsolete. Especially with implementation of the radio location GPS/NAVSTAR and GLONAS global systems and the forthcoming operation of Low Earth Orbit (LEO) communication satellites [8]. b). the number of footnotes in the ITU regulations is

confusing and are recommended to be reduced [17 ; 24]. c). the allocation procedure of satellite Geostationnary orbit slots locations has to be revised in order to prevent saturation, deterioration of global communication and unfair commercial activities [24 ; 30 ; 31].

6. ITU main activity in the field of standardization and regulation may be optimized by promoting vigorous standardization steps crucial for positive development of telecommunication. Standardization and telecommunication efficient progress require enhancement in the training, quality and quantity of the involved personnel, international cooperation and acceleration of the standardization procedure [24].
7. Changes and improvements may also be beneficial in organizing the library, document center, technical publications edition, computer and quality management center.
8. The future merging effects of telecommunications, computers and broadcasting necessitate an expert discussion.

REFERENCES

1. Bordewijk, J.L., "Telecommunications in a Modern World", Philips Review, June 1972, pp.15-20.
2. Wallenstein, G.D., "Setting Global Telecommunications Standards: The Stakes, the Players and the Process", Artech House 1990.
3. Berger, R., "The Future is in Telecommunications", Telcom Report International, Vol.16, No.3, 1993, pp.6-7.
4. Macpherson, A., "International Telecommunications : Standards Organizations", Artech House 1990.
5. Tarjanne, P.J., "The Challenges Facing the ITU", New Breeze, Vol.6, No.2, April 1994, pp.3-5.
6. Bellchambers, W.H., "New World-New Mind-New ITU", ITU Newsletter February 1994, pp.9-15.
7. Kirby, R.C., "Radio Communication in the New ITU", ITU Radio Communication Info., No.1, June 1993, p.1.
8. Struzak, R.G., Kirby, R.C., "Evolution of Radio Communications and Future ITU", The International Wroclaw Symposium on Electromagnetic Compatibility, September 1992, pp.15-19.
9. Maclean, B.S., "The ITU and the Networks of the Future : Policy Horizon for the Information Age", ITU Telecommunication Journal, August 1993, pp.287-292.
10. Messerchmitt, D.G., "Telecommunications in the 21 Century", IEIC Trans. on Com. (Japan), January 1993, pp.3-11.
11. Bross, D., "The Clarke Interview", Via Satellite, December 1991, pp.40-44.
12. Utsumi, Y., "Pulling it All Together", New Breeze, Vol.6, No.3, July 1994, pp.2, 10-14.
13. ITU "Report on the State of World Telecommunications", ITU Newsletter, January 1994, pp.9-12.
14. Pelton, J.N., Quigley, J., "New Opportunities in Satellite Tele-Education", Space Communications, Elsevier, Vol.9, No.4, April 1992, pp.253-261.
15. Personnel Information from the Late Professor F. Ollendorf at the Technion in Israel, 1963.
16. Welch, S., "PBS5's Educational Satellite Network" Via Satellite, November 1993, pp.34-41.
17. Perez, R. Editor, "Handbook on EMC" - Gavan, J., 'EMC in Radio Systems", Chapters 19 and 20, Academic Press 1995.
18. Ungerer, H, "What is Going On To Take the Barriers Down", World Communications and Technology, 1994, pp.175-177.
19. Struzak, R.G., "Simulation Model for Evaluating Interference Threat to Radio Communication Systems", Telecommunication Journal, Dec 1990, pp.827-839.
20. ITU-CCIR Handbook, "Spectrum Management and Computer-Aided Techniques", Geneva 1987,

Chapters (1-3); 6.

21. Gavan, J., Joffe, E., "An Investigation of the Desensitizing Effects by High Power HF Broadcast Transmitter on HF Airborne Receivers", IEEE Transactions on Electromagnetic Compatibility, Vol.34, No.5, May 1992, pp.61-73.
22. Mazda, F., "Standardizing on Standards", Telecommunications, May 1993, pp.40-50.
23. Tokuo, I., "Changing Standardization environment Prompts Changes in ITSC", New Breeze, Japan, Winter 1993, pp.23-26.
24. ITU General Secretariat, "Radio Regulations" Tomes 1,2,3, Geneva 1990.
25. Spaulding, A.D., "Effects of Noise and Interference on System Performance", NATO AGARD Conf., Summer 1987, pp.10(1-17).
26. Struzak, R.G., Olms, K., "Frequency Management and Software for Spectrum Engineering for Personal Computers", ITU Telecommunication Journal, April 1993, pp.168-174.
27. Gavan, J., "Radio System Interference Effects between Two Collocated Vehicular Transceivers : Analysis, Computation and Mitigation Methods", IEEE Transactions on Vehicular Technology, August 1994, pp.
28. Bykhovsky, M.A., Pavliouk, A.P., "Approach to the Development of a Uniform Methodology for EMC Analysis of Radio Systems", ITU Telecommunication Journal, Vol.60, February 1993, pp.65-80.
29. Pelton, J.N., "Low Orbit Satellites - A New Global Player in Telecommunications", Space Commmunications, Elsevier, Vol.9, No.4, April 1992, pp.241-242.
30. Editorial, "Make ITU a Stronger Referee", Space News 24/30, January 1994, pp.2-3.
31. Degman, K., "Orbital Settlers shoot it out with Space Outlaws", Via Satellite, September 1993, pp.30-41.

- *VII* -

MOBILE COMMUNICATIONS

INTERNATIONAL REGULATION OF THE RADIO FREQUENCY SPECTRUM AND ORBITAL POSITIONS

STEWART D. WHITE

Media and Communications Group, ASHURST MORRIS CRISP, Broadwalk House, 5 Appold Street, London EC2A 2HA, Telephone : 0171-972 7813, Facsimile : 0171-972 7990

INTRODUCTION

Radio frequency spectrum and orbital positions are finite resources. This article outlines the reasons why they require regulation both at an international level and nationally and identifies and describes the role and duties of the most important organisation concerned with international telecommunications policy and regulation. As an example of the current international regulations, the article examines the procedure for the co-ordination of geostationary fixed-satellite services, before briefly discussing enforcement of such regulations and drawing conclusions as to their future form. The author thanks the International Telecommunication Union for its assistance in providing information in support of this article including, in particular, information on the outcome of the 1994 Kyoto Plenipotentiary Conference and for allowing certain flowcharts to be extracted from the Union's materials for reproduction here.

I - IS REGULATION NECESSARY ?

It is generally accepted that outer space is not available for appropriation by any nation or person. In effect, the international community has agreed that the use and exploitation of outer space are subject to international law and that outer space is for the benefit of all mankind [1]. Usage of satellites, that is, stations [2] in outer space used for a variety of purposes, both military and peaceful, is therefore governed by international law. Such purposes include broadcasting and telecommunications usage.

Apart from this broad-based "public trust", there is a particular point in the need for international co-operation or regulation of satellite usage in outer space. Satellites use radio frequency spectrum for a number of purposes principal among which are the sending of signals to and from the satellite for communications purposes. This is known as "up-linking" and "down-linking", respectively. Radio frequency spectrum is also required for such functions as telemetry to and from the satellite as well as control and command functions from the earth to the satellite. The most efficacious frequencies for such purposes are limited, not so much by the laws of man, but by the laws of physics.

For example, for what are known as geostationary satellites [3], that is those satellites which are, in layman's terms, "stationary" at a height of approximately 36,000 km over the equator as a result of revolving at the same speed as the earth, there are a limited number of orbital positions:

(a) into which you can place such satellites, irrespective of technical considerations; and
(b) further limited by technical considerations, such as the need to avoid "harmful interference" between the services carried on one satellite by another.

Thus, as the potential uses of satellites were considered and exploitation of them began, the question which needed to be addressed was: How should these finite resources be shared amongst all those in the world who wished to exploit satellite technology and who wished to do so without "interference" ? In other words, how can such resources be used efficiently and so as to enable the introduction of as many services as possible?

To date, technology has made it possible to use these finite physical resources more and more efficiently. However, technology will not deal with issues such as "equal access" that the less developed countries now entering into the satellite communications industry are claiming from the leading developed countries many of which have established their positions sometimes on a "first-come, first-served" basis. Thus there is a tension in the existing international regulations as they try to juggle efficient usage of these finite resources against their equitable usage. More recently the plight of the developing countries has been recognised by the recommendation that developed countries take into account requests for favourable treatment made by developing countries [4]. Ultimately, the members of the international community will have to work together to solve these difficulties: it is and has always been a question of political will.

II - THE INTERNATIONAL TELECOMMUNICATION UNION ("ITU")

International co-operation in the communications field is not new. As far back as 1865, the founding members of the International Telegraph Union recognised that there was a need to govern international communications, although at that time not by reason of the scarcity of radio frequency spectrum, but arguably more to place such communications firmly under the control of governments rather than the private sector. However, by 1927, the usage of radio frequency spectrum had got to the stage that the then International Radiotelegraphy Union agreed in Washington a table for use of frequencies so as to avoid services "causing harmful interference" to each other [5].

The successor to these bodies, the International Telecommunication Union (ITU), is probably the most important organisation concerned with international telecommunications policy and regulation. It has approximately 180 member countries (Members) and is the oldest of the international bodies concerned with telecommunications.

A new Constitution and a new Convention were adopted at the Additional Plenipotentiary Conference in Geneva in 1992. Together with certain other Administrative Regulations they set out the structure, role and function of the ITU. They are part of international treaty law and execution of them imposes on the Members the obligation to abide by them and, importantly, the Administrative Regulations made under them include the obligation on Members to take such steps as are necessary to oblige observance of them on those operating as authorised by Members [6]. Whilst both the instruments adopted in Geneva will come into force on 1st July 1994, all their provisions relating to the new structure and methods of the ITU have been applied as if in full force since 1st March, 1993.

In addition to the ITU, the United Nations has been active in this field, particularly through the General Assembly and its Committee on the Peaceful Uses of Outer Space. The ITU's relationship with the UN is defined by an agreement entered into between the two organisations [7]. The ITU provides information on its activities concerning telecommunication and the peaceful use of outer space to the UN annually [8].

The two relevant primary functions of the ITU are set out in Article 1.2 of the ITU's Constitution. Article 1.2 provides:-

(a) effect allocation of bands of the radio frequency spectrum, the allotment of radio frequencies and the registration of radio frequency assignments and any associated orbital positions in the geostationary satellite orbit in order to avoid harmful interference between radio stations of different countries; and

(b) co-ordinate efforts to eliminate harmful interference between radio stations of different countries and to improve the use made of the radio frequency spectrum and of the geostationary satellite orbit for radio communication services.

In addition to them, Articles 44 and 45 of the ITU's Constitution are relevant. Article 44 provides:-

1. Members shall endeavour to limit the number of frequencies and the spectrum used to the minimum essential to provide in a satisfactory manner the necessary services. To that end, they shall endeavour to apply the latest technical advances as soon as possible.
2. In using frequency bands for radio services, Members shall bear in mind that radio frequencies and the geostationary satellite orbit are limited natural resources and that they must be used rationally, efficiently and economically in conformity with the provisions of the Radio Regulations, so that countries or groups of countries may have equitable access to both, taking into account the special needs of the developing countries and the geographical situation of particular countries."

Interference, in particular harmful interference, is dealt with in Article 45 of the ITU's Constitution:

1. All stations, whatever their purpose, must be established and operated in such a manner as not to cause harmful interference to the radio services or communications of other Members or of recognised operating agencies, or of other duly authorised operating agencies which carry on a radio service, and which operate in accordance with the provisions of the Radio Regulations.
2. Each Member undertakes to require the operating agencies which it recognises and the other operating agencies duly authorised for

this purpose to observe the provisions of No. 197 [ie No.1. above] above.

3. Further, the Members recognise the necessity of taking all practicable steps to prevent the operation of electrical apparatus and installations of all kinds from causing harmful interference to the radio services or radio communications mentioned in No. 197 [ie No.1. above] above."

Interference is defined in the RR's as:

"Interference: the effect of unwanted energy due to one or a combination of emissions, radiations or inductions upon reception in a radiocommunication system, manifested by any performance degradation, misinterpretation or loss of information which could be extracted in the absence of such unwanted energy."

Harmful interference is defined in the RR's as :

"Harmful interference: interference which endangers the functioning of a radionavigation service or of other safety services or seriously degrades, obstructs or repeatedly interrupts a radiocommunication service operating in accordance with the Radio Regulations."

As mentioned above, since 1st March 1993 the ITU comprises:-

(a) a Plenipotentiary Conference, which is the supreme authority of the ITU;
(b) the Council, which acts on behalf of the Plenipotentiary Conference in the interim between conferences;
(c) a World Conference on International Telecommunications;
(d) the Radio Communication Sector (which includes World and Regional Radio Communication Conferences, Radio Communication Assemblies and the Radio Regulations Board);
(e) the Telecommunication Standardisation Sector (which includes World Telecommunication Standardisation Conferences);
(f) the Telecommunications Development Sector (which includes World and Regional Telecommunication Development Conferences); and
(g) the General Secretariat.

For the purposes of international regulation of the frequency spectrum and orbital positions, the Radio Regulations Board (RRB) and the Radiocommunications Bureau are the relevant bodies. These bodies apply the RR's as adopted from time to time by the Members in Plenipotentiary Conference. The RR's deal with operational, practical and technical matters. It is the RR's to which reference is now made to provide an example of the international regulation of the radio frequency spectrum and orbital positions.

III - THE RADIO REGULATIONS

The Radio Regulations deal with all matters touching upon the transmission, emission and reception of radio waves. Thus, there are provisions relating to maritime mobile services, radio astronomy, land mobile services, aeronautical mobile services, radionavigation services, space research services, meteorological services and all forms of other such services. There are also provisions dealing with the regulation of safety and distress signals.

The Radio Regulations have developed into a cumbersome 3-volume looseleaf edition which incorporates all the revisions to previous editions in the form of appendices and also incorporates all the Resolutions and Recommendations made pursuant to the various World Administrative Radio Conferences. Fortunately, the ITU has embarked on a programme of simplification of the Radio Regulations. A Voluntary Group of Experts is currently working on, inter alia, the simplification of the Radio Regulations with a view to submitting a report for consideration at the World Radiocommunication Conference in 1995 [9].

To try and make more simple the verbal complexity of the Radio Regulations, flowcharts are included [10]. Set out in the following pages are copies of flowcharts "AW" and "AX". These flowcharts outline, respectively:-

(a) the advance publication procedure applicable to stations in geostationary or non-geostationary satellite networks; and
(b) the co-ordination procedure applicable to space or earth stations in a geostationary satellite network.

It should be noted that the flowcharts are intended by the ITU to act as an aid to understanding and that they do not form part of the Radio Regulations. There are at least two reasons for this, firstly the Radio Regulations contain much more detail than the flowcharts contain and, secondly, the Radio Regulations can, in certain circumstances, be more flexible than the flowcharts suggest.

Set out below is a more detailed consideration of the procedure for dealing with the co-ordination, notification and registration of frequency assignments for Geostationary Fixed Satellite Services. Parts of this procedure may be followed on the flowcharts. As mentioned above, flowchart "AW" covers advance publication and flowchart "AX" covers co-ordination. Where there are references to "Boxes", these are references to the box numbers on the relevant flowcharts.

The relevant Articles of the Radio Regulations are Articles 11 and 13. These Articles divide the process into four stages:

(a) advance publication;
(b) co-ordination;
(c) notification; and
(d) registration.

(a) **Advance publication** : Requirements are imposed on any Administration which intends to bring into use a satellite network within a satellite system. Such an Administration (Box 1) must no earlier than six years and no later than two years before the date of bringing into service a proposed satellite network, provide the RRB with certain basic information (RRs 1041-1044 and Box 2).

Once the RRB has received this information, it reviews it in accordance with the technical standards set out in the RRs (boxes 3 and 6). The information which is sent is known as the "advance publication" information. Once the information is received in proper form, the RRB then publishes it in a document call the Weekly Circular (Box 4). The Weekly Circular is received by all of the Administrations and puts them on notice as to any proposed satellite network. The RRB can request further information if the information supplied is incomplete (RR 1044 and Box 6).

If modifications are subsequently made to the proposed network such that its character is significantly changed, the advance publication procedure may have to be recommenced (Box 16).

Within four months of the date of publication in the Weekly Circular any Administration which believes that unacceptable interference may be caused to its existing or planned satellite networks by the proposed assignments, must send to the Administration proposing the new satellite network its comments on the particulars of such anticipated interference (Boxes 11, 10, 9, 12 and 19). RR 1047 states that:

> "If, after studying the information published...any Administration is of the opinion that interference which may be unacceptable may be caused to assignments of its existing or planned satellite networks, it shall, within four months after the date of the weekly circular...send the Administration concerned its comments on the particulars of the interference to its existing or planned satellite systems. A copy of

Figure 1

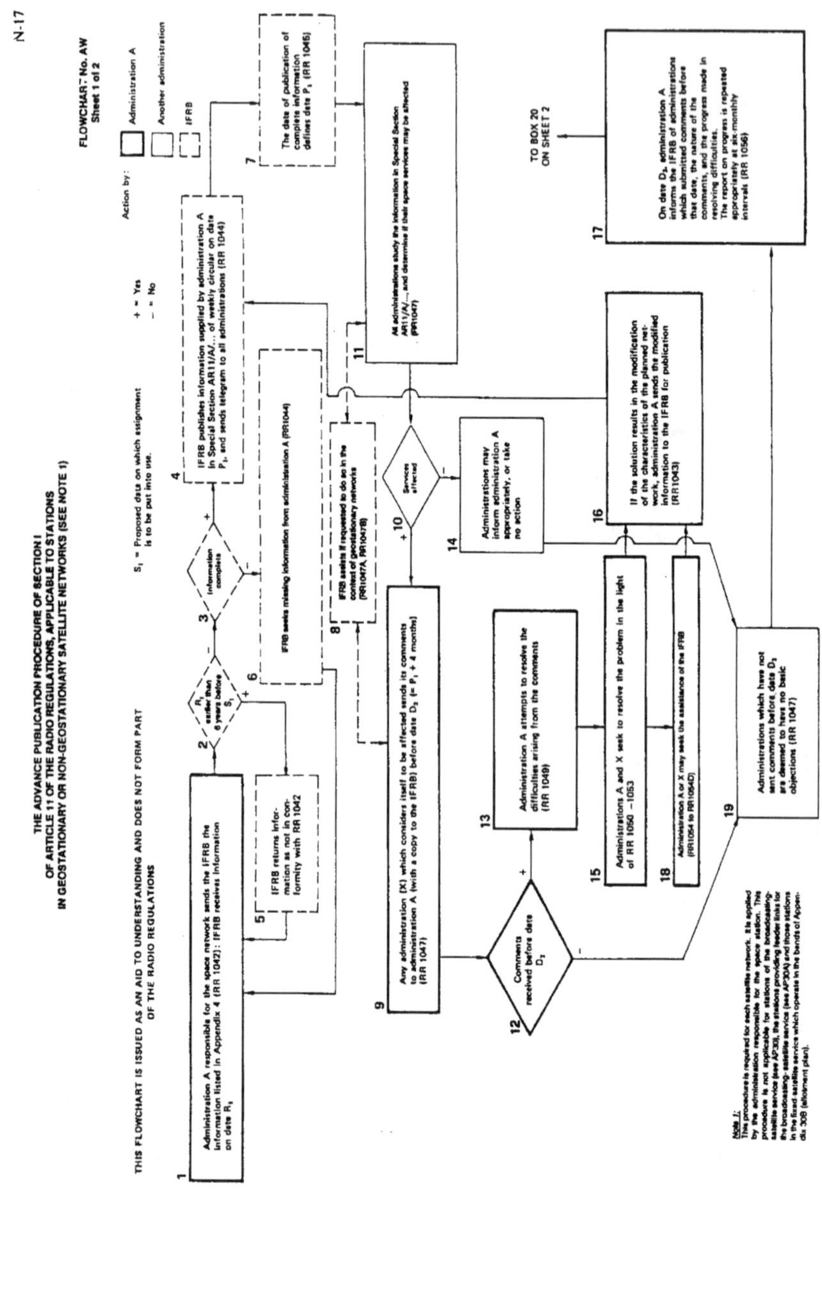
N-17
THE ADVANCE PUBLICATION PROCEDURE OF SECTION I OF ARTICLE 11 OF THE RADIO REGULATIONS, APPLICABLE TO STATIONS IN GEOSTATIONARY OR NON-GEOSTATIONARY SATELLITE NETWORKS (SEE NOTE 1)
THIS FLOWCHART IS ISSUED AS AN AID TO UNDERSTANDING AND DOES NOT FORM PART OF THE RADIO REGULATIONS
FLOWCHART No. AW Sheet 1 of 2
Action by: Administration A; Another administration; IFRB
+ = Yes
– = No
1 Administration A responsible for the space network sends the IFRB the information listed in Appendix 4 (RR 1042); IFRB receives information on date R₁
2 R₁ earlier than 6 years before S₁
3 Information complete
4 IFRB publishes information supplied by administration A in Special Section AR11/A/... of weekly circular on date P₁, and sends telegram to all administrations (RR 1044)
5 IFRB returns information as not in conformity with RR 1042
6 IFRB seeks missing information from administration A (RR1044)
7 The date of publication of complete information defines date P₁ (RR 1045)
8 IFRB assists if requested to do so in the context of geostationary networks (RR1047A, RR1047B)
9 Any administration (X) which considers itself to be affected sends its comments to administration A (with a copy to the IFRB) before date D₂ (= P₁ + 4 months) (RR 1047)
10 Services affected
11 All administrations study the information in Special Section AR11/A/..., and determine if their space services may be affected (RR1047)
12 Comments received before date D₂
13 Administration A attempts to resolve the difficulties arising from the comments (RR 1049)
14 Administrations may inform administration A appropriately, or take no action
15 Administrations A and X seek to resolve the problem in the light of RR 1050 – 1053
16 If the solution results in the modification of the characteristics of the planned network, administration A sends the modified information to the IFRB for publication (RR 1043)
17 On date D₂, administration A informs the IFRB of administrations which submitted comments before that date, the nature of the comments, and the progress made in resolving difficulties. The report on progress is repeated appropriately at six-monthly intervals (RR 1056)
18 Administration A or X may seek the assistance of the IFRB (RR1054 to RR1054D)
19 Administrations which have not sent comments before date D₂ are deemed to have no basic objections (RR 1047)
TO BOX 20 ON SHEET 2
Note 1: This procedure is required for each satellite network. It is applied by the administration responsible for the space station. This procedure is not applicable for stations of the broadcasting-satellite service (see AP30), the stations providing feeder links for the broadcasting-satellite service (see AP30A) and those stations in the fixed satellite service which operate in the bands of Appendix 30B (allotment plan).

Figure 2

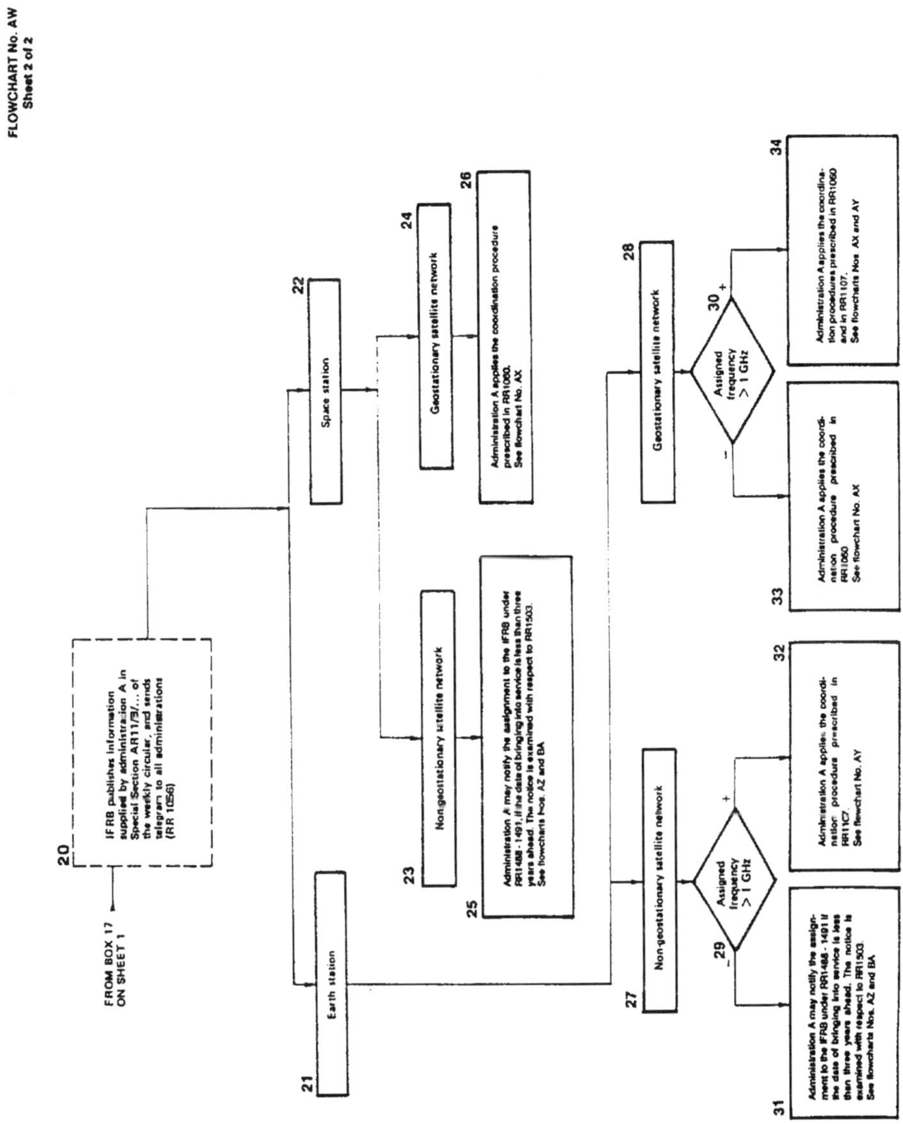
N-18
FLOWCHART No. AW
Sheet 2 of 2
FROM BOX 17 ON SHEET 1
20
IFRB publishes information supplied by administration A in Special Section AR11/B/... of the weekly circular, and sends telegram to all administrations (RR 1056)
21
Earth station
22
Space station
23
Non-geostationary satellite network
24
Geostationary satellite network
25
Administration A may notify the assignment to the IFRB under RR1488 - 1491, if the date of bringing into service is less than three years ahead. The notice is examined with respect to RR1503. See flowcharts Nos. AZ and BA
26
Administration A applies the coordination procedure prescribed in RR1060. See flowchart No. AX
27
Non-geostationary satellite network
28
Geostationary satellite network
29
Assigned frequency > 1 GHz
30
Assigned frequency > 1 GHz
31
Administration A may notify the assignment to the IFRB under RR1488 - 1491 if the date of bringing into service is less than three years ahead. The notice is examined with respect to RR1503. See flowcharts Nos. AZ and BA
32
Administration A applies the coordination procedure prescribed in RR1107. See flowchart No. AY
33
Administration A applies the coordination procedure prescribed in RR1060 See flowchart No. AX
34
Administration A applies the coordination procedures prescribed in RR1060 and in RR1107. See flowcharts Nos. AX and AY

these comments shall also be sent to the [RRB]. If no such comments are received from an Administration within the period mentioned above, it may be assumed that the Administration has no basic objections to the planned satellite network(s) of that system on which details have been published."

Provision is made for resolution of any difficulties which may have been identified as a result of advance publication (Boxes 13, 15 and 16). When the RRs were redrafted following World Administrative Radio Conference -71 (WARC-71) there appeared to be an order of priorities whereby the onus of resolving difficulties was placed initially upon the Administration proposing the new satellite network. Thus, such Administration has to explore all possible solutions without considering the possibility of adjusting other networks (RR 1051). It is only after this process has been completed that a request can be forwarded to the Administration whose satellite network is likely to be affected, to investigate any modifications to such other Administration's system which may resolve the problem (RR 1052). Finally, the Administrations should seek to work together to resolve the difficulties identified as a result of advance publication by mutually acceptable means. This is an important provision often forgotten where "Administrations" have competing objectives, especially if one thinks it has "priority" over the other.

As a result of various disputes which were not resolved and the absence of any powers of sanction available to the RRB, the RRB (then the International Frequency Registration Board or IFRB) issued on 26 February 1993 a rule of procedure (Rule H40) regarding priority of dates in the application of Articles 11 and 13 of the Regulations (Radio Regulation Board Rules of Procedure, Rule no H40 (Rev 1)). Pursuant to Rule H40, no Administration obtains any particular priority as a result of being the first to start either the advance publication procedure or request co-ordination.

The RRB can be requested to assist in resolving any disputes (Box 18). However, as RR's 1054 to 1054D make clear, it only has "advisory" and not "enforcement" powers. Once relevant problems have been mutually resolved, the RRB publishes in the Weekly Circular the results of the advance publication procedure (RR 1056 and Boxes 17 and 20). If all necessary agreements have been obtained the Administration intending to bring the network into use may then proceed to co-ordination of the frequency assignments pertaining to the network (Box 26).

(b) **Co-ordination** : When an Administration sends the information for advance publication, it may at the same time, or thereafter, request the RRB to forward to any Administration which it thinks would be affected by the proposed satellite network (RR 1058) technical details that are required for the network co-ordination of any frequency assignments pertaining to its proposed satellite network.

Frequency assignments which are to be taken into account (Boxes 1 and 3) in the application of co-ordination procedures are those:

(i) in the same frequency bands as that of the planned assignment and which belong to the same service or to any other service which has that band allocated to it with equal rights or higher category allocation (that is, a secondary service assignment must be co-ordinated with a permitted or primary service but not vice versa) (RR 1061); and

(ii) which are in conformity with the RRs and which are either recorded in the Master International Frequency Register (MIFR) or co-ordinated under the provisions of the RRs or, where co-ordination is not required, have been notified to the RRB (RRs 1061-1065).

Once the Administrations with which co-ordination will be required have been identified the Administration proposing the new satellite network must forward all relevant data to the Administrations which are affected with copies to be supplied to the RRB (RR 1073 and Boxes 2 and 7). As with advance publication, the RRB can request further information (Boxes 9, 11, 16, 15, 14, 10 and 17). When full information has been provided, the RRB examines the information supplied for conformity with the RR's and sends the information to affected Administrations together with the results of their examination. The RRB will then publish the information in the Special Section of the Weekly Circular (Box 13).

As with advance publication, if modifications are subsequently made to the information that has been published in the Weekly Circular, the co-ordination procedure may have to be recommenced (Boxes 32,33, 37, 41 and 19).

There are detailed provisions relating to time limits within which the Administrations must forward their comments to the Administration proposing the new satellite network (Boxes 19-31). Importantly, within four months of the publication of the Weekly Circular containing the co-ordination information, any Administration which has any objection to the technical specifications of the proposed satellite network must

Figure 3

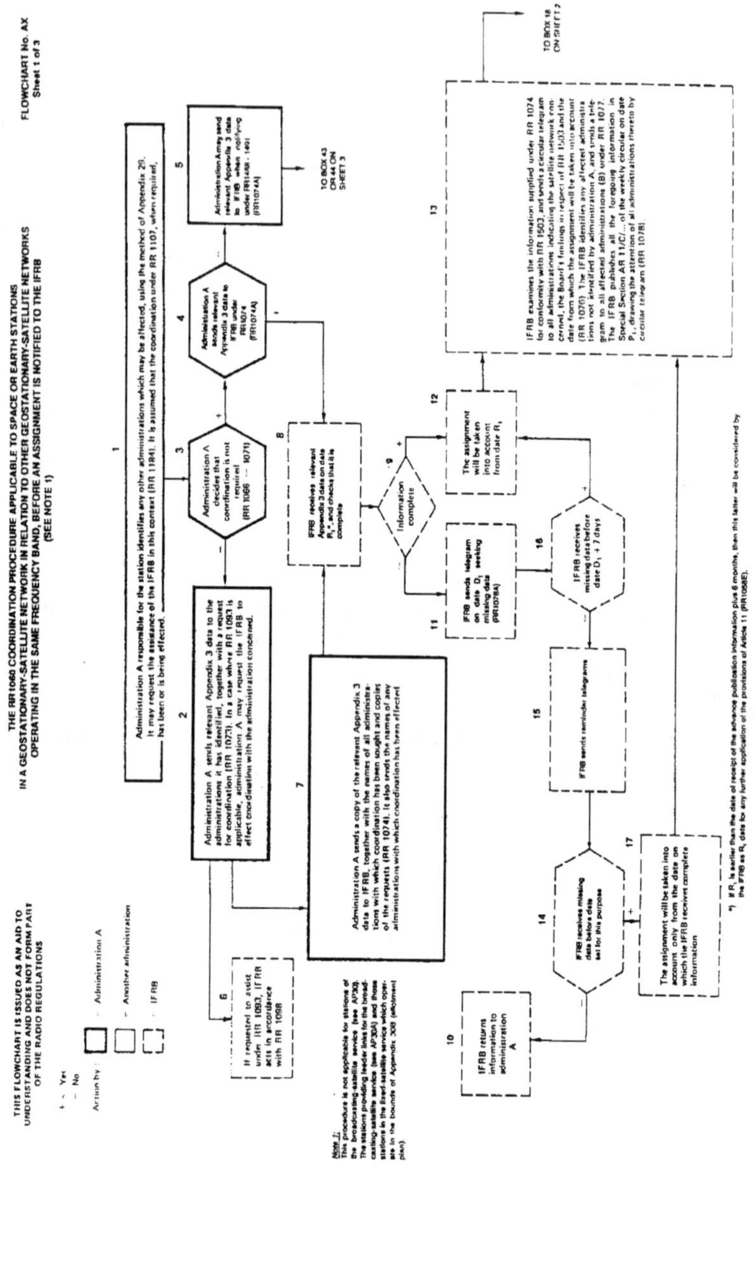
N-19
THIS FLOWCHART IS ISSUED AS AN AID TO UNDERSTANDING AND DOES NOT FORM PART OF THE RADIO REGULATIONS
THE RR1060 COORDINATION PROCEDURE APPLICABLE TO SPACE OR EARTH STATIONS IN A GEOSTATIONARY-SATELLITE NETWORK IN RELATION TO OTHER GEOSTATIONARY-SATELLITE NETWORKS OPERATING IN THE SAME FREQUENCY BAND, BEFORE AN ASSIGNMENT IS NOTIFIED TO THE IFRB (SEE NOTE 1)
FLOWCHART No. AX
Sheet 1 of 3
+ – Yes
– No
Action by: – Administration A
– Another administration
– IFRB
1
Administration A responsible for the station identifies any other administrations which may be affected, using the method of Appendix 29. It may request the assistance of the IFRB in this context (RR 1184). It is assumed that the coordination under RR 1107, when required, has been or is being effected.
2
Administration A sends relevant Appendix 3 data to the administrations it has identified, together with a request for coordination (RR 1073). In a case where RR 1093 is applicable, administration A may request the IFRB to effect coordination with the administration concerned.
3
Administration A decides that coordination is not required (RR 1066 — 1071)
4
Administration A sends relevant Appendix 3 data to IFRB under RR1074 (RR1074A)
5
Administration A may send relevant Appendix 3 data to IFRB when notifying under RR1488 - 1491 (RR1074A)
TO BOX 43 OR 44 ON SHEET 3
6
If requested to assist under RR 1093, IFRB acts in accordance with RR 1098
7
Administration A sends a copy of the relevant Appendix 3 data to IFRB, together with the names of all administrations with which coordination has been sought and copies of the requests (RR 1074). It also sends the names of any administrations with which coordination has been effected
8
IFRB receives relevant Appendix 3 data on date R₁*, and checks that it is complete
9
Information complete
10
IFRB returns information to administration A
11
IFRB sends telegram on date D₁ seeking missing data (RR1078A)
12
The assignment will be taken into account from date R₁
13
IFRB examines the information supplied under RR 1074 for conformity with RR 1503, and sends a circular telegram to all administrations indicating the satellite network concerned, the Board's findings in respect of RR 1503 and the date from which the assignment will be taken into account (RR 1076). The IFRB identifies any affected administrations not identified by administration A, and sends a telegram to all affected administrations (B) under RR 1077. The IFRB publishes all the foregoing information in Special Section AR 11/C/... of the weekly circular on date P₁, drawing the attention of all administrations thereto by circular telegram (RR 1078).
TO BOX 18 ON SHEET 2
14
IFRB receives missing data before date set for this purpose
15
IFRB sends reminder telegrams
16
IFRB receives missing data before date D₁ + 7 days
17
The assignment will be taken into account only from the date on which the IFRB receives complete information
Note 1:
This procedure is not applicable for stations of the broadcasting-satellite service (see AP30). The stations providing feeder links for the broadcasting-satellite service (see AP30A) and those stations in the fixed-satellite service which operate in the bounds of Appendix 30B (allotment plan).
*) If R₁ is earlier than the date of receipt of the advance publication information plus 6 months, then this latter will be considered by the IFRB as R₁ date for any further application of the provisions of Article 11 (RR1058E).

Figure 4

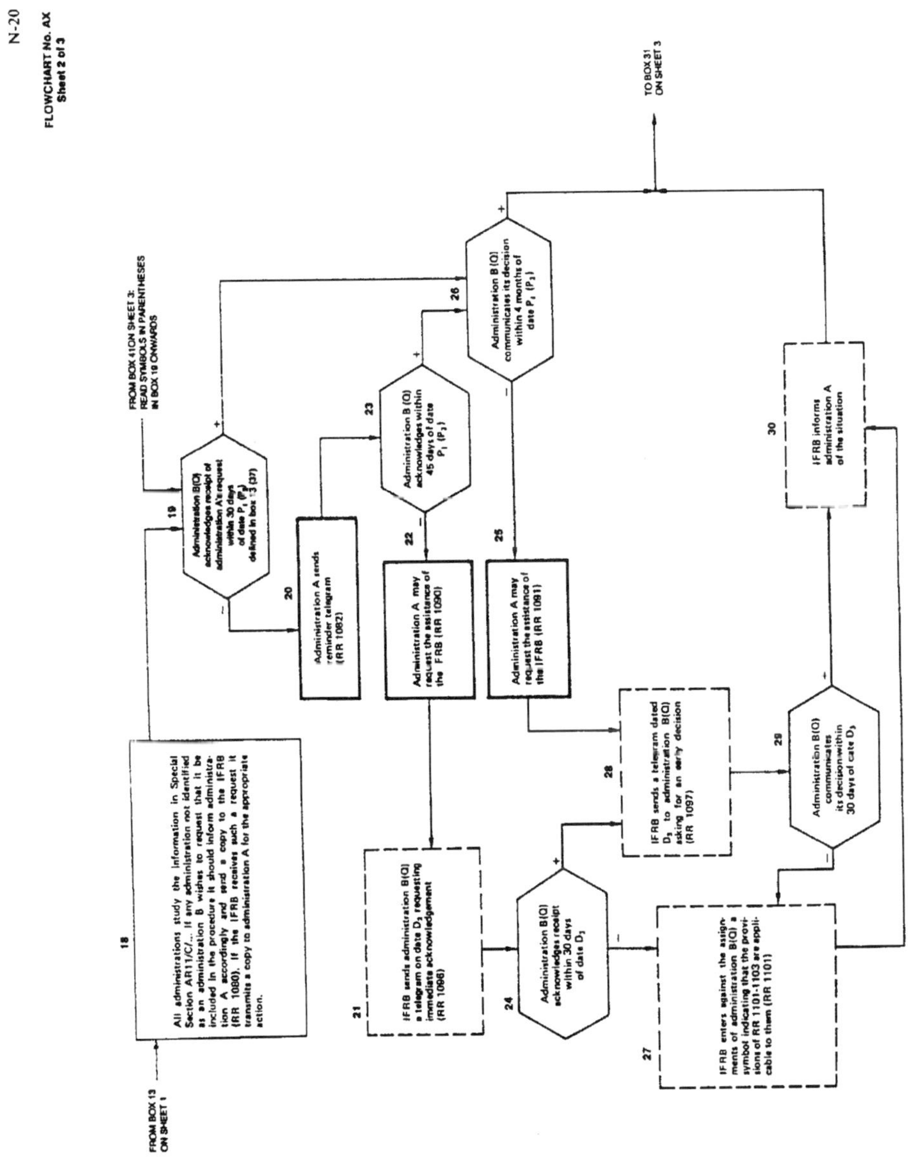

Figure 5

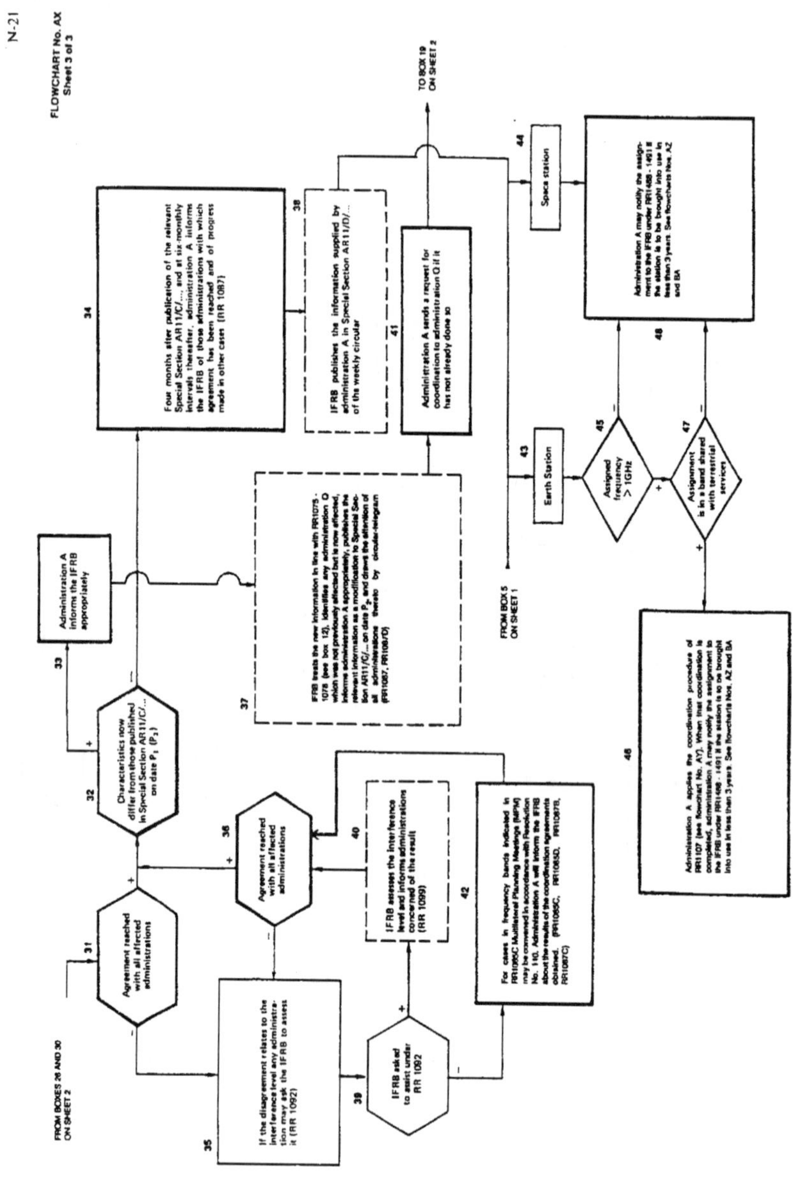

forward its objections to the Administration proposing the new satellite network. It also has to provide copies to the RRB of any suggestions or solutions to the problem. In practice, however, Administrations do not always manage to meet the four month deadline. In the past political objections have been raised to having objections accepted outside such a period. When notifying the Administration proposing the new satellite network that there are objections, the technical details upon which the disagreement is based must be identified. Suggestions should also be made by the objecting Administration as to how to resolve such problems (RR 1084).

The RRs also state that the affected Administrations shall make all possible mutual efforts to overcome the difficulties in a manner acceptable to the parties (RR 1085A). This is the principle behind Rule H40 whereby there is no priority and the onus on resolving any difficulty is shared equally between those Administrations which control satellite systems and which have already been co-ordinated, as well as the Administration proposing the new satellite network.

Once the co-ordination procedure has begun, after the four months' time period, the Administration proposing the new satellite system informs the RRB of the Administrations with which an agreement has and has not been reached (Box 34). Thereafter, it reports to the RRB at six-monthly intervals as to what progress has been made in securing the consent of other relevant Administrations (RR 1087).

When a disagreement cannot be resolved, the RRB can be requested to provide assistance to try and resolve such difficulties. There are specific instances when the RRB's assistance may be sought. These include, where an Administration identified under the advance publication procedure as being affected by the proposed satellite system does not acknowledge receipt of the co-ordination data (Box 22), or, if it does acknowledge receipt, does not provide an answer as to whether or not it has agreed to the proposed satellite network (Box 25). The RRB also assists where there is a continued disagreement between the Administration seeking co-ordination and the Administration with which the co-ordination is sought or if co-ordination is not possible for any other reason (RRs 1088-1094 and Boxes 35, 39, 40 and 42).

Once the RRB becomes involved, it takes all necessary steps to facilitate the reaching of an agreement, including the setting up of meetings between the relevant Administrations (Box 42). It also assesses the extent of interference and evaluates the technical grounds upon which the Administration is objecting to the proposed satellite network (Box 35).

A useful feature of this process is the "30-day 'cure' period". This is where the RRB gives an Administration with whom co-ordination is sought 30 days within which it must announce its decision as to whether it will be affected by the proposed satellite network (Box 29). If that Administration fails to act within this period (Box 27), it is deemed to have undertaken that:

a) no complaint will be made in respect of any harmful interference affecting the services rendered by its space radiocommunication stations which may be caused by the use of the assignment to a station of the satellite network for which co-ordination was requested; and
b) its space radiocommunication stations will not cause harmful interference to the satellite network assignments for which co-ordination was requested.

In the absence of any response the RRB lacks any powers of sanction. In practice, to avoid this deemed undertaking, the objecting Administration can simply provide unhelpful responses (RRs 1095-1103).

A recent resolution of the ITU Plenipotentiary Conference held in Kyoto [9] instructed the Director of the Radiocommunication Bureau to initiate a review of issues concerning satellite networth co-ordination with the objectives of ensuring equitable access, co-ordination procedures meet administrations' needs and flexible and efficient use of the radio frequency spectrum and orbital positions. A preliminary report is to be made to the World Radiocommunication Conference in 1995 with a final report to be made to the said Conference in 1997. Thus the position on co-ordination may change to reflect the stated objectives.

(c) **Notification**: Frequency assignments must be notified to the RRB in the following circumstances:

(i) if use of the frequency is capable of causing harmful interference to the services of any other Administration; or
(ii) if the frequency is to be used for international radiocommunications; or
(iii) if it is desired to obtain international recognition of the use of the frequency (RRs 1488-1491).

The RRB is then under an obligation to publish the notification in the Weekly Circular within 40 days of receiving the notice. The Weekly Circular has to contain all particulars of the notice. The publication in the Weekly Circular amounts to notification to all Administrations of the complete notice. There are, as before, provisions for the RRB to request further information if the information provided is not complete (RRs 1498-1500).

The RRB is under an obligation to examine each notice to ensure its conformity with the Table of Frequency Allocations and with other provisions in the RRs (RRs 1502-1512). Apart from a few technical exclusions, the RRB must examine the proposed notification against all the existing co-ordinated or notified frequency assignments which are likely to be affected. There are different standards against which the notice must be examined depending on whether there has been a complete co-ordination procedure or whether co-ordination was not required (see paragraph 2 of "(b) Co-ordination" above).

If the RRB arrives at an unfavourable finding based upon the possibility of harmful interference with an assignment which has been entered in the MIFR for a space station which it believes is not in use, the RRB can consult with the Administration responsible for the assignment and if it is discovered that the recorded assignment has not been in use for two years it can disregard such recorded assignment. If it is discovered that this is the case, the recorded assignment becomes suspended and if it is brought back into use, it must go through the co-ordination and notification procedures again.

(d) **Registration**: Once the examination of the notice is completed, the RRB can proceed to make an entry in the MIFR. If the examination of the RRB is favourable, the RRB enters it as such in the MIFR.

Any registration can also include a reference to the fact that the assignment will operate in accordance with the provisions of RR 342, whereby the Administration agrees that the assignment is made on the basis that it will operate only so long as harmful interference is not caused and as soon as such harmful interference is caused to a service which is operating in accordance with the RRs, the assignment must cease.

When the RRB makes an unfavourable finding with respect to an Administration's notification, the RRB will return the notice for the purpose of it being changed and resubmitted, if appropriate (RRs 1506 and 1542). Even though an Administration resubmits a notice with insufficient changes to permit a favourable finding (RR 1543) nevertheless, pursuant to Rule H40, the notice may be recorded in the MIFR under the following conditions:

(i) "provisionally" (RRs 1544 and 1556) in a case where an earlier notified assignment which was at the origin of the unfavourable finding is not in operation; or

(ii) "definitely" (RR 1544) in a case where both assignments have already been in operation for at least four months without any harmful interference being reported.

Both of the above cases are placed under the general provision of RR 1559 which stipulates that stations using frequencies recorded with a finding under RR 1544 must immediately cease interference upon receipt of a complaint relating to harmful interference caused to stations which have been recorded with favourable findings.

As mentioned above, the flowcharts and the procedures detailed above give only an outline of the relevant RR's and describe only certain procedures relating to Geostationary Fixed-Satellite Services, respectively. They do not, therefore, include details of all the various procedures provided for by the RRs, which are exceptionally complicated and make distinctions for different services. It is for this reason that it is very much hoped that the simplification which the Voluntary Group of Experts is carrying out will be put into effect as soon as possible. The burden of responsibilities on the RRB is so vast that there is an enormous backlog of co-ordination to be effected especially as more and more Administrations are requesting frequency assignments to be recorded in the MIFR.

IV - ENFORCEMENT OF THE RADIO REGULATIONS

Chapter 5 of the RRs contains provisions designed to identify harmful interference and to deal with it. Article 20 (RRs 1872-1886) sets up a system of international monitoring whereby Administrations agree to assist with the monitoring of the use of radio waves to ensure efficient and economical use of the radio frequency spectrum and to help in the prompt elimination of harmful interference. The Secretary General nominates monitoring stations across the world which conduct such monitoring as may be required of them by the RRB or by the Designated Monitoring Station (DMS) of other Administrations.

Further to the monitoring of the use of the radio frequency spectrum, Article 21 (RRs 1915-1917) provides that infringements of the Constitution, the Convention or the RRs are to be reported to the Administration of the country having jurisdiction over the station committing the infringement. There is a duty on such Administration to "...ascertain the facts, fix the responsibility and take the necessary action"

(RR 1917). The lack of any express provision as to what action the Administration is to take is very much in line with the lack of enforcement powers possessed by the RRB and the ITU under the RRs [11]. In other words, enforcement is a national obligation under appropriate domestic law.

An interesting question then arises: What if the body responsible for domestic enforcement disagrees with any finding in the cause of harmful interference or in the taking account of its responsibilities under the Constitution, the Convention and the Radio Regulators to take practicable steps to avoid harmful interference and use efficiency and technological developments to make services available to the public (e.g., Constitution Article 1.1(c) and (d)) where that Administration believes that there are technical solutions to allow services to co-exist without harmful interference?

In this context, Article 22 (RRs 1943-1966) sets out the procedures to be used in cases of harmful interference. The reliance placed by Article 22 on agreement between parties further emphasizes the consensual nature of treaty arrangements. In particular Article 22 (RR 1943) states:

> "It is essential that Members exercise the utmost goodwill and mutual assistance in the application of the provisions of Article 35 of the Convention and of this article to the settlement of problems of harmful interference."

Furthermore, Article 35 of the Convention (which here means the International Telecommunication Convention, Nairobi, 1982) states:

> "1. All stations, whatever their purpose, must be established and operated in such a manner as not to cause harmful interference to the radio services or communications of other Members or of recognised private operating agencies, or of other duly authorised operating agencies which carry on radio service, and which operate in accordance with the provisions of the Radio Regulations.
>
> 2. Each Member undertakes to require the private operating agencies which it recognises and the other operating agencies duly authorised for this purpose, to observe the provisions of No. 158 [ie No. 1. above].
>
> 3. Further, the Members recognise the desirability of taking all practicable steps to prevent the operation of electrical apparatus and installations of all kinds from causing harmful interference to the radio services or communications mentioned in No. 158 [ie No. 1. above]."

The consensual nature of the RRB is further reinforced by the subsequent RR's which encourage Administrations to deal with harmful interference by agreement. The only obligation of the Administrations is to provide the necessary information to help establish the extent and nature of the interference.

In cases of harmful interference occurring as a result of emissions from space stations, the Administration which has jurisdiction over the stations experiencing the interference may request from the Administration which controls such space station the necessary data to allow determination of the positions of such space stations (RR 1955). If rapid action is required, the RRs permit information to be exchanged directly between the relevant Administrations' DMS's, subject to the authorisation of the relevant Administrations (RR 1956).

If harmful interference persists, the Administration having jurisdiction over the transmitting station whose service is being interfered with may address to the Administration having jurisdiction over the interfering station a report of the irregularity or infraction in accordance with Article 22 (RR 1959). It is only after all these steps have been taken that an Administration can forward details to the RRB for its information.

The relevant Administration may also request the RRB to provide technical support by conducting a study and the RRB can then make recommendations for the solution of the problem (RRs 1576-1584 and 1962). Once the RRB reaches its conclusions it makes recommendations, a copy of which is forwarded to the Administration reporting the harmful interference. The conclusions and recommendations are also forwarded by telegram to the Administration believed to be responsible for the source of harmful interference, together with a request for prompt action (RR 1966).

It is clear that the RRB has no power to impose sanctions against the Administration whose station is causing the interference. This goes to the very heart of the international regulatory regime - enforcement is a matter of domestic law. The members have not yet granted any regulatory powers to the RRB involving enforcement of the RRs. Furthermore, the principles of international public law do not enable enforcement through the RRB.

Therefore the question arises as to whether the RRB should remain competent only to give recommendations without any regulatory power or whether the development of telecommunications might be enhanced by the ITU being granted regulatory powers, in the absence of the Members of the ITU having recourse to enforcement of international obligations by other Members (and their Administrations or recognised operating agencies) in the International Court of Justice.

It is outside the scope of this article to comment on this other than to point out that if the Members actually fulfilled their obligation to the ITU by having appropriate domestic law to reflect their obligations, there would be no question of needing a "super" regulator. As the preamble to the Radio Regulations states "the application of the provisions of [the Radio Regulations]... does not imply the expression of any opinion whatsoever on the part of the ITU concerning the sovereignty or the legal status of any country, territory or geographical area". Given the different state of technological development between the 180 or so Members of the ITU, it is unlikely that this principle would or indeed should ever change. Technological developments should not be hampered by the pace of development of the slowest Member of the ITU.

CONCLUSION

The finite resources of the radio frequency spectrum and orbital positions will continue to need regulation at an international level to ensure their efficient, orderly and equitable use. Given that technology changes rapidly it is difficult to see how the tension between "equal access" and "first-come, first-served" can be resolved. The regulations will have to combat these difficulties alongside the need to be easily understood if an efficient, orderly and equitable use of these resources is to be maintained. Furthermore, Administrations need to recognise that regulations, particularly the Radio Regulations, in themselves are not "black letter" law. Over and above them are obligations on members of the international community to work together to find solutions to problems. Given that these resources are scarce there is even more reason for the usage of the radio frequency spectrum and geostationary orbits to be efficient and to take advantage of technological developments that have taken place since the WARC-77 Plans.

APPENDIX

"Administration"	any governmental department or service responsible for discharging the obligations of a member country of the ITU;
"allocation"	entry in the Table of Frequency Allocations of a given frequency band for the purpose of its use by one or more terrestrial or space radiocommunication services or the radio astronomy service under specified conditions;
"allotment"	entry of a designated frequency channel in an agreed plan, adopted by a competent conference, for use by one or more Administrations for a terrestrial or space radiocommunication service in one or more

identified countries or geographical areas and under specified conditions;

"assignment" the authorisation given by an Administration for a radio station to use a radio frequency or radio frequency channel;

"Fixed Satellite Service" a radiocommunication service between earth stations at given positions, when one or more satellites are used; the given position may be a specified fixed point or any fixed point within specified areas; in some cases this service includes satellite-to-satellite links, which may also be operated in the inter-satellite service; the fixed satellite service may also include feeder links for other space radiocommunication services;

"harmful interference" interference which endangers the functioning of a radionavigation service or of other safety services or seriously degrades, obstructs or repeatedly interrupts a radiocommunication service operating in accordance with the Radio Regulations;

"interference" the effect of unwanted energy due to one or a combination of emissions, radiations or inductions upon reception in a radiocommunication system, manifested by any performance degradation, misinterpretation or loss of information which could be extracted in the absence of such unwanted energy;

"ITU" International Telecommunication Union;

"MIFR" Master International Frequency Register;

"RRB" Radio Regulations Board; "services" space radiocommunication services including geostationary fixed-satellite services, broadcasting satellite services, mobile satellite services, all of which may be further distinguished as "primary", "permitted" or "secondary" services; and "UN" United Nations.

REFERENCES

1. See Chapter 1 of "Satellite Communications in Europe: Law and Regulation" by Stewart White, Stephen Bate and Timothy Johnson.
2. A "satellite" is a body which revolves around another body of preponderant mass and which has a motion primarily and permanently determined by the force of attraction of that other body.
3. A "geostationary satellite" is a geosynchronous satellite whose circular and direct orbit lies in the plane of the earth's equator and which thus remains fixed relative to the earth; by extension, a satellite which remains approximately fixed relative to the earth.
4. Recommendation COM4/A of the Final Acts of the ITU Plenipotentiary Conference (PP-94), Kyoto 1994.
5. Article 5 of the Radio Telegraphy Regulations, Washington 1927.
6. ITU Constitution (Geneva 1992), Articles 6.1, 6.2 and 54.2.
7. ITU Constitution (Geneva 1992), Article 49.
8. E.g. see 31st Report by the ITU, Booklet No. 40 (Geneva 1992) (ISBN 92-61-04631-2).
9. Resolution COM4/10 of the Final Acts of the ITU Plenipotentiary Conference (PP-94), Kyoto 1994.
10. These flowcharts are included in this article by kind permission of the Secretary General of the ITU.
11. See final paragraph of "(b) Co-ordination" above.

REGULATORY LIBERALISATION OF MOBILE COMMUNICATIONS IN THE EUROPEAN UNION

ELEANOR PAUL

S J Berwin & Co, Solicitors, 222 Grays Inn Road, London, WC1X 8HB, United Kingdom Tel:(44) 0171 837 2222 ; Fax (44) 0171 833 2860

INTRODUCTION

A licence is required to operate any mobile communications network in the European Union. Licences are issued nationally by the telecommunications regulators in each of the Member States. The availability of licences, the licensing procedure, licence fees, licence terms and licensed frequencies have all been left to the discretion of each of the Member States. Generally, mobile licences are held by the monopoly (mostly state-owned) post, telegraph and telephone companies (PTTs). There are not many private operators and as a result competition is limited. Only the supply of mobile terminal equipment is open to full competition in all the Member States.

Until the Green Paper on Mobile and Personal Communications which was adopted in April 1994 [1] there was no European level of harmonisation of mobile communications licensing.

This article reviews the current regulatory framework for the major mobile communications technologies and considers the proposals contained in the Green Paper and the subsequent refinement of those proposals by the European Commission.

I - CURRENT REGULATORY FRAMEWORK

In the European Union, mobile communications licences have historically been granted by the Member States in relation to a specific technology, for instance, radio paging or cellular telephony. The licensed frequency depends upon the technology involved and its availability for use in mobile communications in the particular Member State.

This has meant that different Member States have historically allocated different frequencies to the same mobile technology and, in addition, some have allowed licensees to set their own technical standards for the technology. The result is incompatibility of networks and terminal equipment.

Member States, network operators and equipment manufacturers have recognised the need for common technical standards and international standards have been and are continuing to be developed. These are addressed in more detail below.

Subject to compliance with such international standards, Member States are free to impose restrictions on the number of licences they issue and the terms of those licences. Licensing procedures vary enormously and the criteria for the award of licences are unclear in some countries. Take, for example, the criteria for the issue in 1994 of the PCN licence in France. The award of the licence was made following a closed selection process and some observers consider that the award was made on political grounds rather than on technical proposals and investment plans. [2]

In some countries certain operators are restricted from applying for mobile licences. For example, BT was prohibited from tendering for a PCN licence in the United Kingdom.

Mobile communications liberalisation was expressly excluded from the watershed European Commission Green Paper on telecommunications, published in 1987. [3] The following is a summary of the current regulatory framework for the major mobile communications technologies in the European Union.

1.1. Analogue Cellular Telephony

Except in France, the United Kingdom, and Sweden analogue cellular telephony is provided by the monopoly PTT's who were automatically licensed under their monopoly powers. The private operators in France and the United Kingdom were selected by competitive tender. Until only recently these private operators were restricted from providing their own infrastructure, that is, the lines between the base station and the local switches. They were forced, therefore, to lease lines in France from France Telecom and in the United Kingdom from BT or Mercury. In Sweden, the analogue NMT systems are run by the Telia AB subsidiary Telia Mobitel, Telia AB being the state owned limited company (previously Televerket).

In addition, because most calls from mobile phones terminate on the fixed networks, mobile networks need to interconnect to the fixed networks. The cost of interconnection is high because there is generally only limited or no competition in the fixed networks. Also the choice of access points is normally limited. Access points are important because they determine how far the call has to be carried on the fixed network and hence the cost. Interconnection is generally not regulated but left to commercial negotiation, subject to regulator intervention in the event of a dispute.

The same problems arise in the case of the new digital technologies, such as GSM, where more private mobile operators have been licensed but in many cases are required to use PTT fixed networks to connect between base stations and switches. GSM calls originate or terminate on the fixed networks. Digital technologies are discussed below.

There are no European wide standards for analogue cellular telephony. Standards are set nationally which has resulted in incompatibility of the networks so that analogue equipment purchased for use in one Member State cannot be used in another.

1.2. Radio Paging

The regulatory framework for analogue radio paging is similar to that of analogue cellular telephony in that the monopoly PTT's were licensed to provide the service without any selection process. Only six countries have licensed private operators - the United Kingdom, France, Portugal, Spain, The Netherlands and Sweden. Most of the private operators are restricted from providing their own fixed infrastructure (the connections between base stations and local switches) and must therefore lease lines. The duration of licences varies from ten years in Spain to twenty five years in the United Kingdom.

Digital radio paging is still under development. Some Member States, such as France, have issued digital licences based upon the ERMES standard [4]. Generally, however, radio paging licences do not specify whether the system is analogue or digital.

1.3. Mobile Data

The development of mobile data services in Europe has been slow. There are few dedicated networks operating and limited information is available. Private operators have been licensed in the United Kingdom, The Netherlands and France. Some, but not all, of the PTT's in Europe provide a dedicated mobile data service. The regulatory framework throughout the European Union is not clear, for example, some countries have not yet decided whether to license dedicated mobile data networks.

1.4. Private Mobile Radio

Private mobile radio or PMR, is the closed radio system (normally analogue) operating usually with a single base station and using only a small number of radio channels. Interconnection to the public switched telephone network (PSTN) is prohibited. PMR has developed on a regional basis with the Member States issuing licences on a case-by-case basis. The system is commonly used by taxi companies.

Public Access Mobile Radio (PAMR) permits sharing the network for uses similar to PMR. Shared use allows more efficient use of the available frequency. It also prohibits interconnection with the PSTN. PAMR licences are issued on the basis of competitive tender. Generally, only one national licence is issued but there may be a number of regional licences. PAMR is commonly used by haulage companies.

The regulatory environment for PAMR is still under development in some countries, for example, Belgium and Italy.

1.5. Digital Cellular Telephony - Telepoint, GSM and PCN

a) Telepoint

Telepoint or CT2 is the system by which calls can be made from within about 200 metres of a base station usually located in a public place. Four networks were established in the United Kingdom but all have ceased operating due to the high call prices compared to phone boxes, cost of handsets and the fact that users could not receive incoming calls.

Despite the failure of Telepoint in the UK, Telepoint has been successful in France. Plans for Telepoint in Germany, however, have been postponed. The European Commission, by Council Recommendation [5] has proposed a digital European Cordless Telecommunications standard (DECT) which has yet to be finalised.

b) GSM

GSM stands for Groupe Special Mobile, a working group of Member States and non-European Union countries formed in 1987 to co-ordinate licensed frequencies and technical standards to create a pan-European cellular system for voice and data services. The system is designed to be compatible with general digital transmission technologies emerging from the Integrated Services Digital Network (ISDN).

The GSM standard for mobile cellular telephony was developed outside the European Commission's policy on the liberalisation of mobile telecommunications but the Commission has supported its establishment through various recommendations and directives. A Council Directive [6] required Member States to reserve GSM frequencies and set out the basis for a common digital specification to allow users to "roam", that is, use the equipment in different Member States. The GSM standard has been adopted widely outside the European Union, including in Asia, Australia and South Africa.

GSM licences are not technology specific, which means that an operator can provide cellular telephony, radio paging, mobile data and other services without the need for separate licences.

The GSM regulatory environment in Europe is one of limited competition. In each Member State, one GSM licence is held by the PTT and the other in the private sector following competitive tenders. Greece is an exception in two ways, first because both licences are held in the private sector and secondly because the licences were auctioned. Sweden will also be an exception because it has allocated more than two licences, and Austria has only granted one licence to the PTT.

The award criteria is not always clear and licence terms vary from five years in Denmark to twenty five years in the United Kingdom. Restrictions in the licences vary, for instance, in some Member States direct interconnection with other mobile networks is prohibited.

Notwithstanding competition in the provision of GSM, prices are still high. Due to restrictions on operators from establishing their own fixed infrastructure (except in the United Kingdom, Germany, the Netherlands, Greece and recently France) and the lack of competition in the fixed networks, leased line charges and interconnection costs are high, accounting for up to 50% of the mobile operator's revenues.

c) PCN

Personal Communications Networks are micro cellular networks operating to the GSM standard (known as DCS 1800) but at a much higher frequency. However, by way of contrast with GSM, in connection with which Member States were required to identify (and reserve) a specified frequency range, the PCN frequency of 1,800 MHz is not widely available due to use of the frequency by either the military or television broadcasters. PCN has developed much slower than GSM. In France, a PCN licence was recently issued to Bouygues but the frequency is not yet available throughout France. A national network will depend upon the relinquishment of the frequency by existing users.

The regulatory framework for PCN is still developing in Europe. Licences have been issued only in the United Kingdom, Germany and France to the private sector. No PTTs offer the service, possibly due to the high cost of establishing networks which would, in any case, offer competitive services to the PTTs' GSM services.

II - EUROPEAN STANDARDS ORGANISATIONS AND TYPE APPROVAL

Generally, analogue standards and approval procedures are set by the regulatory authorities in each of the Member States. There are no European-wide analogue standards.

Digital standards are set by European bodies. The European Telecommunications Standards Institute (ETSI) is the leading standards organisation. ETSI's task is to set common standards to link networks and services and to ensure the interoperability of equipment on a European basis and to endeavour to co-operate with standards bodies elsewhere to establish world-wide standards. Standards issued by ETSI must be incorporated into national standards by the Member States.

The European Commission has established the Conformance Testing Services (CTS) programme to develop standards testing facilities and a certification system. The Commission provides funds for the establishment of independent test centres. Member States are required to recognise testing carried out in other Member States by approved testing centres and to recognise the approvals issued in other Member States. [7] That is to say, once a manufacturer has had its equipment tested in one Member State and type approval for that equipment has been issued, it can market it in other Member States and the equipment can be connected to the PSTN in those countries without further testing or approval.

III - THE GREEN PAPER

Rather than being instrumental in the deregulation of mobile communications, the European Commission has supported steps already taken by the Member States in the liberalisation of mobile communications. The Commission has recognised the importance and growth of mobile communications and that the significant regulatory barriers to entry to the mobile market and licensing restrictions which still limit competition, restrict the ability to "roam", and prevent mass price reduction.

The Commission originally proposed the following measures in relation to the regulatory and licensing policies in the European Union :

i) Abolishing the monopoly and special rights granted to operators where the rights held were not gained on objective, proportional, transparent and non-discriminatory grounds.

ii) Ensuring that mobile licences are issued on transparent, objective and non-discriminatory grounds and licence conditions are restricted to those necessary for public service requirements and for effective use of spectrum, network security, data protection, safety and electromagnetic compatibility requirements.

iii) Abolishing restrictions on providing combination services, for example, radio paging and mobile voice telephony and abolishing restrictions on sales so that operators can sell direct or through service providers and suggesting that network operators should have the obligation to contract with service providers except in limited circumstances.

iv) Full freedom for operators to develop their networks, for example, by establishing their own fixed links or to share infrastructure.

v) Facilitating roaming and pan-European service provision, for example, to allow French users to subscribe to a German service provider.

vi) Licence conditions should not restrict ownership by Member States but may restrict ownership by nationals of non-Member States.

vii) In order to facilitate pan-European services, mobile terminal equipment should be subject only to type approval according to the principles set out in the relevant Directives, and there should be mutual recognition of mobile licences.

viii) The allocation of frequencies should be co-ordinated by the existing European Radiocommunications Committee and a new body created known as the Radiocommunications Office.

ix) A European numbering office should be established to coordinate numbering and make proposals for the portability of numbers (personal numbering).
x) In time, restrictions on offering services on the fixed and mobile networks should be abolished so that mobile operators can resell services on the fixed networks and fixed operators can bid for mobile licences.
xi) The Commission believes personal communication services will ultimately be carried via a single technology - the Universal Mobile Telecommunications System (UMTS) and wishes to create the necessary conditions for development of UMTS. Accordingly, the Commission proposes to co-ordinate licensing frequency, standards and numbering decisions via the ITU andits funded research programmes.

In late November 1994 the Commission published a Communication [8] on the consultation process on the Green Paper. Following heavy lobbying from network operators (mostly the monopoly PTTs), the Commission revised some of the more fundamental proposals in the Green Paper. It retracted the implied obligation upon network operators to deal with service providers, that is to say, there would have been no grounds to refuse to appoint a service provider other than for instance, because of doubts as to its financial stability. The "implied obligation to deal" was argued against most strongly by the new Scandinavian Member States, particularly Sweden where the uptake of mobile services is very large and there are a large number of service providers but where there is no obligation to deal.

Further, due to broad support for early action on liberalisation and direct interconnection, the Communication states that the right for mobile network operators to establish their own fixed infrastructure should coincide with the liberalisation of infrastructure generally and the approach should be integrated in the global approach set out in the Green Paper on infrastructure liberalisation [9]. The Green Paper on infrastructure proposes that liberalisation should be linked to the full liberalisation for telecommunications services, including those already open to competition and those to be opened by 1 January 1998. That is to say, where services are open to competition, there should be free choice in the underlying infrastructure for their delivery, such as the provision of links (including microwave links) within mobile networks for the provision of mobile communication services.

In addition, the Commission recognised that its proposal on the mutual recognition of licences was not practical and would therefore have limited effect because of physical limitations imposed by frequency spectrum and also the proposed licensing structures, for instance, only two GSM licences are to be issued in each Member State and therefore a Member State should not be required to recognise further licences.

The Communication sets out the priority areas for action by the Commission which are as follows:

i) The remaining monopolies and special rights are to be abolished as soon as possible and not later that 1 January 1996 through the application of European Union competition rules and the amendment to the Directive on the liberalisation of telecommunications which excluded mobile from its remit [10].
ii) The establishment of a voluntary code of conduct for service providers, the framework for which is to be in place before 1 January 1996. In addition service providers should be given the ability to offer a combination of services and in different Member States subject only to European Union competition rules.
iii) The publication of an interconnection Directive by 1 January 1998.
iv) To promote PCN services, Member States should allocate at least one licence using the DCS 1800 standard as soon as possible and at the latest after the European Radiocommunications Committee decision on allocating DCS 1800 frequencies and prior to 1 January 1998. To promote the extension of GSM using the DCS 1800 standard, existing restrictions on GSM operators using the DCS 1800 standards and upon PCN operators using the GSM standard should be abolished by 1 January 1998, but subject to the availability of frequency and competition rules.
v) Member states should allocate licences for public access/Telepoint/cordless (DECT) based systems and restrictions upon the combination of DECT and other mobile technologies should be abolished before 1 January 1996.
vi) Mobile operators should be able to provide voice services over the fixed networks in line with the liberalisation of voice telephony (which is to be, with the exception of Spain, Greece, Ireland, Portugal and Luxembourg) not later than 1 January 1998 subject only to universal service obligations and competition rules. At the same time, fixed operators will be able to offer mobile services, subject to licensing requirements.
vii) New technologies are to be supported by the Commission by way of studies of the digital Terrestrial Flight Telephone System (TFTS) - in-flight services for passengers, the European digital radio paging system (ERMES), the Trans European Trunked Radio

System (TETRA) for PAMR and mobile data and the digital short range radio system (DSRR) for PMR.

viii) The implementation of measures designed to reduce fraud, including an Equipment Identification Register for digital terminals and Physical Serial Numbers (PSN) for analogue terminals.

ix) While licensing of mobile operators will continue to be at nation allevel, satellite based services should be licensed at a European level. This would cover the Iridium, INMARSAT and the Globalstar proposed mobile satellite based systems. The licensing process is to be established by 1 January 1996.

CONCLUSION

The European Commission has supported and followed the steps taken by Member States to liberalise mobile communications rather than leading the liberalisation process. The proposals in the Green Paper will make little difference in the United Kingdom and Sweden and probably also little difference in France and Germany. However, it should significantly speed up the process in other countries of the European Union. The cost of use of the fixed networks is high and mobile operators may have to wait until 1998 before full competition, arrives and to discover whether the Commission will intervene to establish fair interconnection costs and terms, or whether the Commission will maintain its past position of leaving the issue for commercial negotiation.

REFERENCES

1. Towards the Personal Communications Environment: Green Paper on a Common Approach in the Field of mobile and personal communications in the European Union (COM(94) 145 final, 27.04.1994).
2. See Financial Times 5 October 1994
3. Green Paper on the Development of the Common Market for Telecommunications Services and Equipment (COM(87)290 final, 30.06.1987).
4. European Radio Messaging System
5. 91/288/EEC OJ L144/47 8.06.1991
6. 87/372/EEC OJ L196/85 17.07.1987
7. 91/263/EEC OJ L128/1, 23.05.1991; 93/97/EEC OJ L290/01, 24.11.1993
8. Communication to the European Parliament and the Council on the Consultation on the Green Paper on Mobile and Personal Communications in the European Union (COM (94) 492 final 23.11.1994).
9. Green Paper on the liberalisation of telecommunications infrastructure and cable television networks: Part 1 Principle and Timetable, (COM 94 440, 25.10.1994).
10. Commission Directive of 28 June 1990 on competition on the markets for telecommunication services (90/388/EEC; OJ L 192/10 24.07.1990).

- VIII -

CASE STUDY

JOURNALISTIC CULTURE, GLOBAL FREE SPEECH CONCERNS AND SPACE COMMUNICATION ISSUES : A CASE STUDY

DR. KERRIC HARVEY

National Center for Communication Studies ; The George Washington University ; 812 20th Street NWWashington, D.C. 20052, USA.

Abstract : *A serious barrier to the responsible implementation of a worldwide procedure for press coverage of space-related issues is found in the enormous pressure of both U.S. and international law as they apply to the freedom of the press. Legal decision, historical precedent, and the conventions of newsgathering routines all combine with the dictates of journalistic culture to challenge the fundamental viability of any international agreement regarding the timely and responsible flow of public information about space research topics. Internal pressures and demands in the media environment are reinforced by cultural expectations surrounding the global public's "right to know". The expectations articulate on both the national and international levels of planetary goverance, as the First Amendment to the U.S. Constitution and in similar language contained in several of the human rights instruments of the United Nations. A case in point may be found in the circumstances surrounding news coverage of one of the most volatile space research issues, the Search for Extra-Terrestial Intelligence, also know as SETI. Rather than impose access restrictions on space communication issues in general or to force censorship devices on SETI-related information, in particular, the scientific community should consider developing a non-governmental, university-based Space Communications Institute, which would have the sole purpose of establishing a responsible, credible, and consistent liason mechanism between those who conduct all space-related research, including SETI, and those who report upon it.*

I - COMMUNICATION ISSUES IN SPACE EXPLORATION

Much of the current work in the space communication policy literature addresses the question of how the press will react to coverage situations which require local reporting of planetarily important events, and has focused on the need, the ways, and the means of achieving successful and effective international cooperation in these types of journalistically collaborative situations. Tarter (1991) has provided some insight regarding press attitudes about culturally volatile space issues and their potential news value on the global scene, but from the communications perspective, it is a grave mistake to think about most truly important space-related news stories as events which will be reported upon by the science press, in general, or as an information flow which can be effectively monitored for its scientific accuracy, in particular. News treatment of the Search for Extra-terrestial Intelligence, or SETI, supplies a vivid, if somewhat extreme, example of just how delicate the press/government/science dynamics are likely to be in any space-related issue carrying dramatic consequences for the global population. This paper uses SETI as a case study for exploring the challenges and complexities faced by all types of space scientists and policy-makers as they try to interact with the global press to supply informative, yet responsible, news coverage of space-related issues. Although the paper focuses on SETI, it illustrates some principles of journalistic culture and several points of international media law which have ramifications for anyone involved in the communications aspect of space law and policy.

II - SETI AS A CASE STUDY

The prospect of establishing even limited communication with sentient beings not of earthly origin — that endeavour which space scientists and astronautical researchers have come to call the "Search for Extra-terrestial Intelligence, or "SETI" — demands from policy makers and communication scholars a new way of thinking about international communication flow and the domestic media policies which accompany it. Michaud (1990) and Tarter (1990; 1991) point out that First Contact will be a general media event — it will be compelling human news, a front-page story if ever there was one. What this means, however, is that the norms and expectations of journalistic culture, per se, not just science journalism, have to be factored into any viable plan for managing SETI news flow. This is very different from dealing with relatively specialized science media, which are, by the necessity of their trade, better acquainted with the complex nature of the research process and which, thus, are likely to be somewhat more conservative in their approach to unverified announcements.

Concern about how to ensure responsible handling of information about SETI activities also raises, almost immediately, profound questions about freedom of speech, which ramify on both the moral and the pragmatic levels. The proposed international SETI protocol states, clearly and explicitly, that announcements about Contact be withheld from the press until complete scientific verification is achieved to the mutual satisfaction of the agreement signatories. Furthermore, the protocol implies, if not outright suggests, that all information about ETI communication be routed through a central clearinghouse — an internationally credible authority — before it finds its way into the world press and, thence, into the public eye. Responsible science and the demands of public safety, it is argued, dictate this relatively unusual degree of institutional brokering relative to the release of news about SETI.

The scientific community's sense of public responsibility, as illustrated in suggestions regarding SETI public information policy, are justified by the exigencies of the SETI situation. It must be pointed out, however, that the SETI protocol presumes an inordinately high level of international cooperation and a significant degree of consistency among the cultural values of its member groups, particularly as they relate to the notion of free speech.

III - FREE PRESS AND THE SPACE FRONTIER

Although policymakers and news managers can hope for international cooperation in space issues news management, it is not unlikely that optimistic assumptions may prove, eventually, to be problematic. Logdson and Anderson (1990) have already pointed out the frailty of paper agreements when real-world stakes intrude between international agreement signatories. It is highly probable that challenges to the integrity of the SETI media accord will come, perhaps surprisingly, from the more industrially and informationally "advanced" participants in the world community. Cultural relativism, a persistent wrinkle in the smooth and even application of international law, is often unfairly associated with the demands of Third World nations, who are — in the United Nations (U.N.) record — often portrayed as reluctant to conform to the human rights prescriptions that other countries impose upon the aggregate group.

In the SETI situation, the defiant chips are likely to fall out differently. Most specifically, the real opposition to international embargoes on SETI news — even if those embargoes are dictated by both common sense and historical record — will come from the United States media.

This is because any institutional machinery designed to orchestrate, or organize, or manage, or edit the SETI informational flow will, **by definition**, collide with practice, expectation, and outright law regarding the freedom of the press in U.S. cultural history. With the exception of the Desert Storm episode, the American press has relied on Constitutional mandates to effectively repel government interference of any description, and the international SETI protocols, which emphasize the timely release of verified information, are unwittingly setting themselves up for attack by the very institutions they need most desperately to manage — the U.S. news media.

The Western news media, in general, are likely to bristle at having to conform to international agreements which, they will feel, are designed to muzzle them. Both law and history, in this instance, come down heavily on the side of media freedom — in the international as well as the domestic arenas.

In the United States, the Federal Freedom of Information Act — a Congressional mandate — and a battery of state statutes have both upheld and reinforced the media's presumption of the right to relatively unrestricted access to information as an essential part of the free speech process (Borden, 1993). Similarly consistent support has emerged, from a long snarl of sometimes complicated decisions, for the demonization of prior restraint — the a priori, or pre-publication suppression or regulation of information, or the delay of information release through media channels, even when the release of that information may threaten the national security or the social good. I am not, in any way, trying to advocate the reckless and habitual use of prior restraint as a censorship mechanism. But the U.S. court system's virtual expulsion of the prior restraint concept from the list of socially viable information management options will hinder, severely, any attempt by the SETI community to exercise deliberate and responsible control of information about its work which the press sees fit to share with the world, on the media's timetable.

The media's antagonistic attitudes toward government interference with the "free flow of information" have been legitimatized through a series of important U.S. Supreme Court decisions, each of which supplies a significant precedent for subsequent decisions about how much control over information flow the SETI community may have to relinquish to the global press. It should be noted that, in the case of an event as resounding as First Contact, news embargo timeframes would, logically, need to be extended even further than usual, since the cultural importance of the event and the technical complexity involved in signal verification demand a more exhaustive validation protocol than many other scientific "breakthroughs" might require. But the media are impatient creatures, unlikely to wait for the "official" report if they can get a "credible source" to talk first.

The verification stakes for the press are different than they are for SETI researchers: Even if they rush the story, and the signal turns out to be unverifiable, the media will not take the direct blame for this error — the media's sources will. Technically, that is how journalism works in the contemporary environment — and the Court, leaning on constitutional principles, has upheld the media's inculpability for "honest" mistakes in its reporting, as long the incorrect information comes from arguably "credible" sources. The messengers of world events are seldom liable for the consequences of the tales they report. To prove them liable, even in cases of outright and hurtful untruth, injured parties who are public officials or public figures must be able to prove "actual malice" — must be able to prove that the media indulged in a "reckless disregard for the truth," or, at a higher threshold of accountability, that the media deliberately intended to hurt them — both of which are almost impossible tasks for those lacking in psychic gifts.

How the sudden intrusion of new life forms might affect the courtroom environment is, of course, impossible to say. In sheerly human experiences, however, U.S. legal decision has consistently defended the probing media from almost all comers. In terms of prior restraint, within U.S. press history the primary official excuse for impeding the free press has been the occasional emergence of unique and overwhelming national security issues. Of course, "national security" has been, historically, defined in a number of interesting ways. But, as demonstrated repeatedly, only direct and relatively immediate threats to national well-being can permit the institutions of government to intrude into the free flow of information about matters of public concern. This is not, by any means, altogether a bad thing. But such culturally ingrained attitudes about the mass media do present some serious problems for those seeking to promote responsible reporting of ETI events.

IV - MEDIA LAW AND SPACE NEWS FLOW MANAGEMENT

Examining two main categories of U.S. media law, for example, reveals much evidence to support the idea that the very goals of the SETI protocol are likely to be, in the U.S. at least, declared unconstitutional by definition. The first of these categories foreshadows a situation within which scientists might try to muzzle any loose lips, journalistic or otherwise, that might be flapping on about unverified signals.

Media scholar Thomas L. Tedford (1993) uses the phrase "political heresy" to describe the category of legal action in which speech perceived as threatening to the social order has been restricted through some sort of government intervention — typically, by arresting the speaker. The rationale being tested in these instances is the trade-off between an individual's right to expression and the threat to public safety that may result from the exercise of that person's free speech rights. Without going into too much legal detail, he notes the importance of cases like: Schenck v. U.S. (1919), which generated the clear-and-present-danger test first used to decide if expression inches too close to incitement to be tolerated; Gitlow v. New York (1925), protects free speech rights from state interference, under the 14th Amendment ; Whitney v. California (1927), which (despite the fact that it resulted in the speaker being silenced), further "refined" the clear-and-present danger standard to the point where it is almost impossible to apply, and; Brandenburg v. Ohio (1969), which transformed the clear-and-present danger standard into the "imminent lawless action" standard, which is even more impossible to meet (Borden, 1993).

Justice Brandeis, who together with Justice Holmes voted to uphold Whitney's conviction by the lower court, but who felt uneasy about doing so, generated the language which will probably be resurrected in arguments against the scientific community's authority over SETI-related news flow. I would be dishonest not to note a general theoretical agreement with the spirit of Brandeis' language. But it does bode ill for SETI, and as much as one is reluctant to saddle the press with inappropriate restraints, the 21st century might be an important time to re-visit the meaning of the word "appropriate."

According to Brandeis:

> *Fear of serious injury alone cannot justify suppression of free speech and assembly....To justify suppression of free speech there must be reasonable ground to fear that serious evil will result if free speech is practiced. There must be reasonable ground to believe that the danger apprehended is imminent. There must be reasonable ground to believe that the evil to be prevented is a serious one....In order to support finding of a clear and present danger it must be shown either that immediate serious violence was to be expected or was advocated, or that the past conduct (emphasis added) furnished reason to believe*

that such advocacy was then contemplated....Only an emergency can justify repression. (Tedford, p. 62).

The court's decision in Brandenburg, which yielded a more recent refinement of the clear-and present danger test, argues that an "actual incitement leading to imminent lawless action" (Tedford, p. 72) must be present in order to shackle free speech rights. The point of this is to prove that the perceived danger is real, not imaginary...a relatively specious criteria in SETI circumstances, since the theoretical definition and the operational details of reality will doubtless be altered forever with the introduction of alien forms into the social and legal arenas.

One core problem, of course, is that these decisions refer to political speech (seditious libel), and by extension, the social evil which may result from too profound or too physicalized a challenge to the **political** status quo, such as armed insurgence and the like. Even in that regard, however, the specification that restraints on speech must be justified by a history of ill fortune, should that speech be indulged, renders all SETI-related speech free from restraint under the clear and present danger or imminent lawless action tests, since the First Contact event, by definition, will not have any kind of history — harmful or beneficial — behind it.

A rebuttal to the central logic in any of the clear-and-present danger tests might begin by differentiating between a world in which the center of social gravity is defined by political events, and one in which it is shaped, molded, and textured primarily by the parameters of scientific consciousness. That is the key difference between the 18th century, when the idea of a free press hardened into legal reality, and the 21st century, in which legal ground, increasingly, gives way before technological imperatives.

But the other main category of media law likely to have a resounding impact on SETI information management plans resurrects the notion of prior restraint — of stifling speech before it ever reaches the public purview. If anything, the Supreme Court's body of decision in this area is even more persistently resistant to the type of extended news embargo required for SETI signal verification and related scientific processes.

Near v. Minnesota (1931) did identify a series of circumstances under which prior restraint could be invoked, including threats to the national security during times of war, incitement to violence, and assaults on public decency. But even as it articulated these conditions, the Court established an impossibly high threshold of risk to the social good which would have to be surpassed before prior restraint could, legitimately, be employed (Francois, 1990). In fact, first in Near and the in more recent cases, the Court insisted that post facto punishment of illegal or improper media actions is the proper instrument of control over press misbehavior — a sentiment which is reinforced internationally by Article 13 ("Freedom of Thought and Expression") of the American Convention on Human Rights, O.A.S. Treaty Series No. 36. But suing the press for irresponsible reporting, after the fact, won't do any good at all. By the time the press may be taken to court for inappropriate action, the damage, if any occurs, is likely to have been done. And, as indicated by Tarter and Michaud's work, cited earlier, there is enormous potential for damage to occur.

Over and over again, we see this abhorrence of government intervention sustained by U.S. Supreme Court decision — even when clear national or international security issues are involved. Again the Court cites the First Amendment as the basis for its perspective, codifying its skittishness about imposing prior restraint in the infamous Pentagon Papers case in 1971, and again, in 1979, in lower court rulings in the Progressive Magazine affair. Both the decisions have tremendous resonance for would-be SETI information gatekeepers.

In an agonizing split vote over whether to prohibit the **New York Times** from publishing a history of the federal decision-making process during the Viet Nam war, the Court developed what is called the "doctrine of preferred freedoms" (Tedford, p. 227), which forces the restraining body — in this case, the U.S. government — to bear the burden of proof regarding the potential gravity of the "speech" in question. This is expensive, difficult, and time-consuming. It is also almost impossible, since it forces one to "prove" the impact of hypothetical events.

In the Pentagon Papers case, a temporary restraining order (TRO), invoked against the Times while the Court debated the issue, was eventually lifted, and the papers were published. But much press and public outrage still lingers today about the 15 days it took for the Court to think things through. And this serves to illustrate yet another quixotic detail in U.S. cultural attitudes regarding the absolute freedom of the press. "Delay" is equated with "denial," despite the fact that these are two very different things. Not only do the U.S. media want absolute and unimpeded access, but they want it NOW. Making them wait is tantamount to making them go without — at least, that's how it is presented to the public and argued in courts of law. Fifteen days is not necessarily a long time to spend soul-searching important issues like those at stake in Pentagon Papers case. But immediacy is a big part, and a defining characteristic, of journalistic culture. SETI information managers must never forget that what they may perceive as an opportunity for reflection will be interpreted by the press as intolerable delay.

A Seventh Circuit Court of Appeals case which closely parallels potential SETI scenarios, since it deals directly with the social impact of technological matters, supplies an interesting coda to this selective review of U.S. media law. In 1979, **The Progressive** magazine announced its intention to publish a set of instructions for building the H-bomb, as a means of illustrating how vulnerable the world had become to nuclear-inspired lunacy. No illegal data collection activities were indulged and no classified material was employed in the writing of the article. But the government said that the legality of the piece didn't matter — merely presenting the component information needed to construct a nuclear weapon constituted a sufficiently serious threat to national security, and the government sought and received an injunction against publication. The government argued that the risks imposed on society by circulating recipes for thermonuclear catastrophe outweighed **The Progressive's** First Amendment right to publish the article.

The Progressive fought back, scheduling an appeal before a higher court, threatening to go all the way to the Supreme Court, if necessary. In the meantime, other journalists began leaking the sanctioned material. The government sought more TROs; more journals began to print nuclear matter. Finally, after about six months, the government dropped its case, since what was essentially a media cabal had rendered the **Progressive Magazine** prior restraint order useless by back-channeling the disputed material into other venues (Pember, 1981).

The point here is not to defend the U.S. government, or the **Progressive**, or the **New York Times**, for that matter. It's to call attention to certain patterns in the journalistic and the legal historical records which suggest that any federal management of SETI-related news flow is doomed, if it tries to achieve its information management goals using the traditional methods of media control— licencing of electronic media, or news embargoes, prior restraint, access denial, and **post facto** judicial action, in the print media. In fact, there is an entire adjunct of U. S. media law which sets out the "rules" for disregarding directives which the media may feel are unconstitutional constraints on free speech activities, underscoring even more emphatically the deeply entrenched place of privilege that press freedom enjoys in Western — or, at least, U.S. — cultural thought. Far from making the public trust the government's scientific rigor, the creation of an official mechanism for censoring information about scientific endeavors will cause the public to trust the government less, and will further cement the public's sympathetic identification with "crusading" media, even if those media are tilting at the wrong opponents. The American public will side with the media, despite its somewhat gratuitous display of cynicism as it greedily consumes the products of the media's labor.

V - REACTIONS OF THE GLOBAL PRESS

Internationally, the chances of making media embargoes stick on an unruly world press are even more unlikely. In addition to the migration of American press values into international media production routines, even the most vigorous of the U.N. policy-making organs have met with limited and inconsistent success when they try to impose global standards of conduct on diverse local populations.

The U.N. Charter and supporting documents guarantee, on a world level, much the relationship to information collection and freedom of expression as do the U.S. First Amendment and its subsequent interpretations in U.S. media law. Interestingly, the U.N. documents supply a somewhat less absolutist protection on free speech than do their U.S. counterparts; Article 19 of the International Covenant on Civil and Political Rights, essentially, replicates the First Amendment on an international scale, with one important exception; built into the actual language of the protection are the exceptions to it, which include: infringement on the "rights or reputations of others," and threats to the "...protection of national security or of public order, or of public health or morals" (Newman and Weissbrodt, p. 31). This language puts the exceptions to the protections on free speech, and the responsibilities attached to the enjoyment of protection, on an equal footing with free speech protection itself, acknowledging the intense situationality of many free speech disputes, especially those which are characterized by the presence of competing equal rights.

But, as many scholars have noted, U.N. law and policy are binding on member nations only insofar as they permit them to be so. Although, theoretically, U.N. decisions pre-empt those of "lower" national courts, in actual practice, international law is subordinate to domestic custom. As Pasqua et al put it, "One nation cannot make other nations enforce its laws in their countries unless it is willing and able to use the force of arms to do so. Internationally recognized bodies such as the United Nations and its International Court of Justice often succeed at solving disputes, but member nations have never given them the full power to enforce decisions" (1990, p. 345). Newman and Weissbrodt enlarge on this notion, pointing out, "International law is law to the extent that governments are willing to conform their conduct to norms that accept or grudgingly acknowledge the prevailing rules among nations (p. XVII), noting the U.N.'s failure to ensure women's rights on American Indian reservations as well as under Islamic penal codes as examples of sovereign-state ability to reject international recommendations for behavior. And Pierre Marc Johnson, speaking to a group of academics, embassy professionals, politicians and policymakers on the environmental enforcement aspects of the North American Free Trade Agreement, admited that normative international clout extends only to those nations which permit themselves to linger within easy U.N. reach (November 15, 1993).

In most situations, voluntary submission is required for U.N. doctrine to affect the internal affairs of its member nations. In those few situations where compliance can be required by the collective agency, enforcement is the instrument by which it must be achieved, either through economic sanctions, the pressure of world opinion, or, sometimes, by multilateral military intervention. Enforcement is always problematic. But the crucial point for the purposes of this work is the realization that all of the usual enforcement options take place **after the fact**, after transgression has occurred and a record of it has been established. International remedies for irresponsible press behavior are every much as **post facto** as are traditional U.S. methods of balancing free speech rights against other pressing social concerns, like personal safety, national security, and cultural well-being.

Tarter (1992), Logsdon (1990), and Michaud (1990) have all noted that the period immediately following the First Contact announcement will be crucial in establishing enduring human attitudes toward extra-terrestial encounters, noting that it is those first few days — perhaps the first few hours — that will set the tone for centuries of interspecies interaction. **Post facto** action may, indeed, be the most appropriate remedy for irresponsible speech in most cases. But it supplies very little true relief — in fact, it's kind of beside the point — in the cirumstances surrounding SETI news flow.

CONCLUSIONS AND RECOMMENDATIONS

If there is to be any "meeting of the minds" between the space research community and the global media who report upon it, that connection will have to take a vastly different form and be of a radically different character than that which any previous media/information management intersections have assumed.

To achieve this goal, it is necessary to manage SETI information without appearing to manipulate it. One way to accomplish this is through the creation of a permanent Space Communication Institute, the purpose of which would be not to circumvent the First Amendment so much as to make the media's dogged insistence on it functionally irrelevent, by exploiting the routine practices of the newsgathering industries in such a fashion that the world media, of its own accord, would seek out the Institute's official voice in matters relating to signal detection and reply.

Many critics of the "free" mass media (Gitlin, 1980; Tuchman, 1978; Bennett, 1983) have pointed out that production and publication deadlines associated with a commercial press create an insatiable demand for regularly delivered and easily digestible information, which only other institutions have the resources and the wherewithal to satisfy. They are quick to note that the tyranny of production schedules and the pressing demands of the bottom line combine to create commercial media organizations with voracious and insatiable appetites for information, requiring regular feeding at predictable intervals. In other words, the mass media must go on — the paper must be printed and the nightly news must be broadcast — regardless of whether unruly reality has cooperated with commercial media deadlines and schedules. This constant need for immediately usable information is usually cited as a structural weakness of the mass media, since it makes them vulnerable to those institutions that can provide them with regular nourishment at critical moments, and these informational suppliers are usually "official" institutions of one sort or another.

Much as the U.S. Center for Disease Control (CDC) has become this type of organization, the place of first recourse for journalists seeking expert opinion on epidemiology, and the World Health Organization (WHO) has become the arbitrating voice on issues of national well-being, so must the world space exploration elite establish an organization equipped and intended to deal with press and public inquiries regarding ETI communications. It should be a non-partisan, government affiliated, but university based organization, to distance it from the spectre of "official" control that so haunts the Western press's professional sensibility, and staff it with as international a complement as possible. In keeping with post-Cold War budgetary sentiment, this Institute must also be designed as a dual-purpose organization, charged with providing the world press with regular updates on all forms of space-related communications issues, including benign but persistent news releases regarding current SETI efforts.

Care should be taken, however, to avoid repeating the mistakes of the CDC model, which found itself over-embroiled in the AIDS controversy by neglecting to have "on board" people who could act as interpreters between the scientists and the public, and who could translate scientific research into understandable terms. Toward that end, it would be vital to have anthropologists, communications specialists, social psychologists, and an array of humanities people, as well as scientific experts, included in the institute's permanent advisory board.

Again, the strategic objective here is not to muzzle the press regarding SETI information, but to overwhelm them with accurate, responsible, usable material of the kind that media so desperately requires to meet their ever-present deadlines, to dovetail with standard newsgathering practices and industrial routines. In the best-case scenarios for every one involved — press, public, and research community alike — the Institute would make it so easy for the press to access SETI information that the media would be unlikely to go elsewhere in search of less scientifically responsible information. **The press, no more than any other living organism, is unlikely to bite the hand which feeds it, even at the moment of criticizing the owner of that beneficient limb**. But in the case of SETI, and its overwhelming need to maintain an unusually high degree of expert control of scientific newsflow, it may well be that pre-emptive, proactive information management, such as the type which could be provided through an office such as this, is not merely justified, but is, indeed, imperative.

This notion of needing to monitor the scientific quality of ETI newsflow can be, understandably, unsettling to those who believe in the importance and indispensibility of a free flow of global information.But it's worth considering if SETI might just present one of those situations in which there is no perfectly comfortable solution available, because we, as a world culture, have gotten ourselves into a place where our technical capability far outstrips our cultural ability to deal with it...a very 21st century type of pickle to be in.

Which introduces the central dilemma of managing SETI information — or any type of information which the international space community has embargoed for reasons of scientific verification — within a First Amendment media environment. The First Contact announcement is likely to be a national story with enormous cultural impact attached to it, not just a science news story, which means that assumptions about the likely media response must be contextualized within universal journalistic culture, where individual reporter ambition and presumptive norms surrounding the First Amendment conspire against **a priori** restrictions on First Contact news flow. Both U.S. and international law, in the SETI situation, are likely to further complicate the natural complexities introduced by religion and culture. In the U.S., institutions within the scientific community which may be trying to work with federal regulators to ensure responsible SETI coverage will find that they are able to restrict press access only in times of severe national emergency (if then). The lurking element of threat buried within the very notion of First Contact could, mistakenly, be exploited for this very reason — to buy critical time for the national research community, by presenting Contact as a national security issue, exempted from the usual amount of press access.

But once Contact is defined as a crisis, even if only for the purpose of leveraging press restraint mechanisms into place, a certain element of danger is automatically and irrevocably attached to it. This is counterproductive to the goal of allaying public fears about alien contact. The result is a classic double-bind; science needs to control the SETI news flow in order to achieve responsible coverage that keeps people from over-reacting to the Contact announcement, but the only legal means by which it can achieve that control is by creating the very thing it wishes to avoid — public anxiety, triggered by framing Contact as a crisis situation.

It all comes down to a choice between defining First Contact as a national emergency, like war or natural disaster, in order to gain media compliance, or presenting Contact as a scientific or diplomatic event of no special magnitude and no particular threat, thus inviting the unpredictable attentions of an inherently unmanagable press. To subordinate what, in the SETI situation, may be irresponsible First Amendment impulses, one must establish precisely the kind of national attitude toward Contact that one is trying to avoid by monitoring newsflow, in the first place.

A first step in addressing the complex needs of SETI newsflow management revolve around the creation of an international, non-governmental Space Communication Institute, to be assigned proactive rather than responsive duties in shaping press attitude toward SETI through the constant over-provision (as opposed to the idiosyncratic restriction) of scientifically verifiable material. The Institute must be credible to a number of individual and institutional audiences, some of them characterized by highly contrasting world-views and value systems. It must be of long-standing duration, which in this budgetary climate means it must also be dual-purpose. And, like many other types of space exploration and research, SETI news management must incorporate the expertise of a wide variety of scholars, scientists, and information managers, who can help turn the cultural fears surrounding the idea of the deeply alien into public appreciation of this most spectacular episode in all human exploration.

WORKS CONSULTED

Bennett, Lance W. (1983). **News : The politics of illusion**. New York: Longman.

Borden, Diane. (1993). In private conversation, Washington, D.C.

Brandenburg v. Ohio, 395 U.S. 444, 89 S.Ct. 1827, 23 L.Ed.2d. 430 (1969).

Bronowski, J. (1990). **Science and human values**. (Revised edition with a new dialogue). New York : Harper and Row.

Burkett, David Warren. (1973). **Writing science news for the mass media**. (Second edition, revised). Houston, Texas : Gulf Publishing Company.

Cantril, Hadley, with Hazel Gaudet and Herta Herzog. (1940). **The invasion from Mars** : A study in the psychology of panic. Princeton, New Jersey: Princeton University Press.

Davis, David Brion. (1966). **The problem of slavery in Western culture**. Ithaca, New York : Cornell University Press.

Davis, David Brion. (1984). **Slavery and human progress**. New York : Oxford University Press.

Declaration of principles concerning activities following detection of extraterrestial intelligence. In **Acta Astronautica : Journal of the International Academy of Astronautics**. Vol. 21, No. 2. Pp. 153 - 154.

Dennis v. U.S., 341 U.S. 494 (1951).

Douglas, Mary & Aaron Wildavsky. (1983). **Risk and culture : An essay on the selection of technological risk and environmental dangers**. Berkeley and Los Angeles, California : University of California Press.

Francois, William E. (1990). **Mass media law and regulation**. (Fifth edition). Ames, Iowa : Iowa State University Press.

Gitlin, Todd. (1980). **The whole world is watching**. Berkely and Los Angeles, California: University of California Press.

Gitlow v. New York, 286 U.S. 652 (1925).

Greenberg, Bradley S. & Walter Ganz. (Eds). **Desert Strom and the mass media**. (1993). Cresskill, New Jersey : Hampton Press.

Hanke, Lewis. (1959). **Aristotle and the American Indians**. Chicago : Henry Regnery Company.

Harvey, Kerric. (1993). Adam's other rib : The development of an American robot industry. paper presented at the Southern Humanities Council Annual Conference, Huntsville, Alabama, February 11 - 14.

Hornig, Susanna. (1993). Reading risk: public response to print media accounts of technological risk. In Public Understanding of Science, Volume 2. Pp. 95 - 109.

International covenant of economic, social and cultural rights. (1976, January 10). G.A. res. 2200A (XXI), December 16, 1966, 21 U.N. GAOR Supp. (No. 16) at 49, U.N. Doc. A/6316 (1966), 993 U.N.T.S. 3.

Kellner, Douglas. (1992). **The Persian Gulf TV war**. Boulder, Colorado : Westview Press.

Klein, Herbert S. (1986). **African slvery in Latin America and the Carribbean**. New York : Oxford University Press.

Klein, Herbert S. (1967). **Slavery in the AMericas: A comparative study of Virginia and Cuba. Chicago** : University of Chicago Press.

Logsdon, John & C.M. Anderson. (1990). Announcing the first signal : Policy aspects. **In Acta Astronautica**, Vol. 21, No. 2. Pp. 89 - 92.

Marc-Johnson, Pierre. (1993, November 15). Address to the Fulbright Lecture Series. Canadian Embassy: Washington. D.C.

McLuhan, Marshall & Eric McLuhan. (1988). Laws of media : The new science. Canada : McLuhan Associates, Ltd.

Michaud, Michael A.G. (1990). A unique moment in human history. In Ben Bova & Byron Preiss, (Eds.)., **First Contact : The search for extraterrestial intelligence**. Pp. 243 - 261. New York : NAL Books.

Michaud, Michal A.G. (1989, May). Detection of ETI — An international agreement. In **Space Policy**, Vol. 5, No. 2. Pp. 103 - 106.

Near v. Minnesota, 383 U.S. 697 (1931).

New York Times v. U.S.; U.S. v. Washington Post, 713 U.S. 403 (1971).

Newman, Frank & David Weissbrodt. (1990). **Selected international human rights instruments**. Cinncinati, Ohio: Anderson Publishing Company.

Pasqua, Thomas M. Jr., & Others. (1990). **Mass media in the Information** Age. Englewood Cliffs, New Jersey: Simon & Schuster.

Pember, Don R. (1981). **Mass media law**. (Second Edition). Dubuque, Iowa : William C. Brown Company.

Pollack, John Crothers. (1981). **The politics of crisis reporting : Learning to be a foreign correspondent**. New York: Praeger Press.

Rivenburgh, Nancy. (1990). In private conversation. University of Washington: Seattle.

Schneck v. U.S., 249 U.S. 47 (1919).

Singer, Eleanor & Phyllis M. Endreny. (1993). **Reporting on risk : How the mass media portray accidents, diseases, disasters, and other hazards**. New York : Russell Sage Foundation.

Sterling, Christopher H. & John M. Kitross. (1990) **Stay tuned : A concise history of American broadcasting**. (Second edition). Belmont, California : Wadsworth Publishing.

Tarter, Donald E. & Walter Gillis Peacock. (1991). The consequences of contact : Views of the scientific community and the science media. In J. Heidmann & M.J. Klein (eds.)., **Bioastronomy : The search for extraterrestial life**. New York: Springer-Verlag.

Tarter, Donald E. (1992, May). Interpreting and reporting on a SETI discovery. In **Space Policy**, Pp. 137 - 148.

Tarter, Donald E. (1992). SETI and the media : Views from inside and out. in **Acta Astronautica**, Vol. 26, No. 3/4. Pp. 281 - 289.

Tarter, Donald E. (1993). Treading the edge : Practicing safe science with SETI. In tne **Skeptical Inquirer**, Vol. 17., Spring issue. Pp. 288 - 296.

Tarter, Donald E. (1993, February 13) What if? The protocols for alien contact. Paper presented at the Southern Humanities Council Annual Conference, Huntsville, Alabama.

Tuchman, Gaye. (1978). Making news : **A study in the construction of reality**. New York: Free Press.

Universal declaration of human rights. (1948, December 10). G.A. res. 217 A(III), U.N. Doc. A/810, at 71 (1948).

U.S. v. Progressive, 467 F. Supp. 990, (1979).

Watzlawick, Paul. (1976). **How real is real? Confusion, Disinformation, Communication**. New York : Vintage Books.

Weatherford, Jack. (1988). **Indian givers : How the Indians of the Americas transformed the world**. New York : Fawcett Columbine.

Winks, Robin. (Ed.). (1972). **Slavery : A comparative perspective**. New York : New York University Press.

Whitney v. California, 274 U.S. 357 (1927).

Wiseman, Richard L & Jolene Koester. (1993). **Intercultural communication competence**. New York: Sage Publications.

Tedford, Thomas L. (1993). **Freedom of speech in the United States**. (Second edition). New York: McGraw-Hill, Inc.

- *IX* -

APPENDIX

LAW OFFICES SPECIALIZING IN THE TELECOMMUNICATIONS AND/OR SPACE SECTORS

The Telecommunications and Space Journal intends to become a useful instrument at the service of the largest community of all those interested in the two growing sectors of telecommunications and Space. For this purpose, the editorial bureau sent a questionnaire to more than two thousand law offices around the world having announced a specialization in the telecommunications and/or space areas. Although a few of them answered this questionnaire, it was decided nonetheless to publish the following list which is obviously to be completed in a further edition. The same will be done with insurance companies or banking institutions dealing with telecommunications and/or space matters. All information, personal data, address and telephone/fax numbers have to be sent to the Telecommunications and Space Journal (Editorial Bureau . Attention : Professeur Lucien Rapp).

Advokatfirman Vinge, Box 1703,11187 Stockholm, Sweden, Tel : 46-8 614 30 00 Fax : 46-8 611 90 37. Contact : Tengroth Lennart.

Akim, Gump, Strauss, Hauer, Feld & Dassesse, 5 avenue Louise, B.7, B - 1050 Brussels, Belgium Tel : 32 2 535 29 11, Fax 32 2 535 29 00.

Allison & Humphreys, J.A. Ballard, East India House, 109-117 Middlesex Street London EI 7JF UK, Tel : 0171 570 6000, Fax : 0171 570 6060.

Amaro , Stuber & Street, Av. Paulista, 1499 - 19° andar - CEP 01311-928 Sao Paulo - SP - Brasil , Tel : (55) (11) 284-9911.

Bos, Oosterbaan & Van Eeghen, Dinant T.L., Koningslaan 35, 1075 AB Amsterdam, Netherlands Tel : 31 20 6716756, Fax : 31 20 6718669.

Dechert Price & Rhoads, Avenue Louise 65, 1050 Brussels, Belgium, Tel : 32 2 535 54 11, Fax : 32 2 535 54 00. Contact : Amory Bernard.

Fagny Consult, 145 Avenue Notre Dame 1140 Brussel, Belgium Tel : 32 2/215 88 27, Fax : 32 2/245 35 99.

Faugere & Jutheau, 54 Quai Michelet, 92681 Levallois Perret Cédex, France, Tel : (33-1) 41 34 50 00.

Gide Loyrette Nouel, 26 cours Albert 1er, 75008 Paris France, Tel (1) 40 75 60 00, Fax : (1) 43 59 37 79.

Int'1 Institute of Air and Space Law, PO. Box 9520, 2300 RA Leiden, The Netherlands, Tel : 31-71-277724, Fax : 31-71-277600. Contact : Von Der Dunk F.G..

Levin & vance P.A. ,320 Gold Sw, Suite 1400, Albuquerque, New Mexico 87102 USA, Tel : 505 247 1111. Contact : Levin Richard H;.

Mannheimer Swartling Advokatbyra, Box 1650, Stockholm, S 11186 Sweden Tel : (5) 46 8 613 55 00. Contact : Olofson Rolf.

Nabarro Nathanson Sp. zo.o., ul. Senatorska 12, 00-082 Warszawa, Poland, Tel : (48-22) 27 61 44., Fax : (48-22) 27 48 38.

Partner Simonsen & Mosaeus, PO Box 727, Sentrum, O/Os Oslo Norway, Tel : 095 45 22 93 6500, Fax : 095 45 22 93 6550. Contact : Espen Tondel

Serra Michaud & Associes, 2 rue de la Baume 75008 Paris France, Tel : 4421 97 97 Fax 42 89 57 90. Contact : L. Rapp.

Skadden Arps Slate, Meagher & Flom, 523 avenue Louise, 1050 Bruxelles, Belgium Tel : 02/648 76 66 Fax : 02/640 30 32.

Sterns Patricia M., 849 North Third Avenue, Phoenix, Arizona 85003 USA, Tel 602 254 5197, Fax 602 253 7767.

Studio Legale Ughi e Nunziante, Via XX Settembre I, Rome 00187 Italy, Tel : 6 476841, Fax : 6 4870397. Contact : Gioscia Marcello.

Telefonica De Argentina S.A., Av de Mayo 701 Piso 21, Buenos Aires, 1084 Argentina, Tel : (54 1) 334 0073. Contact : Borio Fernando Raul.

Teleglobe canada Inc., Legal Services, 23rd Floor, 1000 de la Gauchetière St West, Montreal Quebec H3B 4X5, Canada, Tel : (515) 869-7431, Fax : (514) 868-7438.

Tennen Leslie I., 849 North Third Avenue, Phoenix, Arizona 85003 USA, Tel : 602-254-5197, Fax : 602-253-7767.

Trenité van Doorne Advocaten, Postbus 190, 3000 AD Rotterdam, Netherlands.

Tucker Turner Kingsley Wood Co. 18 Bedford Rold, London WC1R 4EB UK, Tel : 071 242 3303, Fax : 071 831 1732. Contact : Richard Briggs Scir.

Wolf Maurice, Wolf Arnold, 8 cardoso, PS., 1150 18Th St NW Washington DC 20036 USA, Tel : 202-857-0585, Fax : 202-293-2068.

......To be followed in Telecommunications and Space Journal Volume 3 1996.

PRIORITY ORDER FORM
VOLUME 2, 1995
VOLUME 1, 1994

How to Order

• Post your completed order form to : Rachel Mayne, SERDI Publishing Company, 29 rue de Saint Petersbourg, 75008 Paris, France.
• Telephone your order on (33) 61.75.79.12.
• Fax it on (33) 61.75.94.15.

Please Send Me :

[ ] copy/ies of "Telecommunications & Space Journal", Volume 2, 1995. (FF 450 ; £45 ; US $ 72,00).
[ ] copy/ies of "Telecommunications & Space Journal", Volume 1, 1994. (FF 450 ; £45 ; US $ 72,00).

Please add postage to your order.
For single copy orders please add FF 20 within France, US $ 5.00 outside France. Postage free for two or more copies.

Name : ....
Address : ....
Telephone : .... Fax : ....

Payment by :

☐ Chèque ☐ American Express Card ☐ Master card

☐ Visa Card

Expiry date of card : ....
Number : ....
Signature : ....Date : ....

EC countries only - please complete details below :

I am registered for VAT, my VAT registration number is :

I am exempted from VAT and enclose proof.

Recommendations to Authors

Manuscripts proposed to "Telecommunications & Space Journal" must be submitted in accordance with the standards below :

- *The text should be submitted in the following format : title page, abstract and key words, text, acknowledgements, references, tables, illustrations and captions. It should be typed with single spacing, on one side only on paper measuring 21x29,7 cm, leaving a left hand margin of 4 cm.*
- *The title page includes : 1). the title of the article ; 2). the authors' fornames and surnames ; 3). the name of departments and institutions to which the authors belong ; 4). the main author's name and address ; 5). the funding sources (donation award, grant etc...).*
- *The second page should include the abstract and the key words.*
- *The third page includes a very short title (less than 40 characters) as a header for each edited page.*
- *The text will be eventualy divided into : introduction, material and methods, results and discussion.*
- *References in the text should be cited by numbers, placed in line with the text and numbered in order of appearance. They should be listed in numerical order at the end of the paper in the form prescribed in the Uniform Requirements*
- *Table : each table should be on a separate page : it should have a title and, if desired, an explanation. Each table should be cited in the text and should be numbered according to the order of appearance.*
- *Illustration for diagrams, original artwork : Black ink on white paper is prefered, but glossy prints will usually be acceptable. All illustrations should bear author's name and the number of the illustration on the reverse side, also an indication in soft pencil if only part of the illustration is required. Captions should be typed on separate sheets. It is the responsability of the author(s) to ensure that any requirements of copyright and courtesy are fulfilled in reproducing illustrations and appropriate acknowledgements included with the captions.*

- *Copyright : Authors of accepted manuscripts must transfer copyright to "Telecommunications & Space Journal", which holds the copyright to all articles, comments, reviews and notes published in "Telecommunications & Space Journal".*

- *Submission of manuscripts : Contributions and correspondence should be sent to :*

Monsieur le Professeur Lucien Rapp
Directeur de l'Institut d'Etudes Internationales
Université de Toulouse I
Place Anatole France
F-31042 Toulouse Cédex
France

Volume 3 - 1996 : before 31 March, 1996

- *Two copies must be submitted. Submission of an article is taken to imply that its content has not previously been published and is not being considered for publication elsewhere.*

- *Manuscripts are sent to reviewers for recommendation and comment. Reviewers' comments are provided to authors.*